Practical Digital Signal Processing for Engineers and Technicians

Titles in the series

Practical Cleanrooms: Technologies and Facilities (David Conway)

Practical Data Acquisition for Instrumentation and Control Systems (John Park, Steve Mackay)

Practical Data Communications for Instrumentation and Control (John Park, Steve Mackay, Edwin Wright)

Practical Digital Signal Processing for Engineers and Technicians (Edmund Lai)

Practical Electrical Network Automation and Communication Systems (Cobus Strauss)

Practical Embedded Controllers (John Park)

Practical Fiber Optics (David Bailey, Edwin Wright)

Practical Industrial Data Networks: Design, Installation and Troubleshooting (Steve Mackay, Edwin Wright, John Park, Deon Reynders)

Practical Industrial Safety, Risk Assessment and Shutdown Systems (Dave Macdonald)

Practical Modern SCADA Protocols: DNP3, 60870.5 and Related Systems (Gordon Clarke, Deon Reynders)

Practical Radio Engineering and Telemetry for Industry (David Bailey)

Practical SCADA for Industry (David Bailey, Edwin Wright)

Practical TCP/IP and Ethernet Networking (Deon Reynders, Edwin Wright)

Practical Variable Speed Drives and Power Electronics (Malcolm Barnes)

Practical Digital Signal Processing for Engineers and Technicians

Edmund Lai PhD, BEng; Lai and Associates, Singapore

AMSTERDAM • BOSTON • HEIDELBERG • LONDON • NEW YORK • OXFORD
PARIS • SAN DIEGO • SAN FRANCISCO • SINGAPORE • SYDNEY • TOKYO
Newnes is an imprint of Elsevier

Newnes
An imprint of Elsevier
Linacre House, Jordan Hill, Oxford OX2 8DP
200 Wheeler Road, Burlington, MA 01803

First published 2004

British Library Cataloguing in Publication Data
A catalogue record for this book is available from the British Library

ISBN 07506 57987

Typeset and Edited by Vivek Mehra, Mumbai, India

Printed and bound in Great Britain

Contents

Preface viii

1 Introduction 1

1.1 Benefits of processing signals digitally 1
1.2 Definition of some terms 2
1.3 DSP systems 3
1.4 Some application areas 4
1.5 Objectives and overview of the book 12

2 Converting analog to digital signals and vice versa 14

2.1 A typical DSP system 14
2.2 Sampling 15
2.3 Quantization 24
2.4 Analog-to-digital converts 34
2.5 Analog reconstruction 42
2.6 Digital-to-analog converters 46
2.7 To probe further 48

3 Time-domain representation of discrete-time signals and systems 50

3.1 Notation 50
3.2 Typical discrete-time signals 50
3.3 Operations on discrete-time signals 52
3.4 Classification of systems 54
3.5 The concept of convolution 55
3.6 Autocorrelation and cross-correlation of sequences 57

4 Frequency-domain representation of discrete-time signals 61

4.1 Discrete Fourier series for discrete-time periodic signals 62
4.2 Discrete Fourier transform for discrete-time aperiodic signals 63
4.3 The inverse discrete Fourier transform and its computation 64
4.4 Properties of the DFT 64
4.5 The fast Fourier transform 67
4.6 Practical implementation issues 71
4.7 Computation of convolution using DFT 74
4.8 Frequency ranges of some natural and man-made signals 78

5 DSP application examples 79

5.1 Periodic signal generation using wave tables 80
5.2 Wireless transmitter implementation 83
5.3 Speech synthesis 88
5.4 Image enhancement 91

5.5	Active noise control	94
5.6	To probe further	97
6	**Finite impulse response filter design**	**98**
6.1	Classification of digital filters	98
6.2	Filter design process	99
6.3	Characteristics of FIR filters	102
6.4	Window method	106
6.5	Frequency sampling method	128
6.6	Parks-McClelland method	134
6.7	Linear programming method	141
6.8	Design examples	142
6.9	To probe further	144
7	**Infinite impulse response (IIR) filter design**	**145**
7.1	Characteristics of IIR filters	146
7.2	Review of classical analog filter	147
7.3	IIR filters from analog filters	157
7.4	Direct design methods	165
7.5	FIR vs IIR	169
7.6	To probe further	170
8	**Digital filter realizations**	**171**
8.1	Direct form	171
8.2	Cascade form	179
8.3	Parallel form	181
8.4	Other structures	183
8.5	Software implementation	186
8.6	Representation of numbers	187
8.7	Finite word-length effects	191
9	**Digital signal processors**	**204**
9.1	Common features	204
9.2	Hardware architecture	206
9.3	Special instructions and addressing modes	215
9.4	General purpose microprocessors for DSP	224
9.5	Choosing a processor	224
9.6	To probe further	225

10	Hardware and software development tools	226
10.1	DSP system design flow	226
10.2	Development tools	231
Appendix A		238
Appendix B		242
Index		285

Preface

Digital signal processing (DSP) can be considered simply to be the capture, analysis and manipulation of an analog signal by a digital computer. The integration of DSP software and hardware into products across a wide range of industries has necessitated the understanding and application of DSP by engineers and technicians.

The aim of this book is to introduce DSP from a practical point of view using a minimum of mathematics. The emphasis is on the practical aspects of DSP, implementation issues, tips and tricks and pitfalls and practical applications. Intuitive explanations and appropriate examples are used to develop a fundamental understanding of DSP theory. The coverage in the book is fairly broad from process control to communications.

Some of the DSP techniques included in the book include:

- Digital filtering for cleaning a signal from noise
- Discrete Fourier transforms for finding a particular frequency component
- Correlation techniques to find a signal buried in noise
- Industrial control with digital controllers
- Instrumentation and test for better accuracy
- Vibration analysis for identifying frequency signatures
- Image and video processing for enhancing images
- Communications especially for filtering out noise

At the conclusion of the reading of the book we hope that you will gain the following:

- A clear understanding of digital signal processing (DSP)
- Benefits and application of DSP technology to improve efficiency
- An understanding of frequency analysis of signals and the application of these techniques
- Information about and actual design of digital filters
- Ability to analyse the performance of DSP systems
- A knowledge of the key issues in designing a DSP system
- An understanding of the features and capabilities of DSP applications

Typical people who will find this book useful include:

- Electrical engineers
- Control system engineers
- Communication system engineers
- Electronic engineers
- Instrumentation engineers
- Condition monitoring engineers and technicians
- Design engineers

A basic knowledge of first year college mathematics is essential to grasp the basic principles in this book; but beyond this the contents are of a fundamental nature and are easy to comprehend.

The structure of the book is as follows.

***Chapter 1**: Introduction.* This chapter gives a brief overview of the benefits of processing signals digitally as well as an overview of the book.

***Chapter 2**: Converting analog to digital signals and vice versa.* A review of a typical DSP system, analog to digital converters and digital to analog converters.

***Chapter 3**: Time domain representation.* A discussion on typical discrete-time signals, operations on discrete time signals, the classification of systems, convolution and auto and cross correlation operations.

***Chapter 4**: Frequency domain representation.* A detailed review of the discrete Fourier and inverse Fourier transform operations with an extension to the Fast Fourier transform and implementation of this important algorithm in software.

***Chapter 5**: DSP application examples.* A review of periodic signal generation using wavetables, a wireless transmitter implementation, speech synthesis, image enhancement and active noise control.

***Chapter 6**: FIR filter design.* An examination of the classification of digital filters, the filter design process, characteristics of FIR filters, the window, frequency sampling and Parks-Mclelland methods.

***Chapter 7**: Infinite impulse response (IIR) filter design.* A review of the characteristics of IIR filters, review of classical analog filter approximations, IIR filter derivation from analog filters and a comparison of the FIR and IIR design methods.

***Chapter 8**: Digital filter realizations.* A review of the direct, cascade, parallel forms and software implementation issues together with finite word-length effects.

***Chapter 9**: Digital signal processors.* An examination of common features, hardware architectures, special instructions and addressing modes and a few suggestions on choosing the most appropriate DSP processor for your design.

***Chapter 10**: Hardware and software development tools.* A concluding review on DSP system design flow and development tools.

1

Introduction

Digital signal processing (DSP) is a field which is primarily technology driven. It started from around mid 1960s when digital computers and digital circuitry became fast enough to process large amounts of data efficiently.

When the term 'digital' is used, often it loosely refers to a finite set of distinct values. This is in contrast to 'analog', which refers to a continuous range of values. In digital signal processing we are concerned with the processing of signals which are discrete in time (sampled) and in most cases, discrete in amplitude (quantized) as well. In other words, we are primarily dealing with data sequences – sequences of numbers.

Such discrete (or digital) signals may arise in one of the following two distinct circumstances:

- The signal may be inherently discrete in time (and/or amplitude)
- The signal may be a sampled version of a continuous-time signal

Examples of the first type of data sequences include monthly sales figures, daily highest/lowest temperatures, stock market indices and students examination marks. Business people, meteorologists, economists, and teachers process these types of data sequences to determine cyclic patterns, trends, and averages. The processing usually involves filtering to remove as much 'noise' as possible so that the pattern of interest will be enhanced or highlighted.

Examples of the second type of discrete-time signals can readily be found in many engineering applications. For instance, speech and audio signals are sampled and then encoded for storage or transmission. A compact disc player reads the encoded digital audio signals and reconstructs the continuous-time signals for playback.

1.1 Benefits of processing signals digitally

A typical question one may ask is why process signals digitally? For the first type of signals discussed previously, the reason is obvious. If the signals are inherently discrete in time, the most natural way to process them is using digital methods. But for continuous-time signals, we have a choice.

Analog signals have to be processed by analog electronics while computers or microprocessors can process digital signals. Analog methods are potentially faster since the analog circuits process signals as they arrive in real-time, provided the settling time is fast enough. On the other hand, digital techniques are algorithmic in nature. If the computer is fast and the algorithms are efficient, then digital processing can be performed in 'real-time' provided the data rate is 'slow enough'. However, with the speed of digital logic increasing exponentially, the upper limit in data rate that can still be considered as real-time processing is becoming higher and higher.

The major advantage of digital signal processing is consistency. For the same signal, the output of a digital process will always be the same. It is not sensitive to offsets and drifts in electronic components.

The second main advantage of DSP is that very complex digital logic circuits can be packed onto a single chip, thus reducing the component count and the size and reliability of the system.

1.2 Definition of some terms

DSP has its origin in electrical/electronic engineering (EE). Therefore the terminology used in DSP are typically that of EE. If you are not an electrical or electronic engineer, there is no problem. In fact many of the terms that are used have counterparts in other engineering areas. It just takes a bit of getting used to.

For those without an engineering background, we shall now attempt to explain a few terms that we shall be using throughout the manual.

- **Signals**
  We have already started using this term in the previous section. A signal is simply a quantity that we can measure over a period of time. This quantity usually changes with time and that is what makes it interesting. Such quantities could be voltage or current. They could also be the pressure, fluid level and temperature. Other quantities of interest include financial indices such as the stock market index. You will be surprised how much of the concepts in DSP has been used to analyze the financial market.

- **Frequency**
  Some signals change slowly over time and others change rapidly. For instance, the (AC) voltage available at our household electrical mains goes up and down like a sine function and they complete one cycle in 50 times or 60 times a second. This signal is said to have a frequency of 50 or 60 hertz (Hz).

- **Spectrum**
  While some signals consist of only a single frequency, others have a combination of a range of frequencies. If you play a string on the violin, there is a fundamental tone (frequency) corresponding to the musical note that is played. But there are other harmonics (integer multiples of the fundamental frequency) present. This musical sound signal is said to have a spectrum of frequencies. The spectrum is a frequency (domain) representation of the time (domain) signal. The two representations are equivalent.

- **Low-pass filter**
  Filters let a certain range of frequency components of a signal through while rejecting the other frequency components. A low-pass filter lets the 'low-frequency' components through. Low-pass filters have a cutoff frequency below which the frequency components can pass through the filter. For

instance, if a signal has two frequency components, say 10 hz and 20 hz, applying a low-pass filter to this signal with a cutoff frequency of 15 hz will result in an output signal, which has only one frequency component at 10 hz; the 20 hz component has been rejected by the filter.

- **Band-pass filter**
  Band-pass filters are similar to low-pass filters in that only a range of frequency components can pass through it intact. This range (the passband) is usually above the DC (zero frequency) and somewhere in the mid-range. For instance, we can have a band-pass filter with a passband between 15 and 25 Hz. Applying this filter to the signal discussed above will result in a signal having only a 20 Hz component.
- **High-pass filter**
  These filters allow frequency components above a certain frequency (cutoff) to pass through intact, rejecting the ones lower than the cutoff frequency.

This should be enough to get us going. New terms will arise from time to time and they will be explained as we come across them.

1.3 DSP systems

DSP systems are discrete-time systems, which means that they accept digital signals as input and output digital signals (or information extracted). Since digital signals are simply sequences of numbers, the input and output relationship of a discrete-time system can be illustrated as in Figure 1.1. The output sequence of sample $y(n)$ is computed from the input sequence of sample $x(n)$ according to some rules, which the system (H) defines.

There are two main methods by which the output sequence is computed from the input sequence. They are called sample-by-sample processing and block processing respectively. We shall encounter both types of processing in later chapters. Most systems can be implemented with either processing method. The output obtained in both cases should be equivalent if the input and the system H are the same.

1.3.1 Sample-by-sample processing

With the sample-by-sample processing method, normally one output sample is obtained when one input sample is presented to the system.

For instance, if the sequence $\{y_0, y_1, y_2, \ldots, y_n, \ldots\}$ is obtained when the input sequence $\{x_0, x_1, x_2, \ldots, x_n, \ldots\}$ is presented to the system. The sample y_0 appears at the output when the input x_0 is available at the input. The sample y_1 appears at the output when the input x_1 is available at the input, etc.

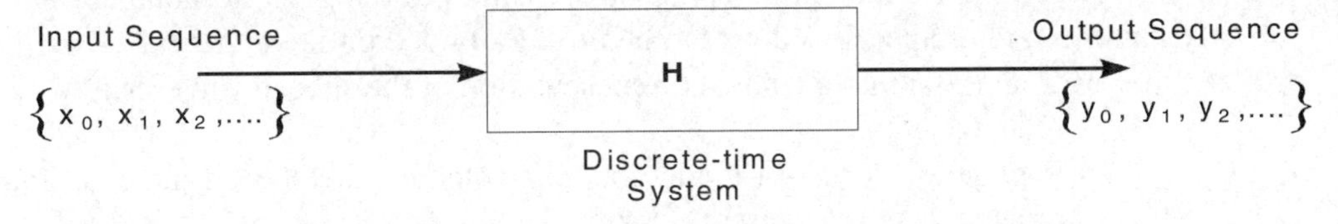

Figure 1.1
A discrete-time system

The delay between the input and output for sample-by-sample processing is at most one sample. The processing has to be completed before the next sample appears at the input.

1.3.2 Block processing

With block processing methods, a block of signal samples is being processed at a time. A block of samples is usually treated as a vector, which is transformed, to an output vector of samples by the system transformation H.

$$\mathrm{x} = \begin{bmatrix} x_0 \\ x_1 \\ x_2 \\ \vdots \end{bmatrix} \xrightarrow{H} \begin{bmatrix} y_0 \\ y_1 \\ y_2 \\ \vdots \end{bmatrix} = \mathrm{y}$$

The delay between input and output in this case is dependent on the number of samples in each block. For example, if we use 8 samples per block, then the first 8 input samples have to be buffered (or collected) before processing can proceed. So the block of 8 output samples will appear at least 8 samples after the first sample x_0 appears. The block computation (according to H) has to be completed before the next block of 8 samples are collected.

1.3.3 Remarks

Both processing methods are extensively used in real applications. We shall encounter DSP algorithms and implementation that uses one or the other. The reader might find it useful in understanding the algorithms or techniques being discussed by realizing which processing method is being used.

1.4 Some application areas

Digital signal processing is being applied to a large range of applications. No attempt is made to include all areas of application here. In fact, new applications are constantly appearing. In this section, we shall try to describe a sufficiently broad range of applications so that the reader can get a feel of what DSP is about.

1.4.1 Speech and audio processing

An area where DSP has found a lot of application is in speech processing. It is also one of the earliest applications of DSP. Digital speech processing includes three main sub-areas: encoding, synthesis, and recognition.

1.4.1.1 Speech coding

There is a considerable amount of redundancy in the speech signal. The encoding process removes as much redundancy as possible while retaining an acceptable quality of the remaining signal. Speech coding can be further divided into two areas:

- Compression – a compact representation of the speech waveform without regard to its meaning.
- Parameterization – a model that characterizes the speech in some linguistically or acoustically meaningful form.

The minimum channel bandwidth required for the transmission of an acceptable quality of speech is around 3 kHz, with a dynamic range of 72 dB. This is normally referred to as telephone quality. Converting into digital form, a sampling rate of 8 k samples per second with a 12-bit quantization (2^{12} amplitude levels) is commonly used, resulting in 96 k bits per second of data. This data rate can be significantly reduced without affecting the quality of the reconstructed speech as far as the listener is concerned. We shall briefly describe three of them:

- **Companding or non-uniform quantization**
  The dynamic range of speech signals is very large. This is due to the fact that voiced sounds such as vowels contain a lot of energy and exhibit wide fluctuations in amplitude while unvoiced sounds like fricatives generally have much lower amplitudes. A compander (compressor–expander) compresses the amplitude of the signal at the transmitter end and expands it at the receiver end. The process is illustrated schematically in Figure 1.2. The compressor compresses the large amplitude samples and expands the small amplitude ones while the expander does the opposite.

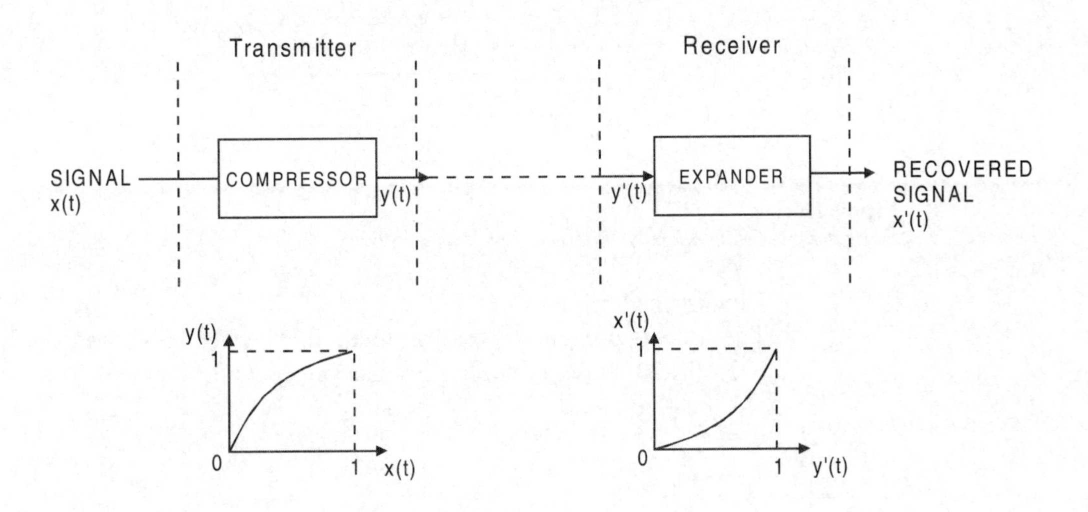

Figure 1.2
Schematic diagram showing the companding process

The μ-law compander (with $\mu = 255$) is a North American standard. A-law companding with $A = 87.56$ is a European (CCITT) standard. The difference in performance is minimal. A-law companding gives slightly better performance at high signal levels while μ-law is better at low levels.

- **Adaptive differential quantization**
  At any adequate sampling rate, consecutive samples of the speech signal are generally highly correlated, except for those sounds that contain a significant amount of wideband noise. The data rate can be greatly reduced by quantizing the difference between two samples instead. Since the dynamic range will be much reduced by differencing, the number of levels required for the quantifier will also be reduced.

  The concept of differential quantization can be extended further. Suppose we have an estimate of the value of the current sample based on information from the previous samples, then we can quantize the difference between the current sample and its estimate. If the prediction is accurate enough, this difference will be quite small.

Figure 1.3 shows the block diagram of an adaptive differential pulse code modulator (ADPCM). It takes a 64 kbits per second pulse code modulated (PCM) signal and encodes it into 32 kbit per second adaptive differential pulse code modulated (ADPCM) signal.

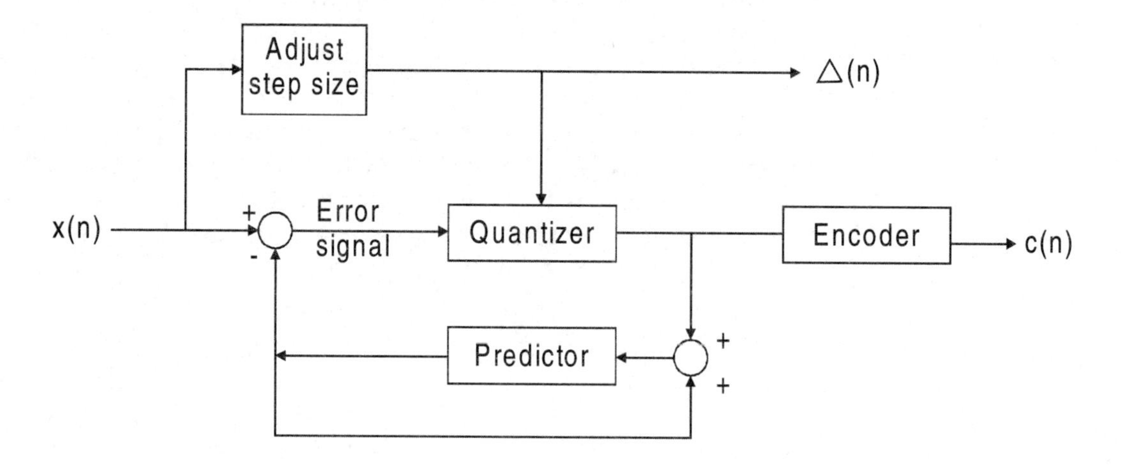

Figure 1.3
Block diagram of an adaptive differential pulse code modulator

- **Linear prediction**
  The linear predictive coding method of speech coding is based on a (simplified) model of speech production shown in Figure 1.4.

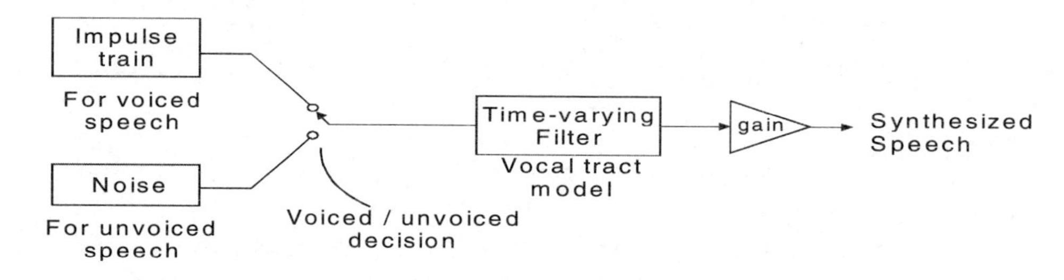

Figure 1.4
A model of speech production

The time-varying digital filter models the vocal tract and is driven by an excitation signal. For voiced speech, this excitation signal is typically a train of scaled unit impulses at pitch frequency. For unvoiced sounds it is random noise.

The analysis system (or encoder) estimates the filter coefficients, detects whether the speech is voiced or unvoiced and estimates the pitch frequency if necessary. This is performed for each overlapping section of speech usually around 10 milliseconds in duration. This information is then encoded and transmitted. The receiver reconstructs the speech signal using these parameters based on the speech production model. It is interesting to note that the reconstructed speech is similar to the original perceptually but the physical appearance of the signal is very different. This is an illustration of the redundancies inherent in speech signals.

1.4.1.2 Speech synthesis

The synthesis or generation of speech can be done through the speech production model mentioned above. Although the duplication of the acoustics of the vocal tract can be carried out quite accurately, the excitation model turns out to be more problematic.

For synthetic speech to sound natural, it is essential that the correct allophone be produced. Despite the fact that different allophones are perceived as the same sound, if the wrong allophone is selected, the synthesized speech will not sound natural. Translation from phonemes to allophones is usually controlled by a set of rules. The control of timing of a word is also very important. But these rules are beyond the realm of DSP.

1.4.1.3 Speech recognition

One of the major goals of speech recognition is to provide an alternative interface between human user and machine. Speech recognition systems can either be speaker dependent or independent, and they can either accept isolated utterances or continuous speech. Each system is capable of handling a certain vocabulary.

The basic approach to speech recognition is to extract features of the speech signals in the training phase. In the recognition phase, the features extracted from the incoming signal are compared to those that have been stored. Owing to the fact that our voice changes with time and the rate at which we speak also varies, speech recognition is a very tough problem. However, there are now commercially available some relatively simple small vocabulary, isolated utterance recognition systems. This comes about after 30 years of research and the advances made in DSP hardware and software.

1.4.2 Image and video processing

Image processing involves the processing of signals, which are two-dimensional. A digital image consists of a two dimensional array of pixel values instead of a one dimensional one for, say, speech signals. We shall briefly describe three areas of image processing.

1.4.2.1 Image enhancement

Image enhancement is used when we need to focus or pick out some important features of an image. For example, we may want to sharpen the image to bring out details such as a car license plate number or some areas of an X-ray film. In aerial photographs, the edges or lines may need to be enhanced in order to pick out buildings or other objects. Certain spectral components of an image may need to be enhanced in images obtained from telescopes or space probes. In some cases, the contrast may need to be enhanced.

While linear filtering may be all that is required for certain types of enhancement, most useful enhancement operations are non-linear in nature.

1.4.2.2 Image restoration

Image restoration deals with techniques for reconstructing an image that may have been blurred by sensor or camera motion and in which additive noise may be present. The blurring process is usually modeled as a linear filtering operation and the problem of image restoration then becomes one of identifying the type of blur and estimating the parameters of the model. The image is then filtered by the inverse of the filter.

1.4.2.3 Image compression and coding

The amount of data in a visual image is very large. A simple black-and-white still picture digitized to a 512×512 array of pixels using 8 bits per pixel involves more than 2 million bits of information. In the case of sequences of images such as in video or television images, the amount of data involved will be even greater. Image compression, like speech compression, seeks to reduce the number of bits required to store or transmit the image with either no loss or an acceptable level of loss or distortion. A number of different techniques have been proposed, including prediction or coding in the (spatial) frequency domain. The most successful techniques typically combine several basic methods. Very sophisticated methods have been developed for digital cameras and digital video discs (DVD).

Standards have been developed for the coding of both image and video signals for different kinds of applications. For still images, the most common one is JPEG. For high quality motion video, there is MPEG and MPEG-2. MPEG-2 was developed with high definition television in mind. It is now used in satellite transmission of broadcast quality video signals.

1.4.3 Adaptive filtering

A major advantage of digital processing is its ability of adapting to changing environments. Even though adaptive signal processing is a more advanced topic, which we will not cover in this course, we shall describe the basic ideas involved in adaptive signal processing and some of its applications.

A basic component in an adaptive digital signal processing system is a digital filter with adjustable filter coefficients – a time-varying digital filter. Changing the characteristics of a filter by a change in the coefficient values is a very simple operation in DSP. The adaptation occurs through an algorithm which takes a reference (or desired) signal and an error signal produced by the difference between the current output of the filter and the current input signal. The algorithm adjusts the filter coefficients so that the averaged error is minimized.

1.4.3.1 Noise cancellation

One example of noise cancellation is the suppression of the maternal ECG component in fetal ECG. The fetal heart rate signal can be obtained from a sensor placed in the abdominal region of the mother. However, this signal is very noisy due to the mother's heartbeat and fetal motion.

The idea behind noise cancellation in this case is to take a direct recording of the mother's heartbeat and after filtering of this signal, subtract it off the fetal heart rate signal to get a relatively noise-free heart rate signal. A schematic diagram of the system is shown in Figure 1.5.

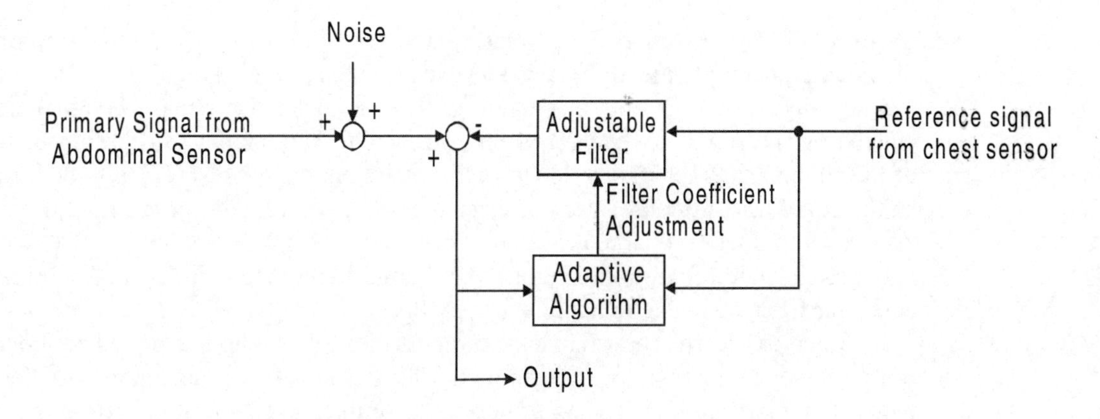

Figure 1.5
An adaptive noise cancellation system

There are two inputs: a primary and a reference. The primary signal is of interest but has a noisy interference component, which is correlated with the reference signal. The adaptive filter is used to produce an estimate of this interference or noise component, which is then subtracted off the primary signal. The filter should be chosen to ensure that the error signal and the reference signal are uncorrelated.

1.4.3.2 Echo cancellation

Echoes are signals that are identical to the original signals but are attenuated and delayed in time. They are typically generated in long distance telephone communication due to impedance mismatch. Such a mismatch usually occurs at the junction or hybrid between the local subscriber loop and the long distance loop. As a result of the mismatch, incident electromagnetic waves are reflected which sound like echoes to the telephone user.

The idea behind echo cancellation is to predict the echo signal values and thus subtract it out. The basic mechanism is illustrated in Figure 1.6. Since the speech signal is constantly changing, the system has to be adaptive.

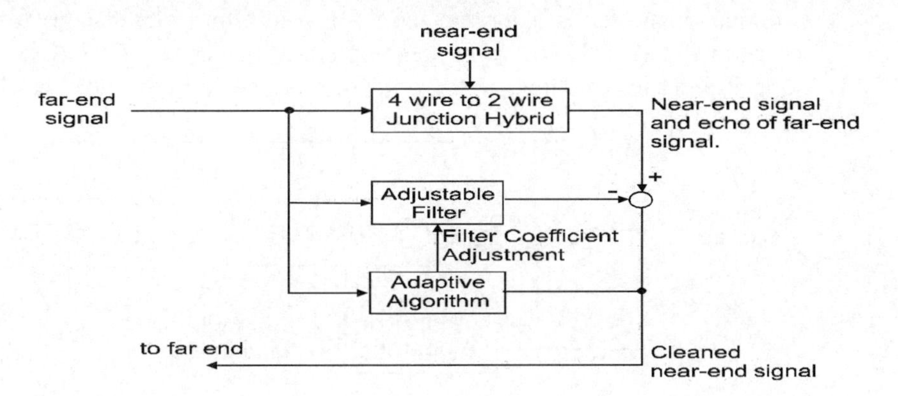

Figure 1.6
An adaptive echo cancellation system

1.4.3.3 Channel equalization

Consider the transmission of a signal over a communication channel (e.g. coaxial cable, optical fiber, wireless). The signal will be subject to channel noise and dispersion caused, for example, by reflection from objects such as buildings in the transmission path. This distorted signal will have to be reconstructed by the receiver.

One way to restore the original signal is to pass the received signal through an equalizing filter to undo the dispersion effects. The equalizer should ideally be the inverse of the channel characteristics. However, channel characteristics typically drift in time and so the equalizer (a digital filter) coefficients will need to be adjusted continuously. If the transmission medium is a cable, the drift will occur very slowly. But for wireless channels in mobile communications the channel characteristics change rapidly and the equalizer filter will have to adapt very quickly.

In order to 'learn' the channel characteristics, the adaptive equalizer operates in a training mode where a pre-determined training signal is transmitted to the receiver. Normal signal transmission has to be regularly interrupted by a brief training session so that the equalizer filter coefficients can be adjusted. Figure 1.7 shows an adaptive equalizer in training mode.

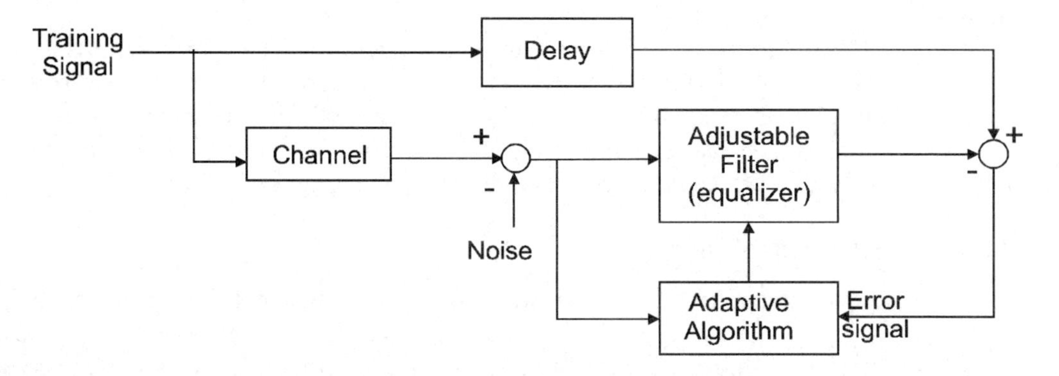

Figure 1.7
An adaptive equalizer in training mode

1.4.4 Control applications

A digital controller is a system used for controlling closed-loop feedback systems as shown in Figure 1.8. The controller implements algebraic algorithms such as filters and compensatory to regulate, correct, or change the behavior of the controlled system.

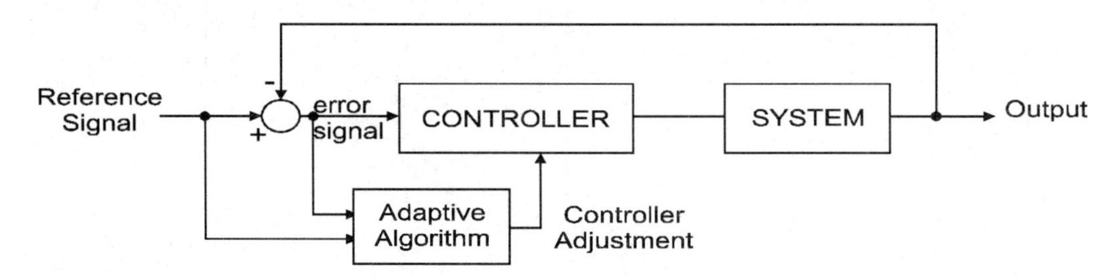

Figure 1.8
A digital closed-loop control system

Digital control has the advantage that complex control algorithms are implemented in software rather than specialized hardware. Thus the controller design and its parameters can easily be altered. Furthermore, increased noise immunity is guaranteed and parameter drift is eliminated. Consequently, they tend to be more reliable and at the same time, feature reduced size, power, weight and cost.

Digital signal processors are very useful for implementing digital controllers since they are typically optimized for digital filtering operations with single instruction arithmetic operations. Furthermore, if the system being controlled changes with time, adaptive control algorithms, similar to adaptive filtering discussed above, can be implemented.

1.4.5 Sensor or antenna array processing

In some applications, a number of spatially distributed sensors are used for receiving signals from some sources. The problem of coherently summing the outputs from these sensors is known as beamforming. Beyond the directivity provided by an individual sensor, a beamformer permits one to 'listen' preferentially to wave fronts propagating from one direction over another. Thus a beamformer implements a spatial filter. Applications of beamforming can be found in seismology, underwater acoustics, biomedical engineering, radio communication systems and astronomy.

In cellular mobile communication systems, smart antennas (an antenna array with digitally steerable beams) are being used to increase user capacity and expand geographic coverage. In order to increase capacity, an array, which can increase the carrier to interference ratio (C/I) at both the base station and the mobile terminal, is required. There are three approaches to maximizing C/I with an antenna array.

- The first one is to create higher gain on the antenna in the intended direction using antenna aperture. This is done by combining the outputs of each individual antenna to create aperture.
- The second approach is the mitigation of multipath fading. In mobile communication, fast fading induced by multipath propagation requires an additional link margin of 8 dB. This margin can be recovered by removing the destructive multipath effects.
- The third approach is the identification and nulling of interferers. It is not difficult for a digital beamformer to create sharp nulls, removing the effects of interference.

Direction of arrival estimation can also be performed using sensor arrays. In the simplest configuration, signals are received at two spatially separated sensors with one signal being an attenuated, delayed and noisy version of the other. If the distance between the sensors is known, and the signal velocity is known, then the direction of arrival can be estimated. If the direction does not change, or changes very slowly with time, then it can be determined by cross-correlating the two signals and finding the global maximum of the cross-correlation function. If the direction changes rapidly, then an adaptive algorithm is needed.

1.4.6 Digital communication receivers and transmitters

One of the most exciting applications of DSP is in the design and implementation of digital communication equipment. Throughout the 1970s and 80s radio systems migrated from analog to digital in almost every aspect, from system control to source and channel coding to hardware technology. A new architecture known generally as 'software radio' is emerging. This technology liberates radio-based services from dependence on hardwired characteristics such as frequency band, channel bandwidth, and channel coding.

The software radio architecture centers on the use of wideband analog-to-digital and digital-to-analog converters that are placed as close to the antenna as possible. Since the signal is being digitized earlier in the system, as much radio functionality as possible can be defined and implemented in software. Thus the hardware is relatively simple and functions are software defined as illustrated in Figure 1.9.

Software definable channel modulation across the entire 25 MHz cellular band has been developed.

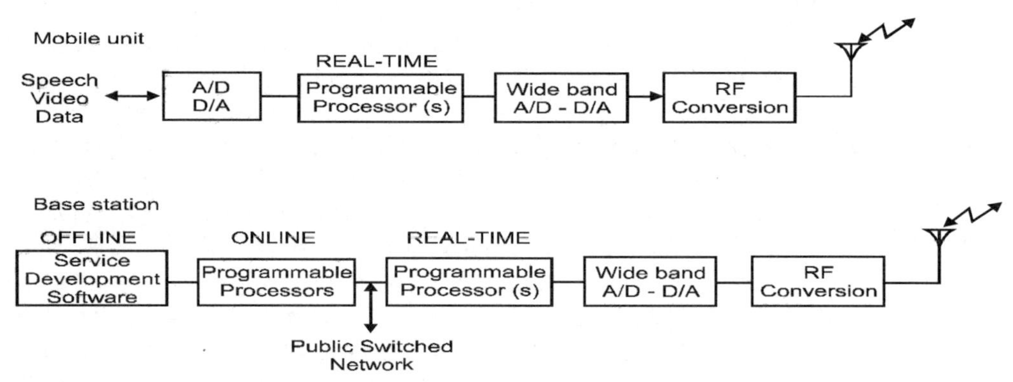

Figure 1.9
Software radio architecture

In an advanced application, a software radio does not just transmit; it characterizes the available transmission channels, probes the propagation path, constructs an appropriate channel modulation, electronically steers its transmit beam in the right direction for systems with antenna arrays and selects the optimum power level. It does not just receive; it characterizes the energy distribution in the channel and in adjacent channels, recognizes the mode of the incoming transmission, adaptively null interferers, estimates the dynamic properties of multipath propagation, equalizes and decodes the channel codes. The main advantage of software radio is that it supports incremental service enhancements through upgrades to its software. This whole area is not possible without the advancements in DSP technology.

1.5 Objectives and overview of the book

1.5.1 Objectives

The main objective of this book is to provide a first introduction to the area of digital signal processing. The emphasis is on providing a balance between theory and practice.

Digital signal processing is in fact a very broad field with numerous applications and potentials. It is an objective of this book to give the interested participants a foundation in DSP so that they may be able to pursue further in this interesting field of digital signal processing.

Software exercises designed to aid in the understanding of concepts and to extend the lecture material further are given. They are based on a software package called MATLAB®. It has become very much the *de facto* industry standard software package for studying and developing signal processing algorithms. It has an intuitive interface and is very easy to use. It also features a visual-programming environment called SIMULINK.

Designing a system using SIMULINK basically involves dragging and dropping visual components on to the screen and making appropriate connections between them.

There are also experiments based on the Texas Instruments TMS320C54x family of digital signal processors which provide the participants with a feel for the performance of DSP chips.

1.5.2 Brief overview of chapters

An overview of the remaining chapters in this manual is as follows:

- Chapter 2 discusses in detail the concepts in converting a continuous-time signal to a discrete-time and discrete-amplitude one and vice versa. Concepts of sampling and quantization and their relation to aliasing are described. These concepts are supplemented with practical analog-to-digital and digital-to-analog conversion techniques.
- Digital signals and systems can either be described as sequences in time or in frequency. In Chapter 3, digital signals are viewed as sequences in time. Digital systems are also characterized by a sequence called the impulse sequence. We shall discuss the properties of digital signals and systems and their interaction. The computation of the correlation of these sequences is discussed in detail.
- The discrete Fourier transform (DFT) provides a link between a time sequence and its frequency representation. The basic characteristics of the DFT and some ways by which the transform can be computed efficiently are described in Chapter 4.
- With the basic concepts in digital signals and systems covered, in Chapter 5 we shall revisit some practical applications. Some of these applications have already been briefly described in this chapter. They shall be further discussed using the concepts learnt in chapters 2 to 4.
- The processing of digital signals is most often performed by digital filters. The design of the two major types of digital filters: finite impulse response (FIR) and infinite impulse response (IIR) filters are thoroughly discussed in chapters 6 and 7.
- The different ways by which these FIR and IIR digital filters can be realized by hardware or software will be discussed in Chapter 8. Chapters 6 to 8 combined gives us a firm understanding in digital filters.
- Finally, in chapters 9 and 10, the architecture, characteristics and development tools of some representative commercially available digital signal processors are described. Some popular commercial software packages that are useful for developing digital signal processing algorithms are also listed and briefly described.

Since this is an introductory course, a number of important but more advanced topics in digital signal processing are not covered. These topics include:

- Adaptive filtering
- Multi-rate processing
- Parametric signal modeling and spectral estimation
- Two (and higher) dimensional digital signal processing
- Other efficient fast Fourier transform algorithms

2

Converting analog to digital signals and vice versa

2.1 A typical DSP system

In the previous chapter, we mentioned that some signals are discrete-time in nature, while others are continuous-time. Most of the signals encountered in engineering applications are analog. In order to process analog signals using digital techniques, they must first be converted into digital signals.

Digital processing of analog signals proceeds in three stages:

- The analog signal is digitized. Digitization involves two processes: sampling (digitization in time) and quantization (digitization in amplitude). This whole process is called analog-to-digital (A/D) conversion.
- The appropriate DSP algorithms process the digitized signal.
- The results or outputs of the processing are converted back into analog signals through interpolation. This process is called digital-to-analog (D/A) conversion.

Figure 2.1 illustrates these three stages in diagram form.

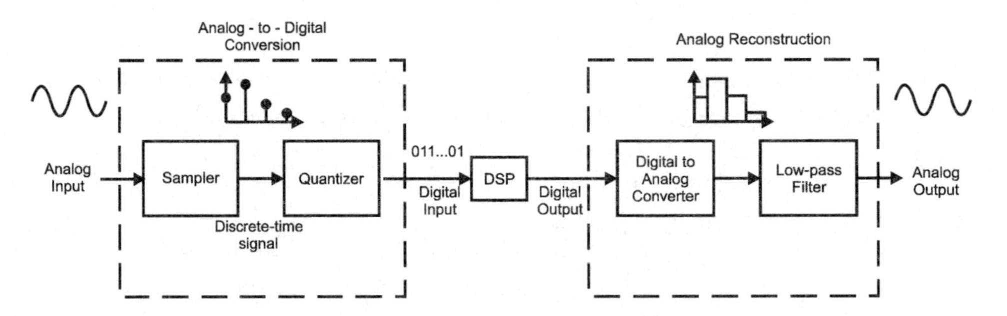

Figure 2.1
The three stages of analog–digital–analog conversions

2.2 Sampling

We shall first consider the sampling operation. It can be illustrated through the changing temperature through a single day. The continuous temperature variation is shown in Figure 2.2. However, the observatory may only be recording the temperature once every hour.

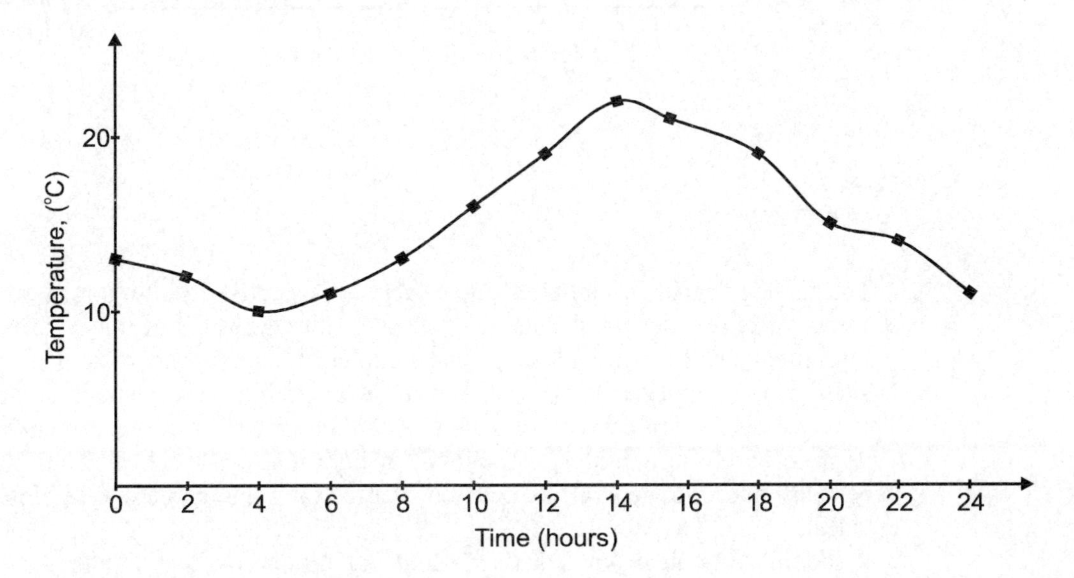

Figure 2.2
Temperature variation throughout a day

The records are shown in Table 2.1. When we plot these values against time, we have a snapshot of the variation in temperature throughout the day. These snapshots are called samples of the signal (temperature). They are plotted as dots in Figure 2.2. In this case the sampling interval, the time between samples, is two hours.

Hour	Temperature
0	13
2	12
4	10
6	11
8	13
10	16
12	19
14	23
16	22
18	20
20	16
22	15
24	12

Table 2.1
Temperature measured at each hour of a day

Figure 2.3 shows the diagram representation of the sampling process.

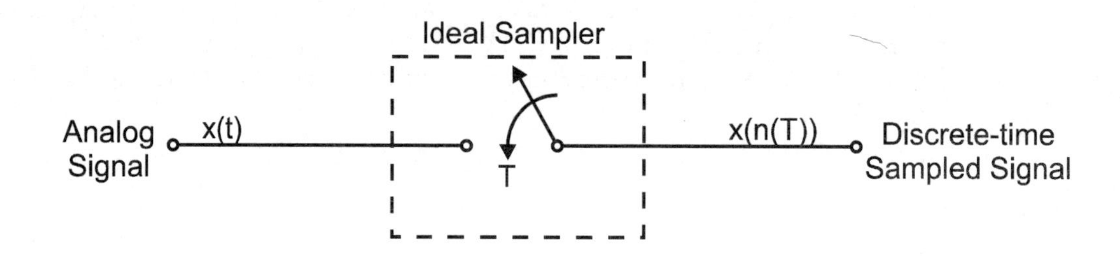

Figure 2.3
The sampling process

The analog signal is sampled once every T seconds, resulting in a sampled data sequence. The sampler is assumed to be ideal in that the value of the signal at an instant (an infinitely small time) is taken. A real sampler, of course, cannot achieve that and the 'switch' in the sampler is actually closed for a finite, though very small, amount of time. This is analogous to a camera with a finite shutter speed. Even if a camera can be built with an infinitely fast shutter, the amount of light that can reach the film plane will be very small indeed. In general, we can consider the sampling process to be close enough to the ideal.

It should be pointed out that throughout our discussions we should assume that the sampling interval is constant. In other words, the spacing between the samples is regular. This is called uniform sampling. Although irregularly sampled signals can, under suitable conditions, be converted to uniformly sampled ones, the concept and mathematics are beyond the scope of this introductory book.

The most important parameter in the sampling process is the sampling period T, or the sampling frequency or sampling rate f_s, which is defined as

$$f_s = \frac{1}{T}$$

Sampling frequency is given in units of 'samples per second' or 'hertz'. If the sampling is too frequent, then the DSP process will have to process a large amount of data in a much shorter time frame. If the sampling is too sparse, then important information might be missing in the sampled signal. The choice is governed by sampling theorem.

2.2.1 Sampling theorem

The sampling theorem specifies the minimum-sampling rate at which a continuous-time signal needs to be uniformly sampled so that the original signal can be completely recovered or reconstructed by these samples alone. This is usually referred to as Shannon's sampling theorem in the literature.

Sampling theorem:
If a continuous time signal contains no frequency components higher than W hz, then it can be completely determined by uniform samples taken at a rate f_s samples per second where

$$f_s \geq 2W$$

or, in term of the sampling period

$$T \leq \frac{1}{2W}$$

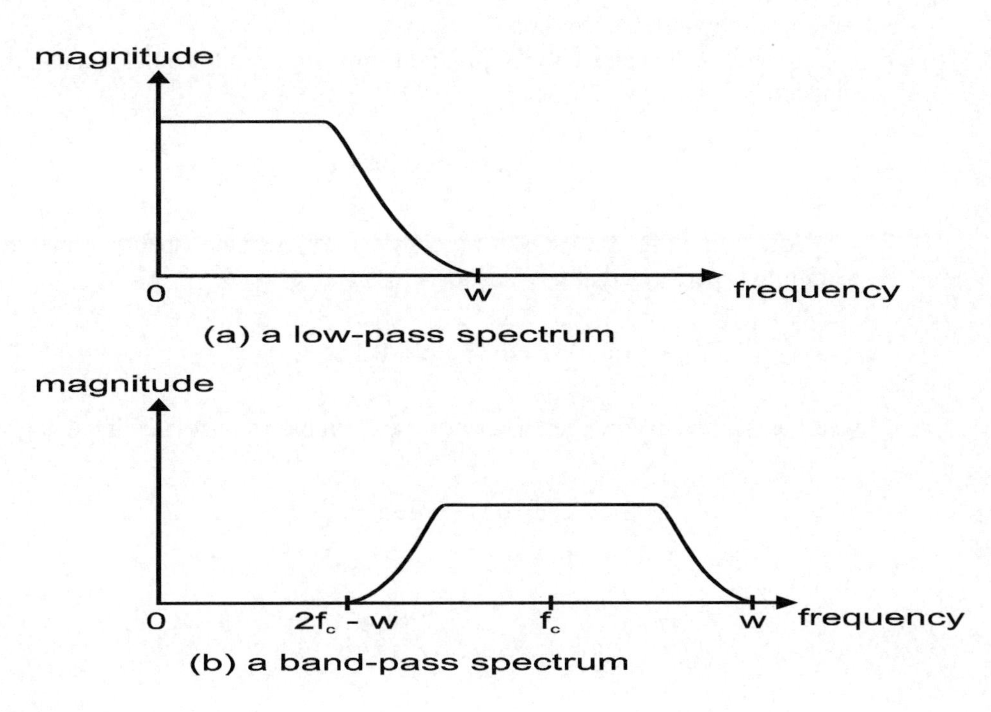

Figure 2.4
Two bandlimited spectra

A signal with no frequency component above a certain maximum frequency is known as a bandlimited signal. Figure 2.4 shows two typical bandlimited signal spectra: one low-pass and one band-pass.

The minimum sampling rate allowed by the sampling theorem ($f_s = 2W$) is called the Nyquist rate.

It is interesting to note that even though this theorem is usually called Shannon's sampling theorem, it was originated by both E.T. and J.M. Whittaker and Ferrar, all British mathematicians. In Russian literature, this theorem was introduced to communications theory by Kotel'nikov and took its name from him. C.E. Shannon used it to study what is now known as information theory in the 1940s. Therefore in mathematics and engineering literature sometimes it is also called WKS sampling theorem after Whittaker, Kotel'nikov and Shannon.

2.2.2 Frequency domain interpretation

The sampling theorem can be proven and derived mathematically. However, a more intuitive understanding of it could be obtained by looking at the sampling process from the frequency domain perspective.

If we consider the sampled signal as an analog signal, it is obvious that the sampling process is equivalent to very drastic chopping of the original signal. The sharp rise and fall of the signal amplitude, just before and after the signal sample instants, introduces a large amount of high frequency components into the signal spectrum.

It can be shown through Fourier transform (which we will discuss in Chapter 4) that the high frequency components generated by sampling appear in a very regular fashion. In fact, every frequency component in the original signal spectrum is periodically replicated over the entire frequency axis. The period at which this replication occurs is determined by the sampling rate.

This replication can easily be justified for a simple sinusoidal signal. Consider a single sinusoid:

$$x(t) = \cos(2\pi f_a t)$$

Before sampling, the spectrum consists of a single spectral line at frequency f_a. Sampling is performed at time instants

$$t = nT, \quad n = 0,1,2,...$$

where n is a positive integer. Therefore, the sampled sinusoidal signal is given by

$$x(nT) = \cos(2\pi f_a nT)$$

At a frequency

$$f = f_a + f_s$$

The sampled signal has value

$$\begin{aligned} x'(nT) &= \cos[2\pi(f_a + f_s)nT] \\ &= \cos[2\pi f_a nT + 2\pi f_s nT] \\ &= \cos[2\pi f_a nT + 2n\pi] \\ &= \cos[2\pi f_a nT] \end{aligned}$$

which is the same as the original sampled signal. Hence, we can say that the sampled signal has frequency components at

$$f = f_a + nf_s$$

This replication is illustrated in Figure 2.5.

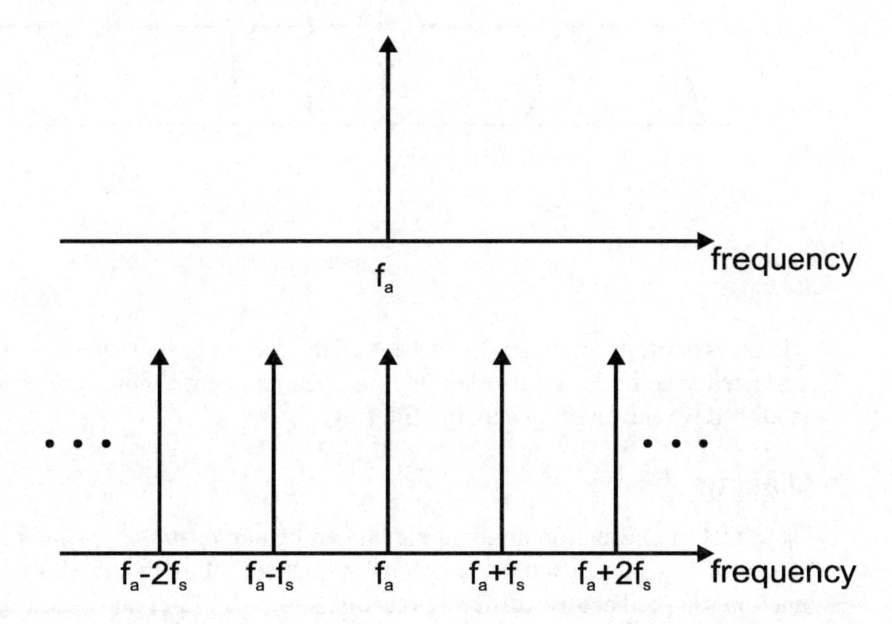

Figure 2.5
Replication of spectrum through sampling

Although it is only illustrated for a single sinusoid, the replication property holds for an arbitrary signal with an arbitrary spectrum. Replication of the signal spectrum for a low-pass bandlimited signal is shown in Figure 2.6.

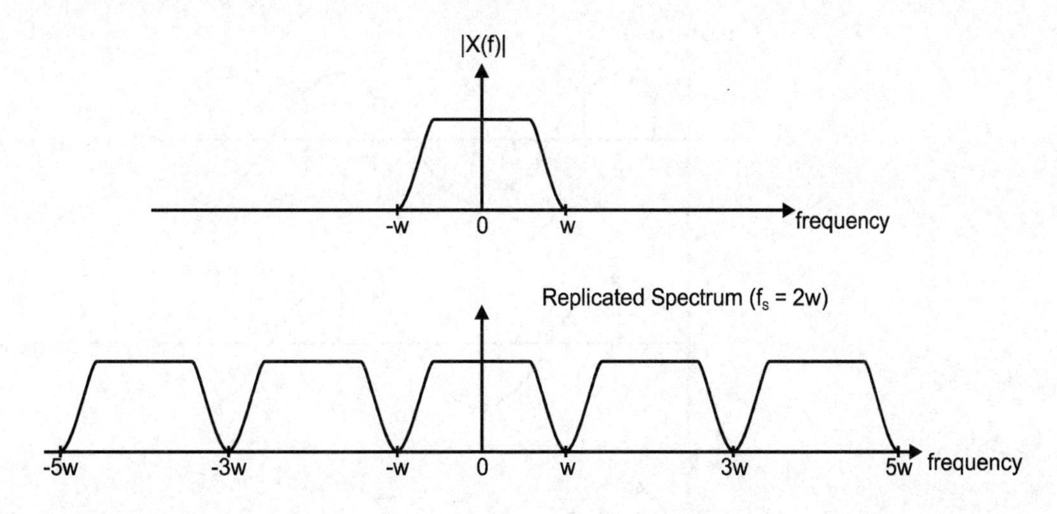

Figure 2.6
The original low-pass spectrum and the replicated spectrum after sampling

Consider the effect if the sampling frequency is less than twice the highest frequency component as required by the sampling theorem. As shown in Figure 2.7, the replicated spectra overlap each other, causing distortion to the original spectrum. Under this circumstance, the original spectrum can never be recovered faithfully. This effect is known as aliasing.

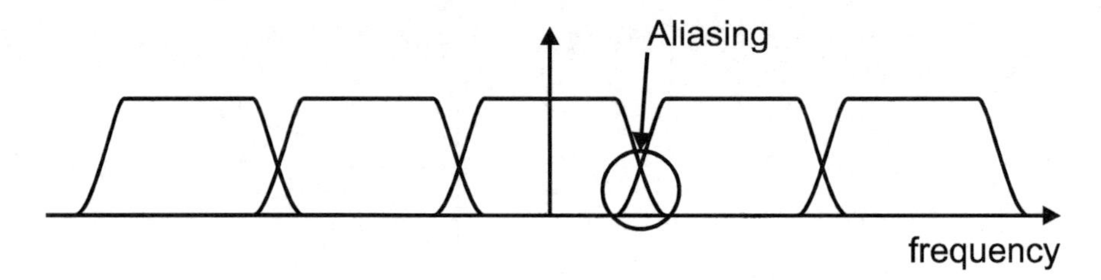

Figure 2.7
Aliasing

If the sampling frequency is at least twice the highest frequency of the spectrum, the replicated spectra do not overlap and no aliasing occurs. Thus, the original spectrum can be faithfully recovered by suitable filtering.

2.2.3 Aliasing

The effect of aliasing on an input signal can be demonstrated by sampling a sine wave of frequency f_a using different sampling frequencies. Figure 2.8 shows such a sinusoidal function sampled at three different rates: $f_s = 4f_a$, $f_s = 2f_a$, and $f_s = 1.5f_a$.

In the first two cases, if we join the sample points using straight lines, it is obvious that the basic 'up–down' nature of the sinusoid is still preserved by the resulting triangular wave as shown in Figure 2.9.

If we pass this triangular wave through a low-pass filter, a smooth interpolated function will result. If the low-pass filter has the appropriate cut-off frequency, the original sine wave can be recovered. This is discussed in detail in section 2.5.

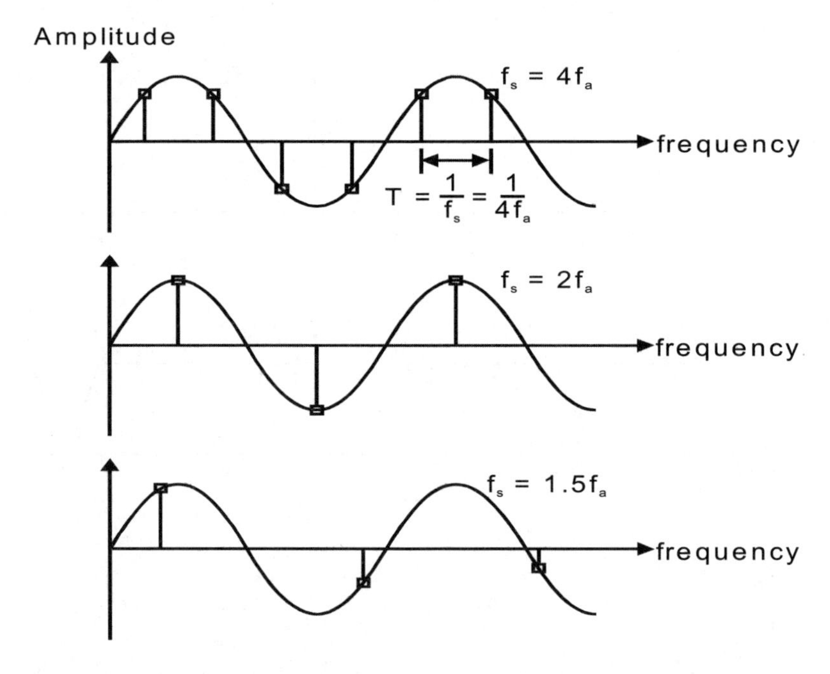

Figure 2.8
A sinusoid sampled at three different rates

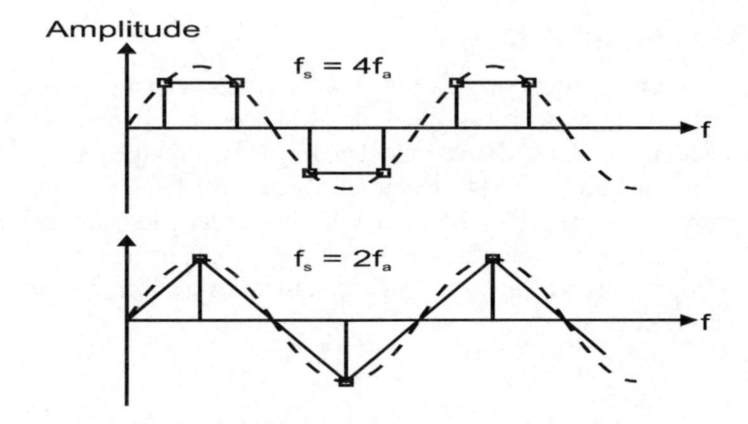

Figure 2.9
Interpolation of sample points with no aliasing

For the last case, the sampling frequency is below the Nyquist rate. We would expect aliasing to occur. This is indeed the case. If we join the sampled points together, it can be observed that the rate at which the resulting function repeats itself differs from the frequency of the original signal. In fact, if we interpolate between the sample points, a smooth function with a lower frequency results, as shown in Figure 2.10.

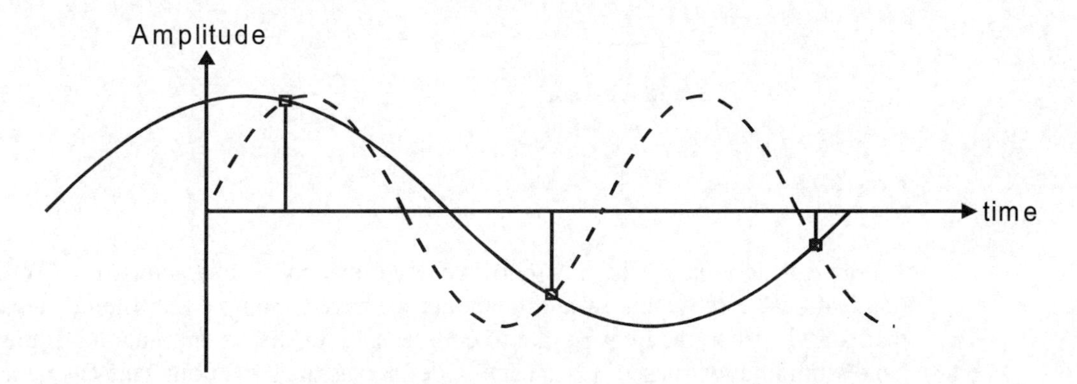

Figure 2.10
Effect of aliasing

Therefore, it is no longer possible to recover the original sine wave from these sampled points. We say that the higher frequency sine wave now has an 'alias' in the lower frequency sine wave inferred from the samples. In other words, these samples are no longer representative of the input signal and therefore any subsequent processing will be invalid.

Notice that the sampling theorem assumes that the signal is strictly bandlimited. In the real world, typical signals have a wide spectrum and are not bandlimited in the strict sense. For instance, we may assume that 20 kHz is the highest frequency the human ears can detect. Thus, we want to sample at a frequency slightly above 40 kHz (say, 44.1 kHz as in compact discs) as dictated by the sampling theorem. However, the actual audio signals normally have a much wider bandwidth than 20 kHz. We can ensure that the signal is bandlimited at 20 kHz by low-pass filtering. This low-pass filter is usually called anti-alias filter.

2.2.4 Anti-aliasing filters

Anti-aliasing filters are always analog filters as they process the signal before it is sampled. In most cases, they are also low-pass filters unless band-pass sampling techniques are used. (Band-pass sampling is beyond the scope of this book.)

The sampling process incorporating an ideal low-pass filter as the anti-alias filter is shown in Figure 2.11. The ideal filter has a flat passband and the cut-off is very sharp. Since the cut-off frequency of this filter is half of that of the sampling frequency, the resulting replicated spectrum of the sampled signal do not overlap each other. Thus no aliasing occurs.

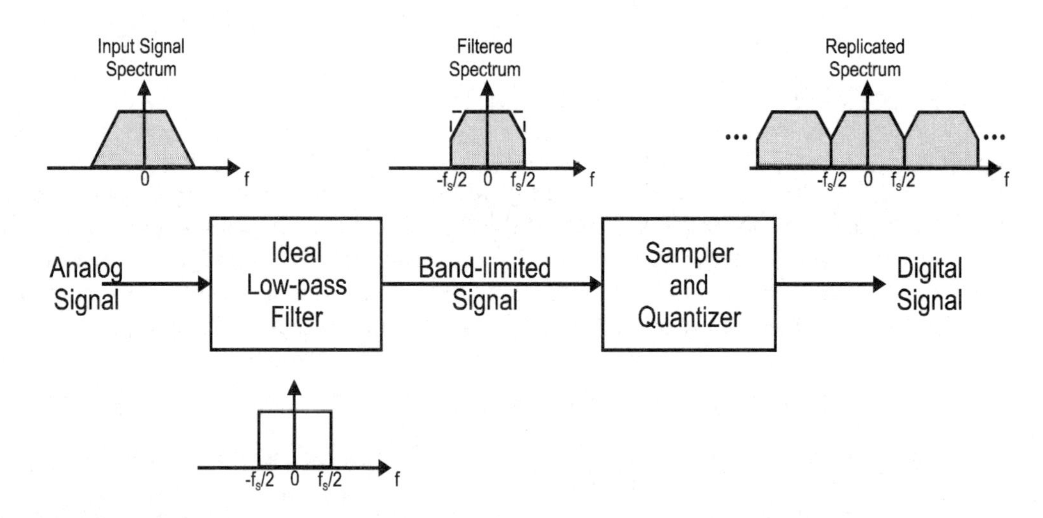

Figure 2.11
The analog-to-digital conversion process with anti-alias filtering

Practical low-pass filters cannot achieve the ideal characteristics. What are the implications? Firstly, this would mean that we have to sample the filtered signals at a rate that is higher than the nyquist rate to compensate for the transition band of the filter. The bandwidth of a low-pass filter is usually defined as the 3 dB point (the frequency at which the magnitude response is 3 dB below the peak level in the passband). However, signal levels below 3 dB are still quite significant for most applications. For the audio signal application example in the previous section, it may be decided that, signal levels below 40 dB will cause insignificant aliasing. The anti-aliasing filter used may have a bandwidth of 20 kHz but the response is 40 dB down starting from 24 kHz. This means that the minimum sampling frequency has to be increased to 48 kHz instead of 40 kHz for the ideal filter.

Alternatively, if we fix the sampling rate, then we need an anti-alias filter with a sharper cut-off. Using the same audio example, if we want to keep the sampling rate at 44.1 kHz, the anti-aliasing filter will need to have an attenuation of 40 dB at about 22 kHz. With a bandwidth of 20 kHz, the filter will need a transition from 3 dB at down to 40 dB within 2 kHz. This typically means that a higher order filter will be required. A higher order filter also implies that more components are needed for its implementation.

2.2.5 Practical limits on sampling rates

As discussed in previous sections, the practical choice of sampling rate is determined by two factors for a certain type of input signal. On one hand, the sampling theorem imposes a lower bound on the allowed values of the sampling frequency. On the other hand, the economics of the hardware imposes an upper bound. These economics include the cost of the analog-to-digital converter (ADC) and the cost of implementing the analog anti-alias filter. A higher speed ADC will allow a higher sampling frequency but may cost substantially more. However, a lower sampling frequency will put a more stringent requirement on the cut-off of the anti-aliasing filter, necessitating a higher order filter and a more complex circuit, which again may cost more.

In real-time applications, each sample is acquired (sampled), quantized and processed by a DSP. The output samples may need to be converted back to analog form. A higher sampling rate will mean that there are more samples to be processed within a certain amount of time. If T_{proc} represents the total DSP processing time, then the time interval between samples T_s will need to be greater than T_{proc}. Otherwise, the processor will not be able to keep up. This means that if we increase the sampling rate we will need a higher speed DSP chip.

2.2.6 Mathematical representation

A mathematical representation of the sampling process (and any other process involved in DSP for that matter) is needed so that we can describe precisely the process and will help us in the analysis of DSP.

The sampling process can be described as a multiplication of the analog signal with a periodic impulse function. This impulse function is also known as the dirac delta function and is usually denoted by $\delta(t)$. It is shown in Figure 2.12.

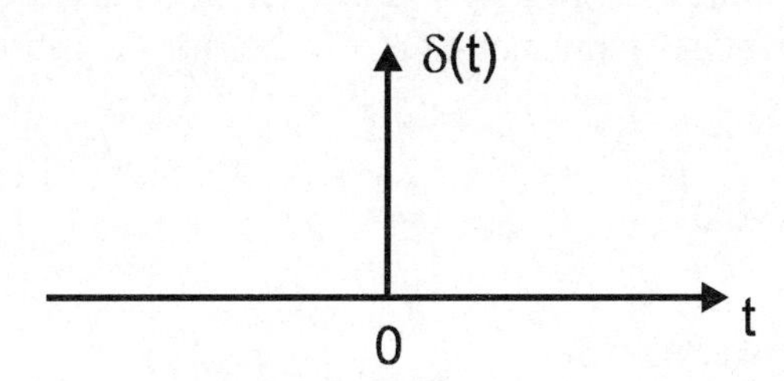

Figure 2.12
The dirac delta function

It can be considered as a rectangular pulse with zero duration and infinite amplitude. It has the property that the energy or the area under the pulse is equal to one. This is expressed as

$$\int_{-\infty}^{\infty} \delta(t)\mathrm{d}t = 1$$

Thus, a weighted or scaled impulse function would be defined as one that satisfies

$$\int_{-\infty}^{\infty} A\delta(t)\mathrm{d}t = A$$

The weighted impulse function is drawn diagrammatically as an arrow with a height proportional to the scaling factor.

The periodic train of impulse functions is expressed as

$$\begin{aligned} s(t) &= \cdots + \delta(t - 2T_s) + \delta(t - T_s) + \delta(t) \\ &\quad + \delta(t + T_s) + \delta(t + 2T_s) + \cdots \\ &= \sum_{n=-\infty}^{\infty} \delta(t - nT_s) \end{aligned}$$

where T_s is the amount of time between two impulses. In terms of sampling, it is the sampling period.

If the input analog signal is denoted by $f(t)$, then the sampled signal is given by

$$\begin{aligned} y(t) &= f(t) \cdot s(t) \\ &= \sum_{n=-\infty}^{\infty} f(t) \cdot \delta(t - nT_s) \end{aligned}$$

or the samples of the output of the sampling process are

$$y(nT_s) = f(nT_s) \cdot \delta(t - nT_s)$$

Sometimes the sampling period is understood and we just use $y(n)$ to denote $y(nT_s)$. This mathematical representation will be used again and again in later chapters of this course.

2.3 Quantization

2.3.1 Sample-and-hold

The next step in the process of converting an analog signal into digital form is the discretization of the sampled signal amplitude or quantization. In practice, because the quantization process takes a finite amount of time, the sampled signal amplitude has to be held constant during this time. The sampling process is usually performed by a sample-and-hold circuit, which can be logically represented as in Figure 2.13. The analog-to-digital converter performs the quantization process.

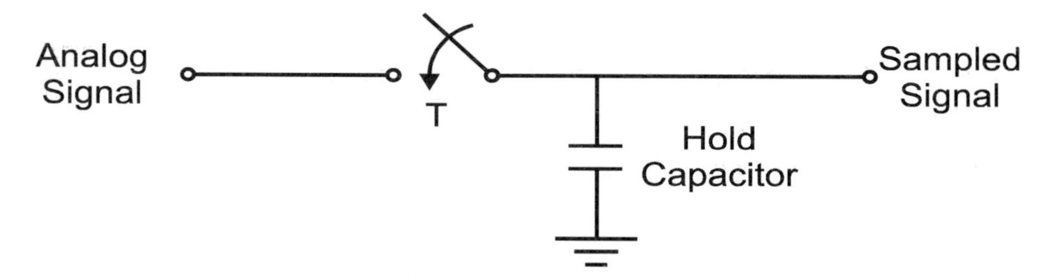

Figure 2.13
Sample and hold circuit

The hold capacitor holds the sampled measurement of the analog signal $x(nT)$ for at most T seconds during which time a quantized value $x_Q(nT)$ is available at the output of the analog-to-digital converter, represented as a B-bit binary number. The sample-and-hold and the ADC may be separate modules or may be integrated on the same chip. Typically, the very fast ADCs require an external sample-and-hold device.

2.3.2 Uniform quantization

The ADC assumes that the input values cover a full-scale range, say R. Typical values of R are between 1 to 15 volts. Since the quantized sampled value $x_Q(nT)$ is represented by B-bits, it can take on only one of 2^B possible quantization levels. If the spacing between these levels is the same throughout the range R, then we have a uniform quantizer. The spacing between quantization levels is called quantization width or the quantizer resolution.

For uniform quantization, the resolution is given by

$$Q = \frac{R}{2^B}$$

The number of bits required to achieve a required resolution of Q is therefore

$$B = \log_2 \frac{R}{Q}$$

Most ADCs can take bipolar inputs, which means the sampled values lie within the symmetric range

$$-\frac{R}{2} \le x(nT) < \frac{R}{2}$$

For unipolar inputs,

$$0 \le x(nT) < R$$

In practice, the input signal $x(t)$ must be preconditioned to lie within the full-scale range of the quantizer. Figure 2.14 shows the quantization levels of a 3-bit quantizer for bipolar inputs.

For a review of the possible binary representations for the quantized output value, see Appendix A.

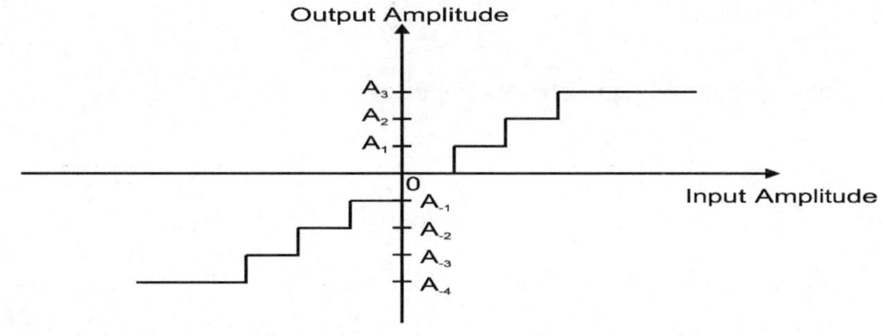

Figure 2.14
A uniform 3-bit quantizer transfer function

Quantization error is the difference between the actual sampled value and the quantized value. Mathematically, this is

$$e(nT) = x(nT) - x_Q(nT)$$

or equivalently,

$$e(n) = x(n) - x_Q(n)$$

If $x(n)$ lies between two quantization levels, it will either be rounded up or truncated. Rounding replaces $x(n)$ by the value of the nearest quantization level. Truncation replaces $x(n)$ by the value of the level below it.

For rounding, the error is given by

$$-\frac{Q}{2} < e < \frac{Q}{2}$$

where as for truncation, it is

$$0 \leq e < Q$$

It is obvious that rounding produces a less biased representation of the analog values. The average error is given by

$$\overline{e} = \frac{1}{Q}\int_{-Q/2}^{Q/2} e\mathrm{d}e = 0$$

which means that on average half the values are rounded up and half rounded down.

The mean-square value of the error gives us an idea of the average power of the error signal. It is given by

$$\begin{aligned}\overline{e^2} &= \frac{1}{Q}\int_{-Q/2}^{Q/2} e^2\mathrm{d}e \\ &= \frac{Q^2}{12}\end{aligned}$$

The root-mean-square error is therefore

$$\begin{aligned}e_{\mathrm{rms}} &= \sqrt{\overline{e^2}} \\ &= \frac{Q}{\sqrt{12}}\end{aligned}$$

The signal-to-quantization-noise ratio is

$$
\begin{aligned}
SQNR &= 20\log_{10}\left[\frac{R}{Q}\right] \\
&= 20\log_{10}\left(2^{B}\right) \\
&= 20B\log_{10}2 \\
&= 6B \ \ \text{dB}
\end{aligned}
$$

Thus if we increase the number of bits of the ADC by one, the signal to quantization noise ratio improves by 6 dB. The above equation gives us the dynamic range of the quantizer.

Example: 2.1
The dynamic range of the human ear is about 100 dB. If a digital audio system is required to match this dynamic range, it will require

$$100/6 = 16.67 \ \text{bits}$$

A 16-bit quantizer will achieve a dynamic range of 96 dB.
If the highest frequency the human ear can hear is 20 kHz, then a sampling rate of at least 40 kHz is required. If the actual sampling rate is 44 kHz, then the bit rate of this system will be

$$16.44 = 704 \ \text{kbits/sec}$$

This is the typical bit rate of a compact disc player.
Since the quantization error is a random number within the range given, it is usually modeled as a random signal (or noise) with a uniform distribution as shown in Figure 2.15.

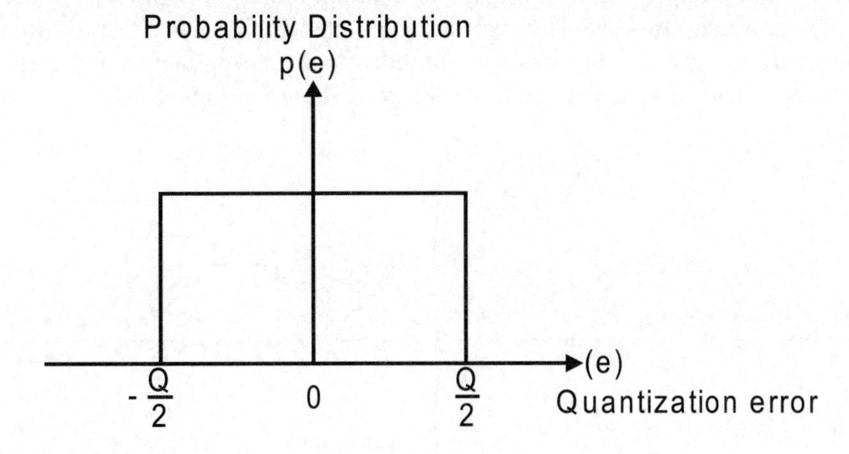

Figure 2.15
Uniform distribution of quantization error

The quantized signal is then modeled as the analog sampled signal with additive quantization noise as in Figure 2.16. This quantization noise is generally assumed to be a zero-mean, uniformly distributed white noise that is uncorrelated with the input signal.

This assumption is generally true for signals that vary through the entire full-scale range and the quantizer has a large number of levels.

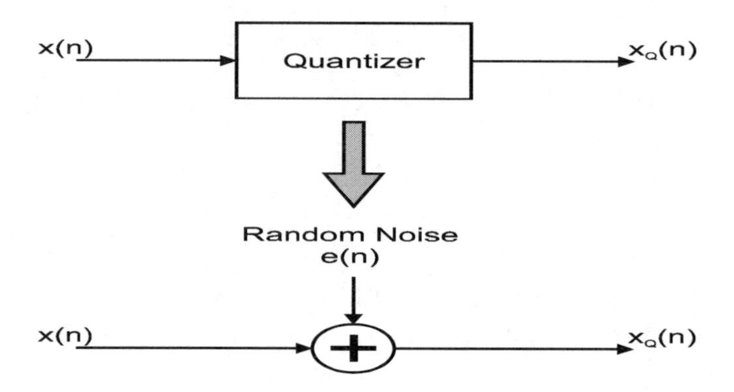

Figure 2.16
Mathematical model of quantization noise

2.3.3 Non-uniform quantization

One of the assumptions we have made in analyzing the quantization error is that the sampled signal amplitude is uniformly distributed over the full-scale range. This assumption may not hold for certain applications. For instance, speech signals are known to have a wide dynamic range. Voiced speech (e.g. vowel sounds) may have amplitudes that span the entire full-scale range, while softer unvoiced speech (e.g. consonants such as fricatives) usually have much smaller amplitudes. In addition, an average person only speaks 60% of the time while she/he is talking. The remaining 40% are silence with negligible signal amplitude.

If uniform quantization is used, the louder voiced sounds will be adequately represented. However, the softer sounds will probably occupy only a small number of quantization levels with similar binary values. This means that we would not be able to distinguish between the softer sounds. As a result, the reconstructed analog speech from these digital samples will not nearly be as intelligible as the original.

To get around this problem, non-uniform quantization can be used. More quantization levels are assigned to the lower amplitudes while the higher amplitudes will have less number of levels. This quantization scheme is shown in Figure 2.17.

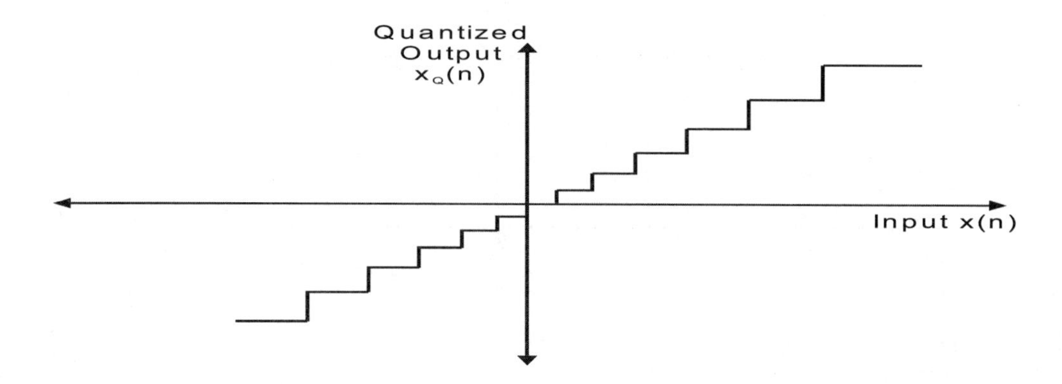

Figure 2.17
Non-uniform quantization

Alternatively, a uniform quantizer can still be used, but the input signal is first compressed by a system with an input–output relationship (or transfer function) similar to that shown in Figure 2.18.

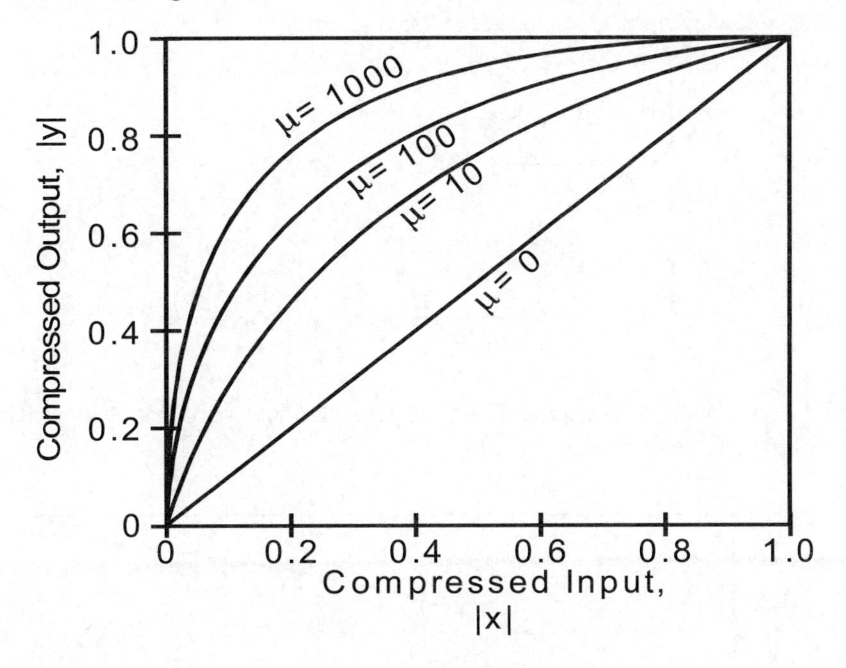

Figure 2.18
μ-law compression characteristics

The higher amplitudes of the input signal are compressed, effectively reducing the number of levels assigned to it. The lower amplitude signals are expanded (or non-uniformly amplified), effectively making it occupy a large number of quantization levels. After processing, an inverse operation is applied to the output signal (expanding it). The system that expands the signal has an input–output relationship that is the inverse of the compressor. The expander expands the high amplitudes and compresses the low amplitudes. The whole process is called companding (COMpressing and exPANDING).

Companding is widely used in public telephone systems. There are two distinct companding schemes. In Europe, A-law companding is used and in the United States, μ-law companding is used.

μ-law compression characteristic is given by the formula:

$$y = y_{\max} \frac{\ln\left[1+\mu\left(\frac{|x|}{x_{\max}}\right)\right]}{\ln(1+\mu)} \operatorname{sgn}(x)$$

where

$$\operatorname{sgn}(x) = \begin{cases} +1, & x \geq 0 \\ -1, & x < 0 \end{cases}$$

Here, x and y represent the input and output values, and $x_{\max}$ and $y_{\max}$ are the maximum positive excursions of the input and output, respectively. μ is a positive constant. The

North American standard specifies μ to be 255. Notice that $\mu = 0$ corresponds to a linear input–output relationship (i.e. uniform quantization). The compression characteristic is shown in Figure 2.18.

The A-law compression characteristic is given by

$$y = \begin{cases} y_{\max} \dfrac{A\left(\dfrac{|x|}{x_{\max}}\right)}{1+\ln A}\operatorname{sgn}(x), & 0 < \dfrac{|x|}{x_{\max}} \le \dfrac{1}{A} \\[2ex] y_{\max} \dfrac{1+\ln\left[A\left(\dfrac{|x|}{x_{\max}}\right)\right]}{1+\ln A}\operatorname{sgn}(x), & \dfrac{1}{A} < \dfrac{|x|}{x_{\max}} < 1 \end{cases}$$

Here, A is a positive constant. The European standard specifies A to be 87.6. Figure 2.19 shows the characteristic graphically.

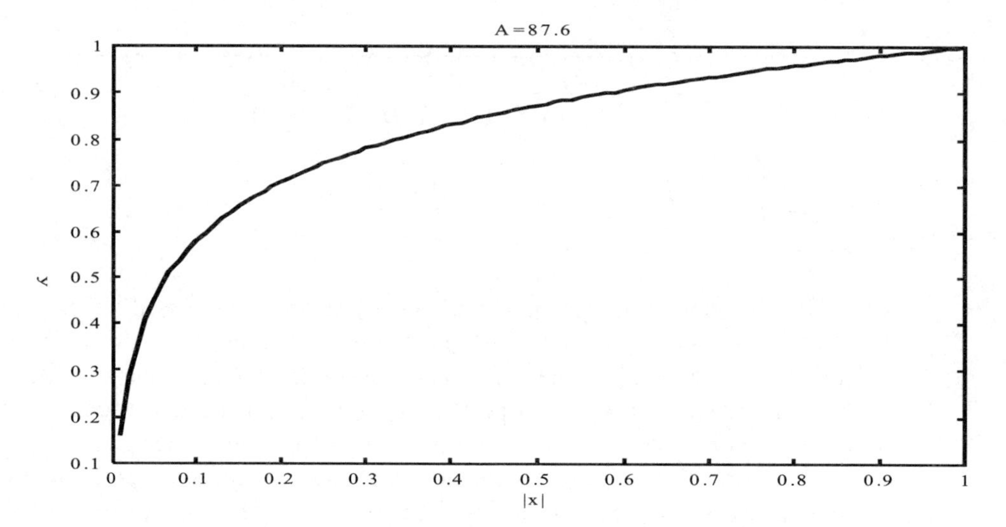

Figure 2.19
The A-law compression characteristics

2.3.4 Dithering

Another assumption we have made in analyzing quantization noise is that, it is assumed to be uniformly distributed over the quantization width. If the noise is not uniformly distributed quantization distortion results.

We shall illustrate quantization distortion through an example. A low amplitude sinusoid is being sampled and quantized. The samples of the sinusoid are given by

$$x(n) = A\cos(2\pi f_0 n)$$

where A is less than the quantization resolution. Let

$$f_s = 40 \text{ samples per cycle}$$

and

$$A = 0.75Q$$

So for a 1 kHz sinusoid, the actual sampling rate is 40 kHz. Figure 2.20(a) shows the original and the quantized signals.

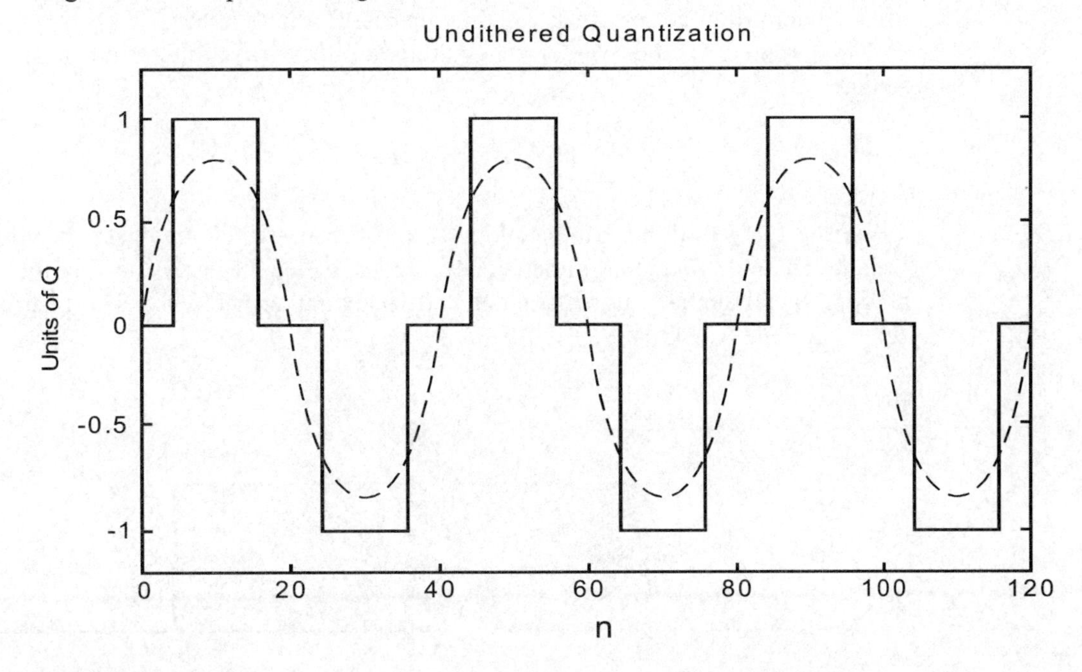

Figure 2.20 (a)
The original and quantized signal

Note that the quantized signal only occupies three of the available quantization levels. The frequency spectrum of this quantized signal is shown in Figure 2.20(b).

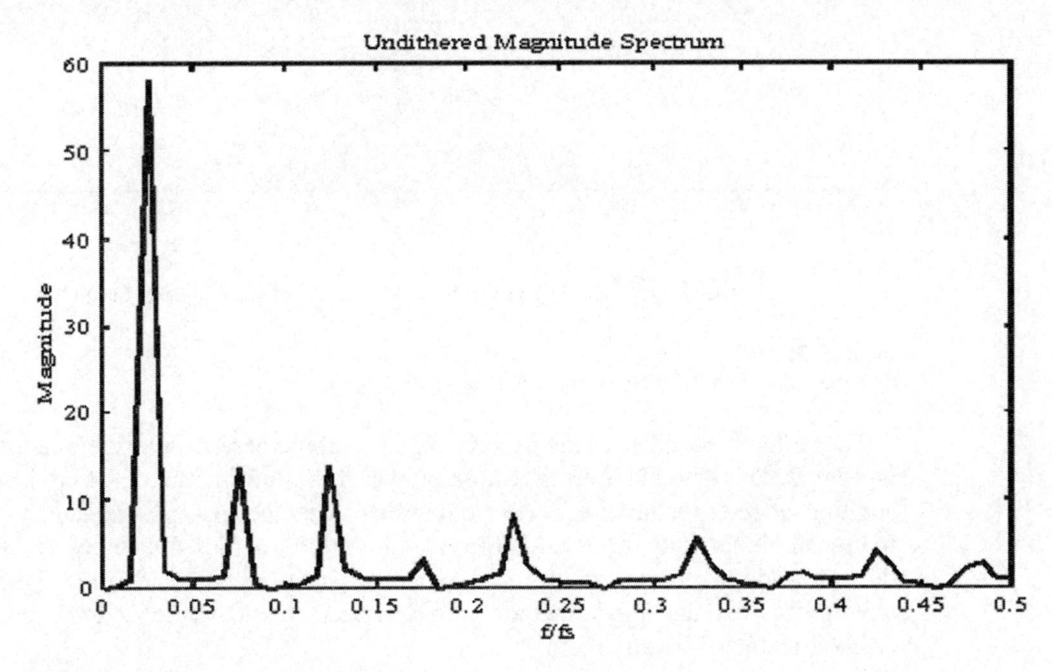

Figure 2.20 (b)
The quantized signal spectrum

It has peaks at f_0, and the odd harmonic frequencies $3f_0$, $5f_0$, etc. Clearly, the odd harmonics are artifacts of the quantization process and can be considered as the spectrum of the quantization noise signal, which in this case, is not white.

This problem can be overcome by adding a dither $v(n)$ to the original sampled signal so that

$$y(n) = x(n) + v(n)$$

Various types of dither can be added. Two of them, which are of practical interest, are rectangular and triangular dither. They are so called because the distribution of the random signal samples is rectangular and triangular respectively. The distributions are shown in Figure 2.21.

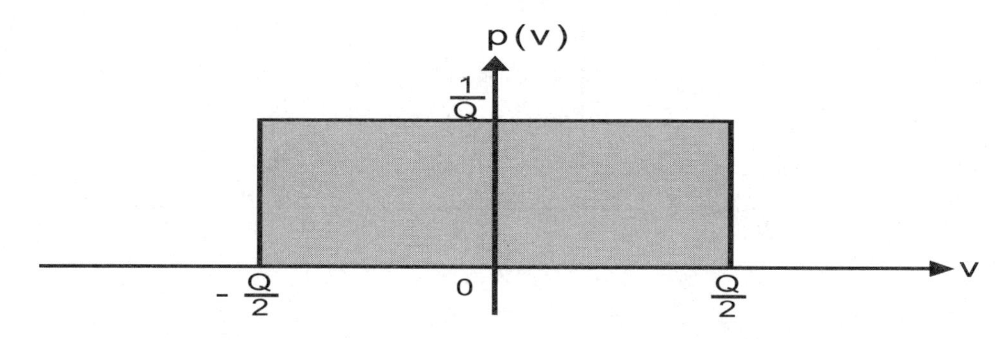

(a) Rectangular dither distribution

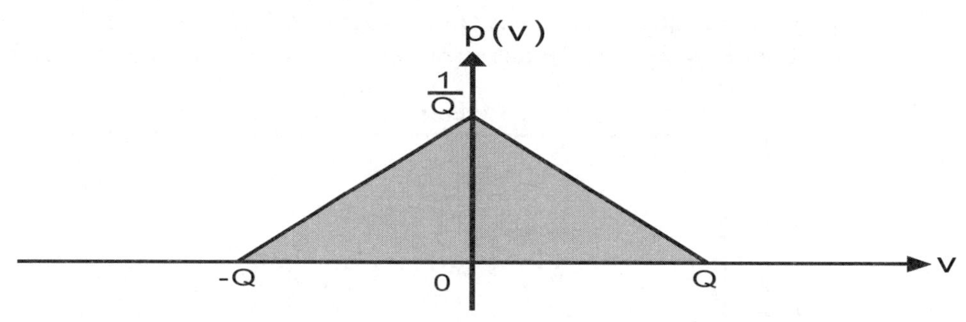

(b) Triangular dither distribution

Figure 2.21
Amplitude distributions of rectangular and triangular dither

The addition of dither to the original signal will increase its average quantization noise power. Recall that the average noise power for uniform quantization is $Q^2/12$. The addition of rectangular dither will double this average noise power and the addition of triangular dither will triple it. However, if we look at the frequency spectrum of the dithered and quantized signal of the example we have been considering (Figure 2.22), we will notice that the noise spectrum now appears to be white and the odd harmonic artifacts are not there any more.

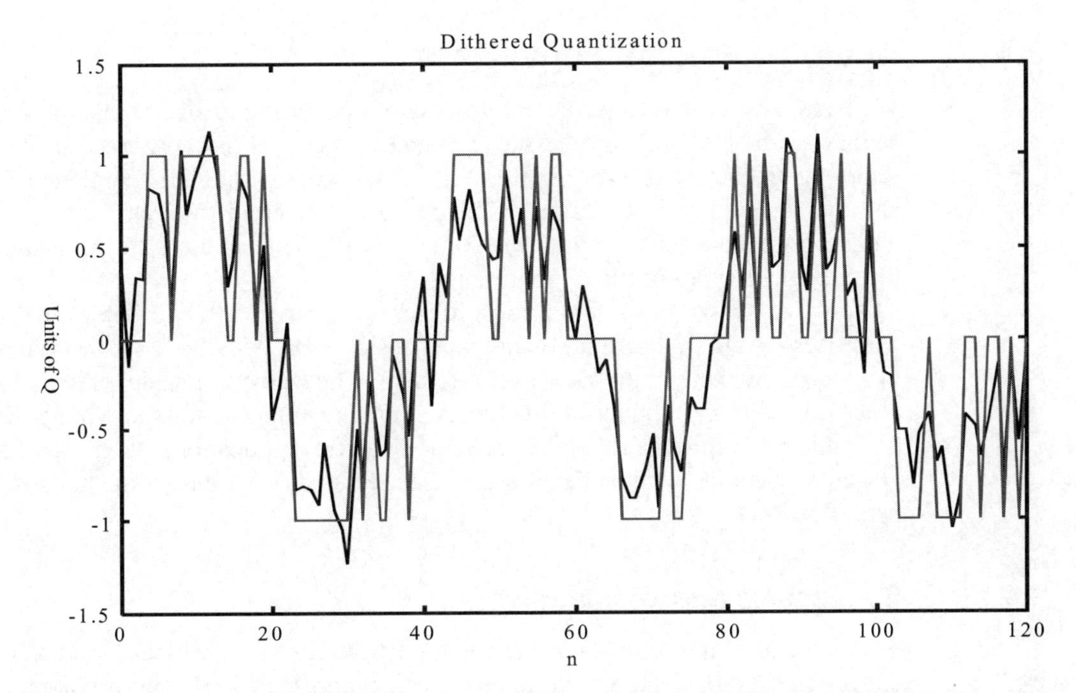

Figure 2.22(a)
The dithered signal and its quantized version

It must be emphasized that in general, the sampling process will cause all the odd harmonics that lie outside of the Nyquist interval (out-of-band harmonics) to be aliased back into the interval (in-band non-harmonic frequencies). Therefore, the overall spectrum will contain peaks at frequencies other than the odd harmonics.

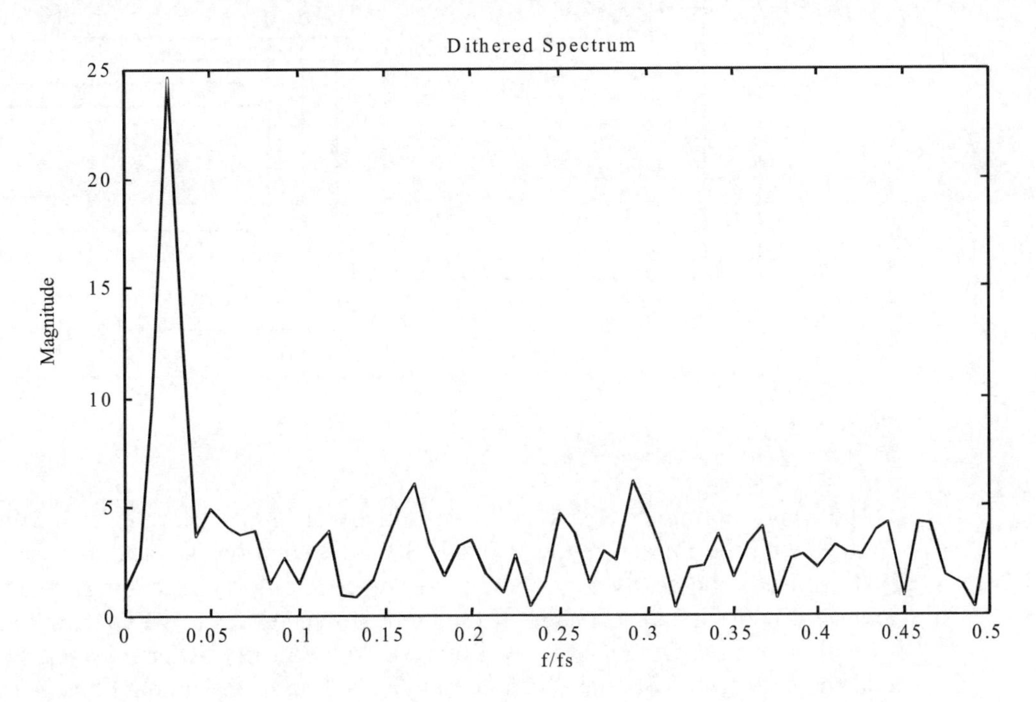

Figure 2.22(b)
Quantization noise spectrum with dithering

2.4 Analog-to-digital converters

We have covered the fundamental process of converting analog signals to digital format so that it can be digitally processed. The analog signal is first low-pass filtered to half the sampling frequency to prevent aliasing. It then goes through a sample-and-hold device and the sampled amplitudes are quantized and converted to binary values. This binary number is represented by n bits where n is typically 8, 10, 12 and 16. Appendix B reviews the three main types of binary representation.

Now we shall take a brief look at some commercially available analog-to-digital converters (ADCs). There are many varieties of ADCs available on the market. Most of them contain the sample-and-hold circuitry. They cover a wide range of conversion speeds, resolution (number of bits representing the output) and input voltage range. Some are general purpose and others are for specific applications such as video signals. Different methods of quantization are used. Four most common methods are discussed here.

2.4.1 Successive approximation

The successive approximation ADC is built from three main blocks: an analog-to-digital converter (DAC), a successive approximation register (SAR) and a comparator. Figure 2.23 shows how these three blocks are connected.

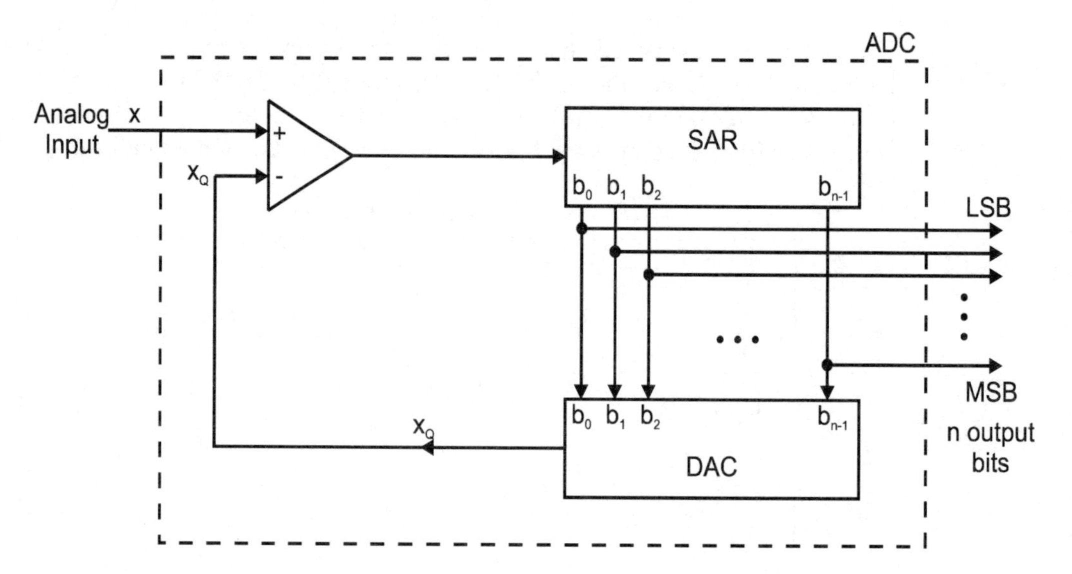

Figure 2.23
Successive approximation converter

The conversion process is as follows. Initially all n bits are reset to zero in the SAR. Starting with the most significant bit (MSB) b_{n-1}, each bit is set to 1 in sequence. The DAC converts the newly formed binary number into a corresponding voltage, which is compared with the input voltage. If the input voltage exceeds the DAC output, then that bit will be left *on*. Otherwise, it will be reset to zero (off). After n cycles, the SAR will hold the correct bit pattern, which is then latched on to the output lines. This technique thus keeps splitting the voltage range in half to determine where the input voltage lies. Alternatively, we can say that the successive approximation algorithm performs a binary search through the quantization levels.

Example: 2.2
Convert analog values $x = 0.2$ and $x = -0.7$ volts to their offset binary representations using successive approximation. Assume a 3-bit quantizer with a range $R = 2$ V.

For 3-bit quantization the conversion will be done in 3 cycles. The bit that is tested, the corresponding quantized value of the intermediate bit pattern and the test results are tabulated below for $x = 0.2$ V.

Cycle	Test bit	$b_2b_1b_0$	x_Q	Test result
1	b_2	100	0.00	1
2	b_1	110	0.50	0
3	b_0	101	0.25	0

For the test result column, a '1' indicates the input is larger than or equal to the DAC output and a '0' otherwise. In cycle 2, the test result is a '0'; thus the SAR will reset b_1 back to zero. Similarly in cycle 3, b_0 is reset to zero, resulting in the output of '100' representing the quantized voltage of 0.00.

The following table reflects the conversion process for $x = -0.7$ V.

Cycle	Test bit	$b_2b_1b_0$	x_Q	Test result
1	b_2	100	0.00	0
2	b_1	010	–0.50	0
3	b_0	011	–0.75	1

The resulting quantized value is –0.75 V and is encoded as '011'.

Notice that in the example, both values are truncated down to the lower level. If rounding to the nearest level is desired, then the input value x must be shifted by half the spacing (resolution) between levels. That is, obtain the shifted value y by

$$y = x + Q/2$$

and quantize y by successive approximation.

Many ADCs also give a two's complement output. If this is the output format required, the successive approximation algorithm has to be slightly modified. This is because the MSB (i.e. the sign bit) must be treated separately from the other bits. If the input value is greater than zero, the MSB must be set to '0'; otherwise, it is a '1'. Note that this is the opposite of the offset binary case. The remaining bits are tested in the usual manner.

Example: 2.3
Perform the quantization as in the previous example but use two's complement representation.

For $x = 0.2$, the process is illustrated in the following table:

Cycle	Test bit	$b_2b_1b_0$	x_Q	Test result
1	b_2	000	0.00	1
2	b_1	010	0.50	0
3	b_0	001	0.25	0

The resulting binary value is '000'.

The following table illustrates the quantization of $x = -0.7$ to two's complement:

Cycle	Test bit	$b_2b_1b_0$	x_Q	Test result
1	b_2	000	0.00	0
2	b_1	110	–0.50	0
3	b_0	101	–0.75	1

The two's complement representation is '101'.

Notice that complementing the MSB (sign bit) will give us the offset binary representation.

A large number of ADCs that operate at sampling frequencies of 1 MHz or less make use of successive approximation.

2.4.2 Dual slope ADC

If high resolution is desired, then the dual slope conversion technique can be employed. A key element of the dual slope ADC is a capacitor. At the start of the conversion cycle, the capacitor is totally discharged (i.e. the capacitor voltage is zero). It is then charged for a certain set time by the input voltage. After this set time the capacitor is switched to a known negative reference voltage and is slowly discharged until the capacitor voltage reaches zero volt. The time taken for the discharge process is recorded using a digital counter. With the counter initially set to zero, the final counter value is proportional to the input voltage. The binary counter value is the converted binary output.

High-resolution conversion can be achieved by simply using a more accurate counter, which is relatively easy to implement. Another advantage of this process is that component value variations will have no effect on the accuracy. For instance, the capacitance may change due to temperature variation. But since the charging and discharging processes are done through the same capacitor, the net effect of this capacitance variation is negligible.

The major disadvantage is that the charging and discharging of capacitors takes a relatively long time. So, this process is normally reserved for high resolution, low sampling frequency ADCs.

2.4.3 Flash ADC

For n-bit quantization, the successive approximation technique requires n cycles. If fast conversion time is required, the comparisons will have to be performed in parallel and at the same time. In flash ADCs, the input voltage is compared with a set of reference voltages at the same time. A ladder of resistors with equal resistances sets these reference voltages.

For an n-bit converter, we need 2^n resistors. The voltages tapped from the terminals of these resistors are then compared with the input voltage and the digital output encoded. Figure 2.24 shows a 2-bit flash ADC.

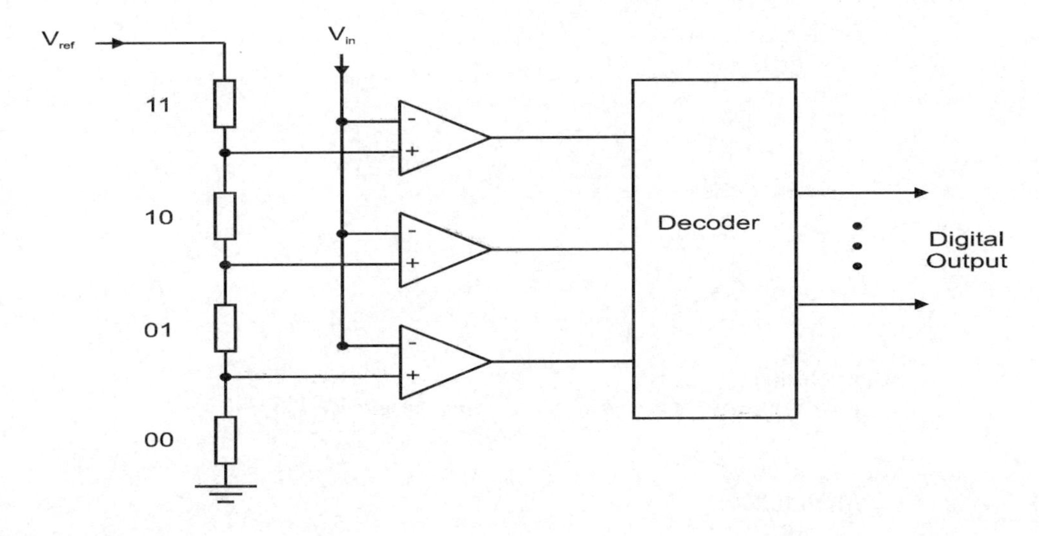

Figure 2.24
A 2-bit flash ADC

In practice, each resistor in the ladder must be matched and laser trimmed to the same value for accuracy. This is a very costly process. Thus flash ADCs are usually expensive and are available only up to 8 bits (with 256 resistors).

2.4.4 Sigma-delta ADC

The concept of sigma-delta converters is significantly different to the above three techniques. Sigma-delta ADC features a very low-resolution quantizer but operates with a sampling rate much higher than the Nyquist rate. Two important techniques, called over sampling and quantization noise shaping, are being used in sigma-delta converters to trade the quantizer resolution with sampling rate. These techniques cleverly apply the theory and concepts we have discussed earlier in this chapter.

2.4.4.1 Oversampling

Oversampling refers to a sampling rate that is higher than the minimum required, which is the Nyquist rate. Usually power of 2 multiples of the Nyquist rates are being used. If the sampling rate used is R times the Nyquist rate:

$$f_s = Rf_{\text{Nyquist}}$$

Then R is called the oversampling ratio. We shall look at the two main advantages of oversampling.

Now consider a particular audio system that uses a sampling frequency of 48 kHz. In this case, we need an anti-aliasing filter that attenuates all frequencies above 24 kHz by, say, 96 dB that is equivalent to a full 16-bit resolution. The cut-off frequency of the filter has to be lower than 24 kHz. Let's assume that we design the filter cut-off to be at 18 kHz.

Figure 2.25(a) shows the frequency response of this filter.

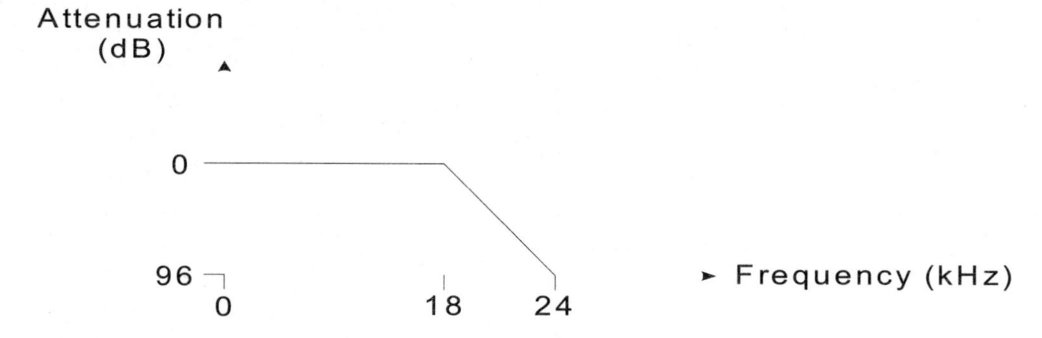

Figure 2.25(a)
Frequency responses of the initializing filters for the critically sampled

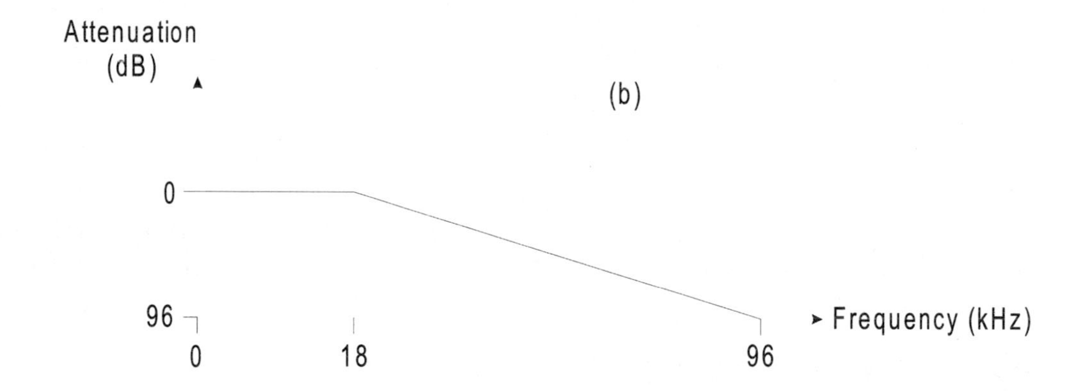

Figure 2.25(b)
Frequency responses of the initializing filters for the 4 times oversampled systems

If we oversample the signal by 4 times, then the sampling frequency becomes 192 kHz. The required attenuation of the anti-aliasing filter is now 96 dB at 96 kHz, instead of 24 kHz. If the cut-off frequency remains at 18 kHz, then the roll-off of this filter is much more gradual than the one before. The frequency response of the filter in this case is shown in Figure 2.25(b).

This means that if the sampling frequency is increased, then the requirements on the anti-aliasing filter are relaxed. In other words, a lower order filter can be used. Recalling that the anti-aliasing filter is an analog filter, a lower order filter translates to less complex analog circuitry and is therefore much easier to maintain.

The second advantage of oversampling is that the quantization noise power is now spread over a much larger frequency range. More precisely, if f_s is the sampling frequency, noise power is spread from 0 to $f_s/2$. Figure 2.26 shows the quantization noise levels of the previous system with and without oversampling. If the numbers of quantization levels remain the same, the same amount of quantization noise is present in both systems. However, in the oversampled system, this amount of quantization noise is spread over a much wider frequency range. Since the input signal occupies a frequency band much narrower than this range, the noise power affecting the input signal is lower. Furthermore, the quantization noise beyond the signal spectrum can now be filtered out using an appropriate digital filter.

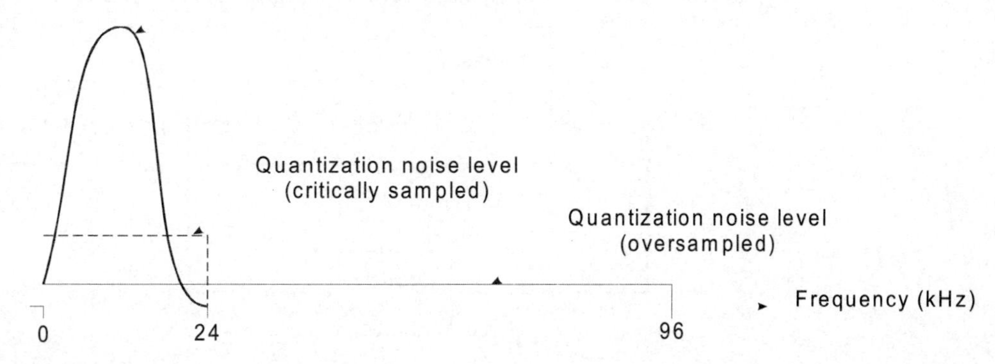

Figure 2.26
Quantization noise levels of the critically sampled and oversampled systems

The amount of quantization noise that remains after digital filtering is now reduced by a factor equal to the oversampling ratio. Thus with 4 times oversampling, the quantization noise is 1/4 of what was before.

$$Q_{\text{os}} = \frac{Q_{\text{Nyquist}}}{4}$$

In decibels, this is

$$Q_{\text{os,db}} = Q_{\text{Nyquist,db}} - 10\log 4$$

and so the quantization noise is reduced by 10log4 = 6.02 dB. Recall that if we increase the quantizer resolution by 1 bit, quantization noise is reduced by about 6 dB. So oversampling by a factor of 4 is equivalent to increasing the quantizer word size by 1 bit.

In the example that we have been considering, the digital filter can be a high-order one with sharp cut-off at 24 kHz. Thus, the quantization noise between 24 kHz and 96 kHz is now filtered out. Since the significant portion of the spectrum is now between 0 and 24 kHz, the output of the digital filter can now be decimated by a factor of 4. This is achieved by retaining only 1 out of 4 samples at the output of the digital filter. In this way, the remainder part of the DSP system will be operating only at the nyquist rate, not at the oversampling rate, reducing the computational requirements.

The use of oversampling method in sigma-delta converter is illustrated in Figure 2.27.

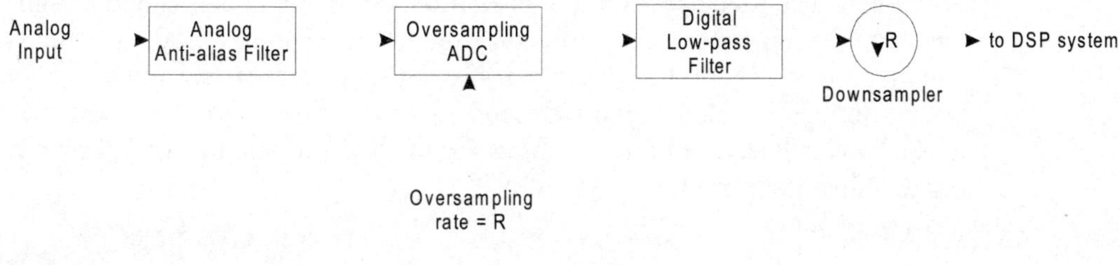

Figure 2.27
Oversampling method in DSP front end

2.4.5 Quantization noise shaping

The second important technique that sigma-delta converters use is quantization noise shaping. The idea is to high-pass filter the quantization noise, so that the required oversampling ratio is reduced for a certain increase in resolution.

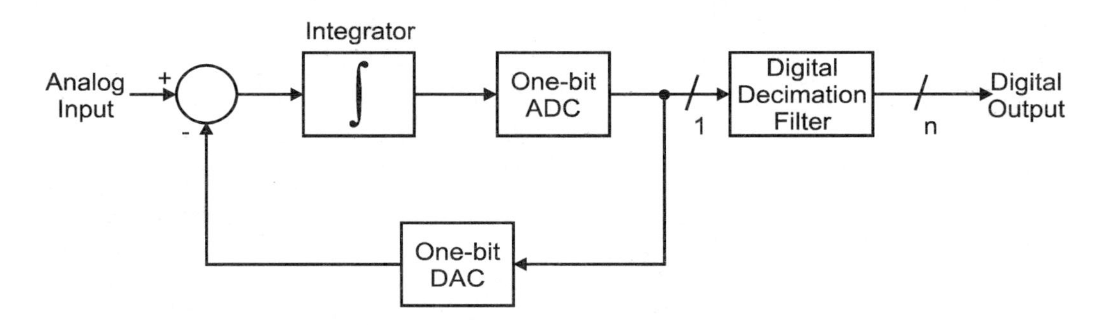

Figure 2.28
A first order sigma-delta ADC

A block diagram of a sigma-delta converter is shown in Figure 2.28. The term 'sigma-delta' comes from the fact that there is a summation point (sigma) and a delta modulator (integrator and 1-bit quantizer). One-bit quantization is assumed here. It is called delta modulation. It quantizes the difference between successive samples of the signal rather than the absolute value of each sample.

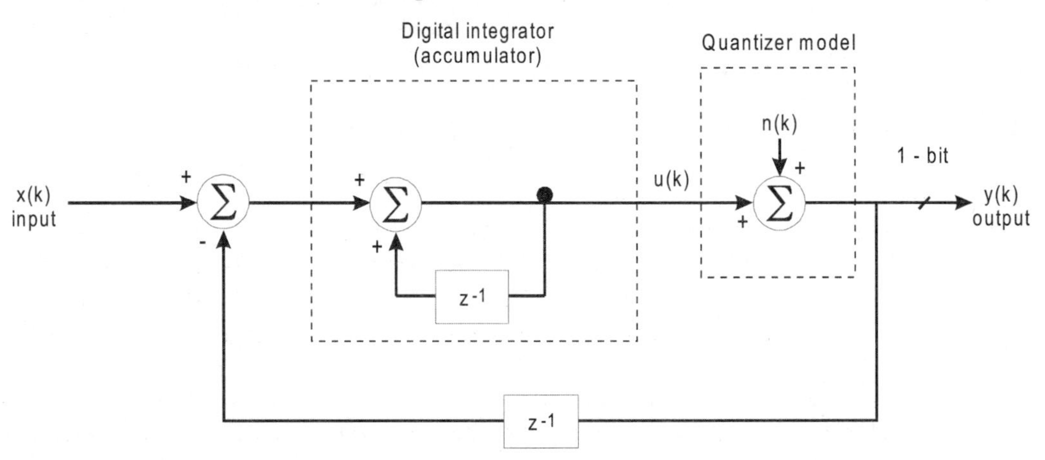

Figure 2.29
Linearize first order sigma-delta converter model

Analysis of this system can be quite involved. The noise performance is frequency dependent. The loop acts as a low-pass filter for the input signal and a high-pass filter for the quantization noise. To appreciate the noise shaping function, the system can be linearized (see Figure 2.29) by replacing the quantizer by the model we have discussed before. In this model, a signal independent white noise source represents the quantization noise. The input and output are now digital and the analog integrator is replaced by a digital integrator such that

$$z(k) = u(k) - z(k-1)$$

Some manipulation of the difference equations gives us the output sample in terms of the input and the quantization noise

$$y(k) = x(k) + n(k) - n(k-1)$$

The noise terms on the right-hand side of this equation are simply a difference between the present and the previous samples of the noise. If this difference is small, which means that the noise signal is changing slowly, the combined noise terms will give a small value. If the difference is large the value of the combined noise terms is large. A slowly changing signal has low frequency components while a rapidly changing signal has high frequency components. Thus $n(k)–n(k–1)$ acts as a high-pass filter of the noise.

In Figure 2.26 we have illustrated the effects of oversampling on the quantization noise level. But so far we have assumed that the noise is uniformly distributed over the frequency range. With high-pass filtering of the noise, the distribution of noise will no longer be uniform. It will now be lower at the low frequency end, with the level increasing towards the high frequency end as shown in Figure 2.30. Recalling that the noise above the Nyquist frequency will eventually be filtered out, the overall quantization noise level within the signal spectrum will be further reduced by the sigma-delta loop.

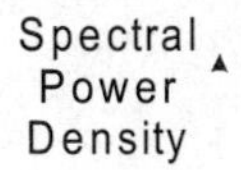

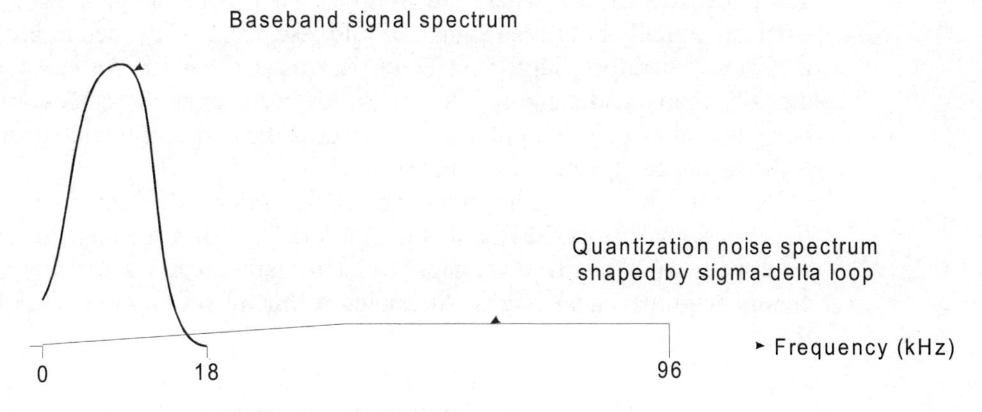

Figure 2.30
Quantization noise spectrum shaped by the sigma-delta technique

As a rule-of-thumb, the first order sigma-delta loop will increase the effective resolution of the system by 0.5 bits.

The 1-bit output of the quantizer is decimated, which is a reduction in sampling rate. This rate reduction is performed by averaging a block of *n*-bits to produce a one-bit output. This averaging process is illustrated in Figure 2.31. This rate reduction is equivalent to filtering in the frequency domain and is discussed at a later chapter.

Output from One-bit ADC :

0 0 1 0 1 1 0	0 1 0 1 0 1 1	1 1 1 0 1 0 1	1 1 0 1 0 1 0
4 x 0 , 3 x 1 = 0	3 x 0 , 4 x 1 = 1	2 x 0 , 5 x 1 = 1	3 x 0 , 4 x 1 = 1

Output from Decimation Filter (÷7) :

0	1	1	1

Figure 2.31
Averaging a block of 7 bits to produce 1 bit

As we have seen, the majority of the processes involved in sigma-delta conversion are digital processes. This means that the chip contains mostly of digital circuitry as opposed to the other three techniques, which have a significant portion of analog circuitry. Thus, sigma-delta ADCs are generally more reliable and stable. It is also possible for this type of ADC to be integrated with the DSP core, reducing the chip count, enhancing system reliability, and reducing overall cost.

2.5 Analog reconstruction

There are applications where the information that we are looking for can be extracted from the digitally processed signal. In this case, there is no need to produce an output that is analog. Therefore, stage 3 in Figure 2.1 does not exist. An example is a receiver for digitally modulated signals. The aim of the receiver is the detection of digital symbols being transmitted. The input to the receiver is the carrier modulated signal and the output is the sequence of detected symbols.

However, there are many applications that require the construction of analog waveforms or signals from the digital signal in the form of a sequence of numbers. Intuitively what we want is to 'fill in the gaps' or interpolating between the sampled values so that a continuous-time signal results. An analog reconstructor as shown in Figure 2.32 performs this.

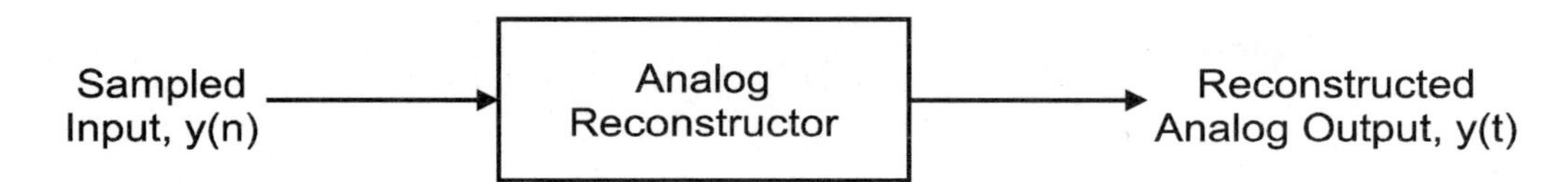

Figure 2.32
Analog signal reconstruction

Generally speaking, any form of interpolation will do the job. However, there are some interpolation methods that are easier to implement and are, in some ways, more desirable than others.

We shall discuss two kinds of reconstructors. The first one is ideal. As the name suggests, this kind of reconstructor is not practical and in fact not physically implementable. But it reflects the ideal situation and will help us to understand the process better. The second type is called staircase reconstructors. They are simple to implement and are in fact most commonly used in practical analog-to-digital converters.

2.5.1 Ideal reconstructor

Let us consider an analog signal $x(t)$ with a frequency spectrum $X(f)$ that has been sampled at the rate of $1/T$ samples per second. The sampled signal $x(n)$ will have a spectrum that consists of replica of $X(f)$ shifted by integer multiples of f_s. Assume that the spectrum $X(f)$ is bandlimited and the sampling rate is sufficiently high so that its replica does not overlap. Then a low-pass filter can recover $X(f)$ with a cut-off frequency of $f_s/2$.

Ideally, this low-pass filter will have frequency characteristics

$$H(f) = \begin{cases} T, & \text{for } |f| \le f_s/2 \\ 0, & \text{otherwise} \end{cases}$$

so that there is no distortion to the spectrum in the Nyquist interval and no frequency component outside this interval is included. $H(f)$ is shown graphically in Figure 2.33(a).

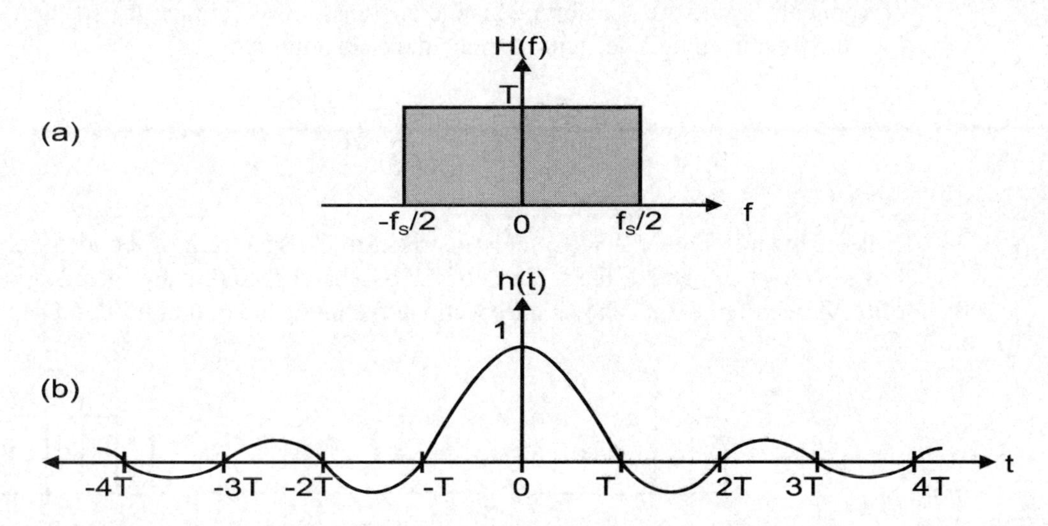

Figure 2.33
An ideal low-pass filter and its impulse response

The time-domain characteristic corresponding to $H(f)$ is given by

$$\begin{aligned} h(t) &= \frac{\sin(\pi t/T)}{\pi t/T} \\ &= \sin c\left(\frac{\pi t}{T}\right) \\ &= \sin c(\pi f_s t) \end{aligned}$$

which is known as the sine function. It is shown in Figure 2.33b.

Notice that $h(t)$ is not physically realizable. This is because it is non-causal. A causal system is one that if excited at $t = 0$ will produce a response starting from $t = 0$. Since $h(t)$ is non-zero in the negative frequency axis, it is non-causal. It means that if this low-pass filter is excited by a single impulse at $t = 0$, the response will have started even before the excitation arrives at the input. Clearly, this is not possible for a real system. Therefore, we cannot implement an ideal reconstructor.

2.5.2 Staircase reconstructor

The reconstructor that is often used in practice is the staircase reconstructor or zero-order holds (ZOH). This reconstructor simply holds the value of the most recent sample until the next sample arrives. So, each sample value is held for T seconds. This is illustrated in Figure 2.34(a).

The ZOH can be characterized by the impulse response:

$$h_{\text{ZOH}}(t) = \begin{cases} 1, & \text{for } 0 \le t \le T \\ 0, & \text{otherwise} \end{cases}$$

This means that if the reconstructor is excited by an impulse at $t = 0$, the output of the reconstructor will be a rectangular waveform with amplitude equal to that of the impulse with duration of T seconds.

It is obvious that the resulting staircase output will contain some high frequency components because of the abrupt change in signal levels. In fact, the spectrum of $h_{\text{ZOH}}(t)$ is a sine function that is decaying exponentially in amplitude.

$$H_{\text{ZOH}}(f) = T\frac{\sin(\pi f T)}{\pi f T}e^{-\text{jpfT}}$$

It is shown in Figure 2.34(b) in comparison to the spectrum of the ideal reconstructor. It is obvious that parts of the replicas of the baseband spectrum are included in the output of the ZOH. Figure 2.35 shows the spectra at the input and output of the ZOH.

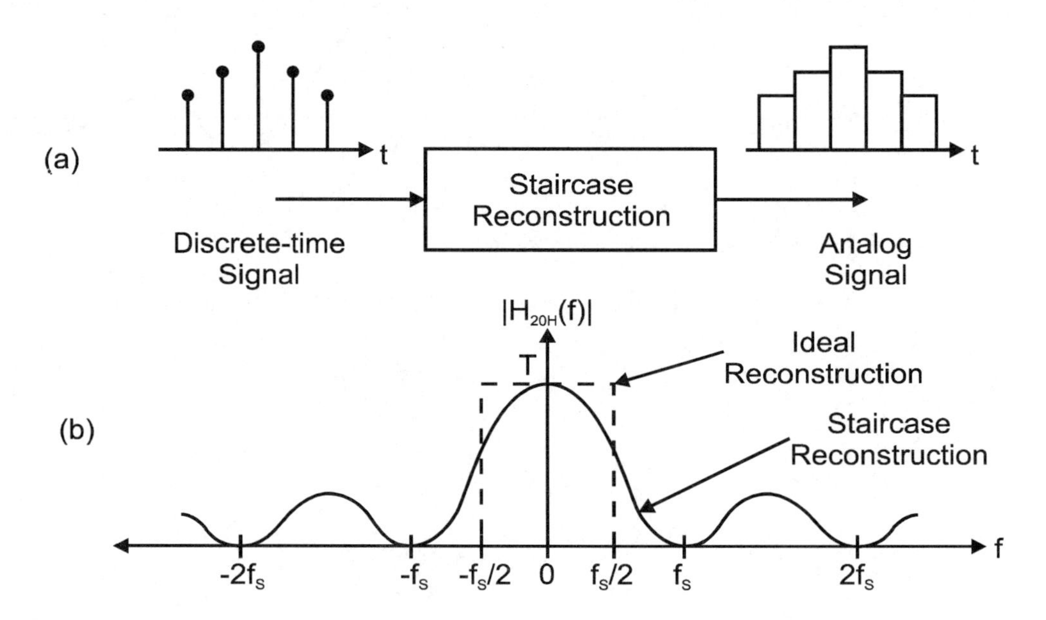

Figure 2.34
Analog reconstruction using zero order holds

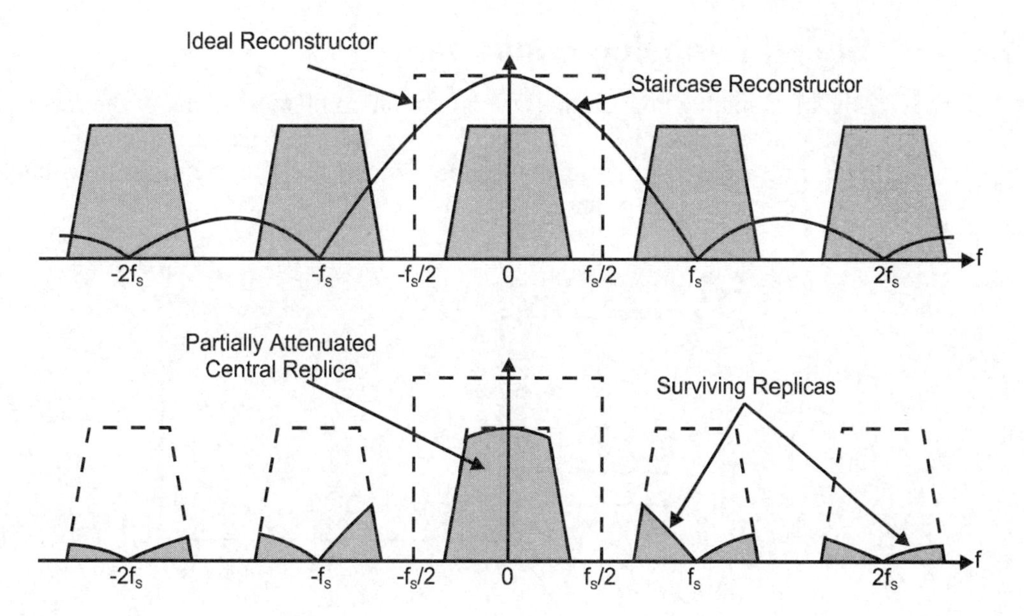

Figure 2.35
Spectra at the input and output of the zero order hold

2.5.3 Image-rejection postfilters

The inclusion of some of the replicas of the baseband spectrum will lead to distortion. It is therefore desirable that they be removed. Since the sampling rate is sufficiently high, low-pass filtering can isolate the baseband spectrum.

A low-pass filter that removes the remaining replicated spectra is also known as an image-rejection filter. The cut-off frequency of this filter should clearly be $f_s/2$. It is very similar to the anti-aliasing filter in characteristics.

Even though the replicated spectra are completely removed (rejected), the baseband spectrum is still slightly distorted by both the ZOH and the image-rejection filter. At the frequency $f_s/2$, the ZOH introduces an attenuation of about 4 dB and the image-rejection filter another 3 dB. While this may be acceptable for some applications, it is highly undesirable for other applications such as high quality digital audio.

This problem can be overcome by equalization. The equalizer has a frequency characteristic $H_{EQ}(f)$ so that the combined frequency response of the equalizer, staircase reconstructor, and the image-rejection filter will be the ideal response $H(f)$ in section 2.5.1. That is,

$$H_{\text{EQ}}(f)H_{\text{ZOH}}(f)H_{\text{post}}(f) = H(f)$$

where $H_{\text{post}}(f)$ is the frequency characteristics of the image-rejection filter.

The advantage of using DSP is that the equalizer can actually be implemented digitally. In other words, the original digital signal samples are first digitally pre-compensated before being converted to analog signal.

2.6 Digital-to-analog converters

Digital-to-analog converters (DACs) are the implementations of the reconstructors. Since DACs are much simpler than ADCs, they are correspondingly cheaper. A digital binary code is converted to an analog output, which may be a current or voltage. Figure 2.36 shows the schematic of an n-bit DAC.

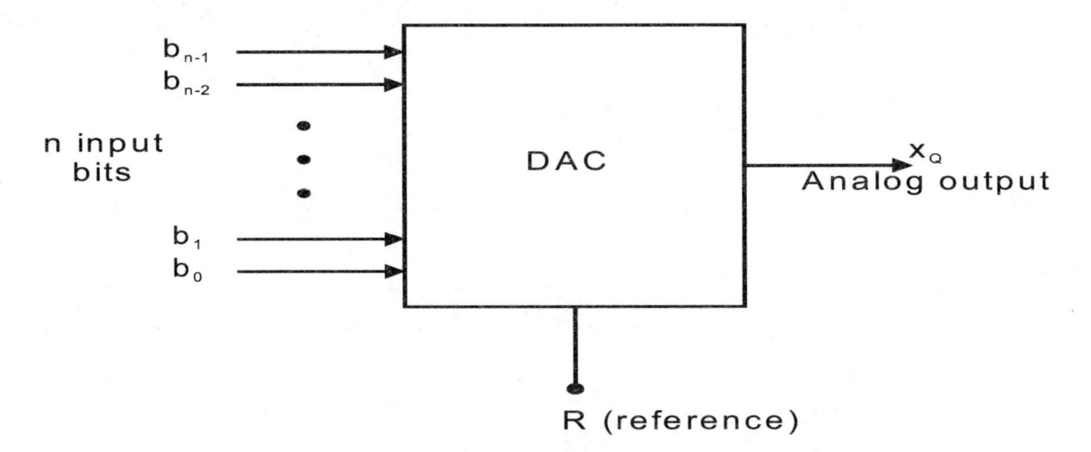

Figure 2.36
An n-bit DAC

An important parameter for ADC is the conversion time – the time it takes for the device to obtain a stable quantized value from the time the conversion starts. For DAC, the corresponding parameter is the settling time – the delay between the binary data appears at the input and a stable voltage is obtained at the output.

2.6.1 Multiplying DAC

The most common DACs are multiplying DACs. It is so called because the output is the sum of the products of the binary code and current sources. Each bit of the binary code turns on or off a corresponding current source. The sum of all the currents available can be converted to a voltage for output or remain as is. Figure 2.37 shows such a current source multiplying DAC. The current sources are normally on and are grounded when not in use.

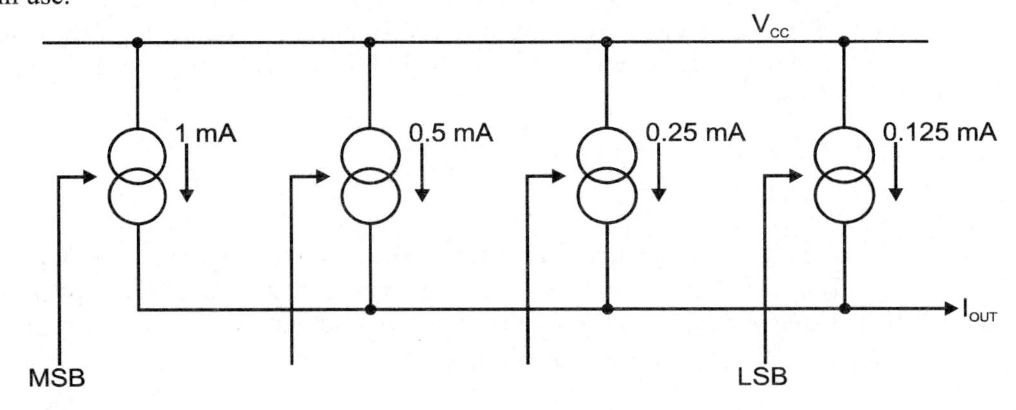

Figure 2.37
A current source multiplying DAC

A voltage source can be used instead of current sources. The voltage source is applied to a series of scaled resistors. The voltages at one end of the resistors are switched either 'on or off' as shown in Figure 2.38. The 'on' voltages are summed. The output is proportional to the weighted sum of the input voltages. Some devices have a built-in reference voltage source. Other ones allow the user to provide an external reference voltage, thereby setting the accuracy of the output.

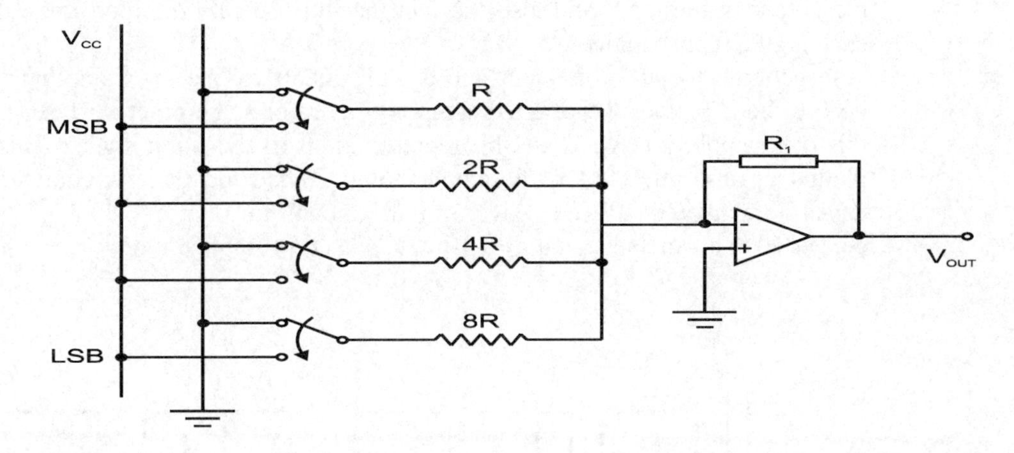

Figure 2.38
A voltage source multiplying DAC

Nearly all commonly used DACs are ZOH devices and therefore image-rejection filters are needed. The settling time of multiplying DACs is short because the conversion is done in parallel.

2.6.2 Bit stream DAC

A disadvantage of multiplying DAC is that the most significant bit (MSB) must be very accurate. This accuracy will be required for the whole range of temperatures specified for the device. Furthermore, it is to be consistent over time.

For an 8-bit DAC, the MSB must be accurate to one part in $2^8 = 256$. The MSB of a 16-bit DAC will need to be accurate to one part in $2^{16} = 65\,536$. Otherwise, some of the least significant bits will be rendered useless and the true resolution of the DAC will diminish. Maintaining voltage and current sources to this level of accuracy is not easy.

One way of overcoming this problem is to use bit stream conversion techniques. The concept is similar to sigma-delta ADCs. In bit stream DACs, a substantially higher sampling frequency is used in exchange for a smaller number of quantization levels.

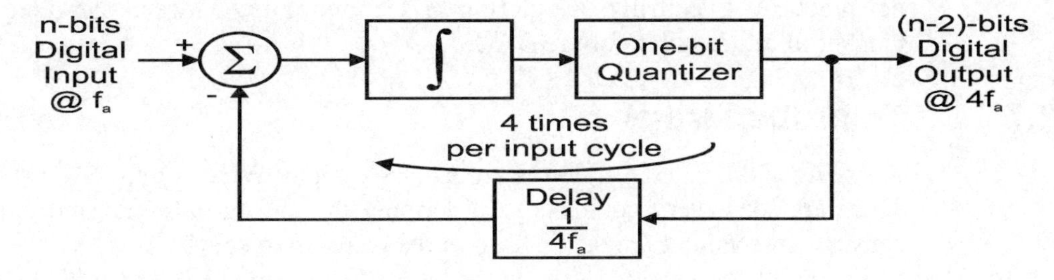

Figure 2.39
The oversampling stage of a bit stream DAC

Figure 2.39 shows the input oversampling stage of a particular bit stream DAC. The input to this stage is an n-bit digital input sampled at a frequency f_a and the output is an $(n-2)$ bit data sequence sampled at $4f_a$. The difference between the current digital input and the digital output is computed. The integrator is digital and simply adds the previous value to the present one. The output of the integrator is quantized into $(n-2)$ bits by truncating the two least significant bits. This loss of resolution is compensated for by the feedback of the output to the input and also the fact that this operation is performed four times for each digital input sample.

In general, for an n-bit input and a q-bit quantizer, the oversampling frequency will need to be 2^{n-q} times the original sampling rate. For some practical DACs, the output of this oversampling stage is a 1-bit representation of the input signal. This bit stream if plotted against time and with sample points joined together, is equivalent to a pulse density modulated (PDM) waveform as shown in Figure 2.40. This bit stream is converted to an analog signal by a 1-bit DAC and subsequently low-pass filtered.

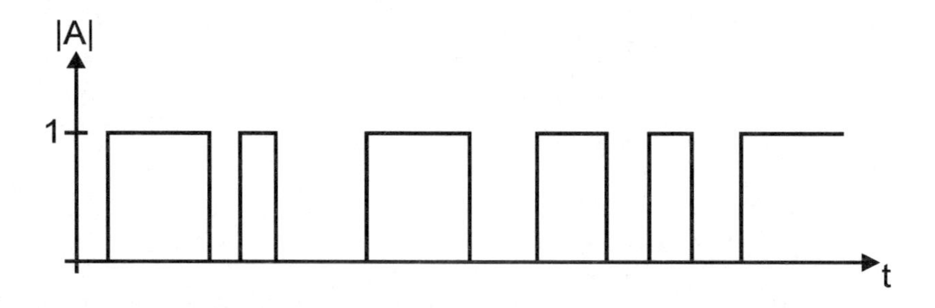

Figure 2.40
A pulse density modulated waveform

Owing to the fact that the original digital signal is being requantized into a small number of levels, the output can sometimes be 'stuck' at an incorrect value. This happens most often when there is a long sequence of the same input value. This 'hang-up' will persist until the next change in input value. The result of this 'hang-up' is that the output will have a substantially different DC (or average) value to that of the input signal.

To overcome this problem, a dithering signal can be added. The effect of dithering has been discussed in section 2.3.4. In this case, dithering lowers the probability of long sequences of any one value.

A further problem with bit stream techniques is the high oversampling frequency. For instance, if we want to resample a CD quality audio to one bit, then we need a frequency of $2^{16} \times 44.1 \times 10^3$ (approximately 3 GHz). This is a very high sampling frequency and is very difficult to implement using present silicon technology. In these cases, 1-bit DAC is not practical. Eventually, the design is a compromise between sampling rate and the number of bits required for the DAC.

2.7 To probe further

Sampling and data conversion is an extensive topic. We can only cover the basics here. There are some very good books describing the analog-to-digital and digital-to-analog conversion techniques in detail. Two of them are listed below:

- D.H. Sheingold (ed.). *Analog-digital conversion handbook*, 3rd edition. Prentice-Hall, 1986.
- G.B. Clayton. *Data converters*. Wiley, 1982.

We have discussed the sampling of a low-pass signal. The minimum sampling frequency is twice that of the bandwidth of the signal. If the signal to be sampled is a bandpass signal, we do not normally want to sample at twice the frequency of the highest frequency component of this signal, since much of the lower frequency components are useless. This leads to the topic of band-pass sampling. Band-pass sampling is of particular interest to applications in carrier modulated communication systems. A good survey of results can be found in the following paper.

- R.G. Vaughan and N.L. Scott, 'The theory of bandpass sampling', *IEEE Transactions on Signal Processing*, Vol.39, no.9, September 1991, pp.1973–1983.

For those who want to understand dithering further, the following two papers are suggested.

- L. Schuchman, 'Dither signals and their effect on quantization noise', *IEEE Transactions on Communications*, vol. COM-12, pp.162.165, 1964.
- S.P. Lipshitz, R.A. Wannamaker and J. Vanderkooy, 'Quantization and dither: a theoretical survey', *Journal of the Audio Engineering Society*, Vol.40, 1992.

Many semiconductor manufacturers produce ADCs and DACs. They come in a variety of configurations. Product information can be obtained from the relevant databooks and the manufacturer's web sites. Some of them provide customers with product information on CDs. They can usually be obtained from the locate distributors.

Below is a very incomplete list of manufacturers and their web sites:

- Analog Devices, Inc. http://www.analog.com
- Motorola, Inc. http://www.motorola.com
- National Semiconductors, Inc. http://www.natsemi.com
- Texas Instruments, Inc. http://www.ti.com

3

Time-domain representation of discrete-time signals and systems

3.1 Notation

A discrete-time signal x consists of a sequence of numbers denoted by $x(n)$, or $x(nT)$, where n is an integer index. The latter notation is usually reserved for sampled data sequences with a uniform sampling period of T seconds. The sampling process is discussed in more detail in a later chapter. If the sampling period is known or is not relevant to the interpretation of results, the subscript notation will be used.

3.2 Typical discrete-time signals

Several discrete-time signals are considered basic and important. More complex signals can be constructed from these elementary ones.

3.2.1 Unit impulse

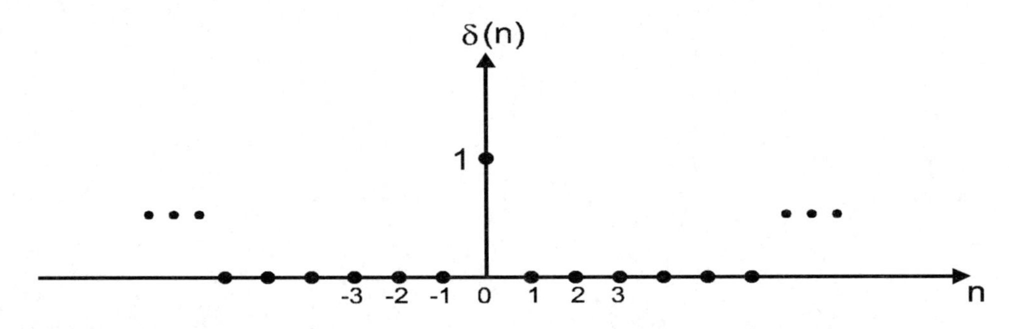

Figure 3.1
The unit impulse sequence

The unit impulse function, usually denoted as δ (n), is shown in Figure 3.1. It consists of a single unit-valued sample at the instant n=0, surrounded on both side by zeros. Mathematically, it is described by the formula

$$\delta(n) = \begin{cases} 1, & n = 0 \\ 0, & n \neq 0 \end{cases}$$

or

$$\delta(n) = \{\ldots, 0, 0, 1, 0, 0, \ldots\}$$

3.2.2 Unit step

The unit step sequence is defined as

$$u(n) = \begin{cases} 0, & n < 0 \\ 1, & n \geq 0 \end{cases}$$

or

$$u(n) = \{\ldots, 0, 0, 1, 1, 1, 1, \ldots\}$$

It is shown graphically in Figure 3.2. The unit step sequence is used to make an arbitrary sequence zero for all indices less than zero by multiplying the arbitrary sequence with the unit step. It can thus indicate the start of an event.

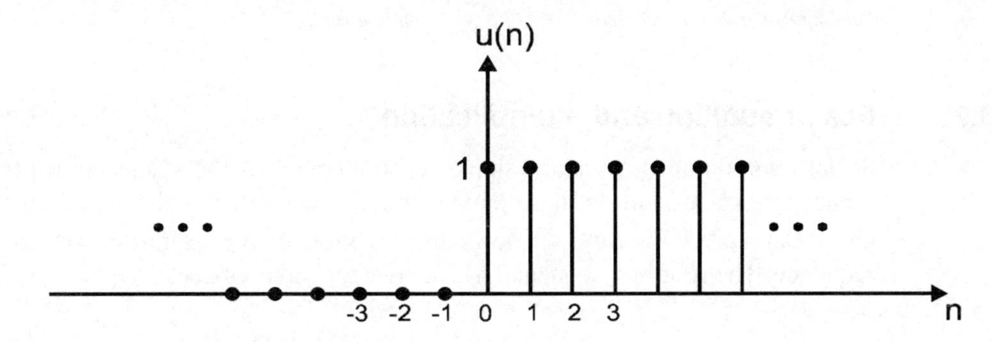

Figure 3.2
The unit step sequence

3.2.3 Random

In some applications it is useful to consider a signal as a random signal rather than a deterministic signal. Random signals are in fact statistical models that obey some specified statistical distribution. Examples of this type of model include white gaussian noise commonly encountered in communication systems analyses. These signal models are very useful for performance analysis. However, we shall focus exclusively on the processing of deterministic signals in this introductory course.

3.3 Operations on discrete-time signals

3.3.1 Delay or shift

An important operation on a sequence is a delay or shift by a number of samples, n_d. For example, Figure 3.3 shows the delayed versions of the unit impulse sequence, $\delta\ (n-1)$ and $\delta\ (n-2)$.

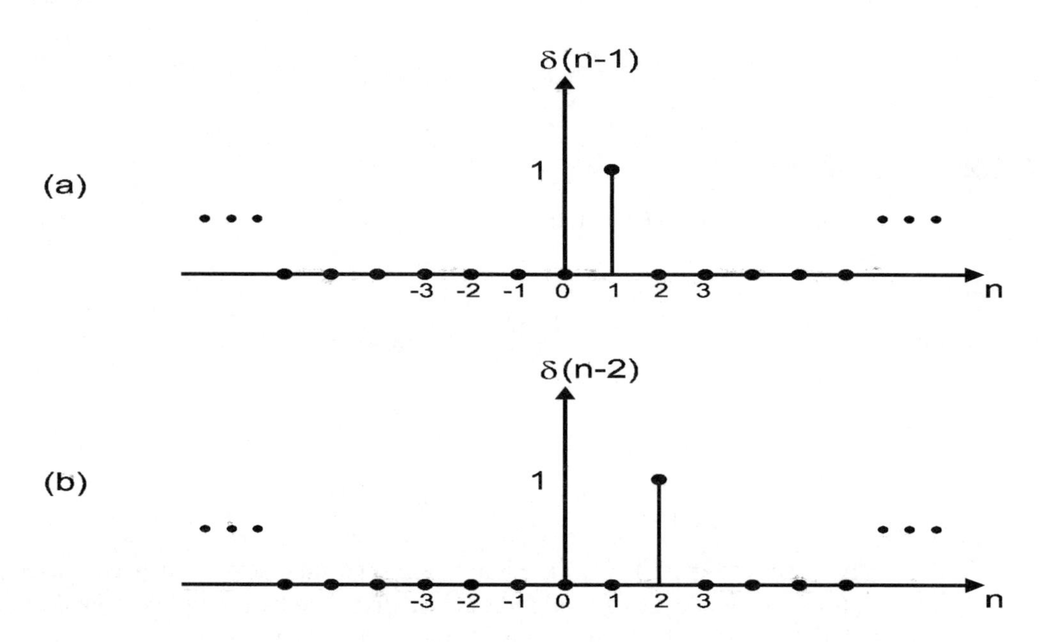

Figure 3.3
The impulse sequence delayed by one and two sample instants

3.3.2 Scalar addition and multiplication

Scalar addition adds a scalar value to each element of the sequence to produce another sequence. Scalar addition thus changes the average value of the signal by the amount of the scalar value. Figure 3.4 shows the addition of a scalar $a = -0.5$ to the unit step sequence. The resulting sequence has an average value of zero.

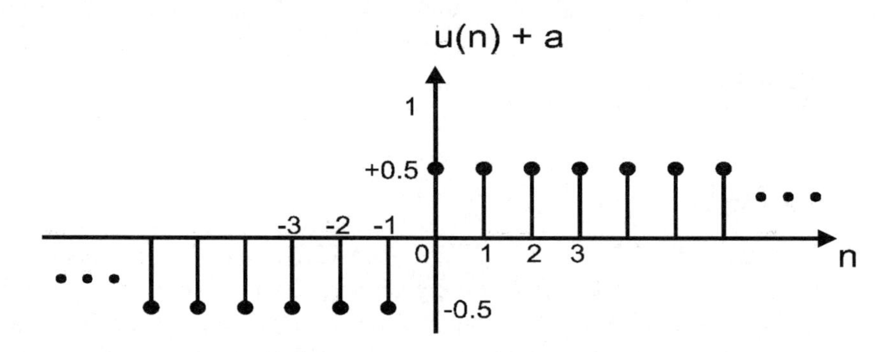

Figure 3.4
Scalar addition of the unit step sequence by $a = -0.5$

Multiplication of a sequence by a scalar results in a sequence that is scaled by that scalar. The unit sequence is multiplied by a scalar $a = -0.5$ in Figure 3.5.

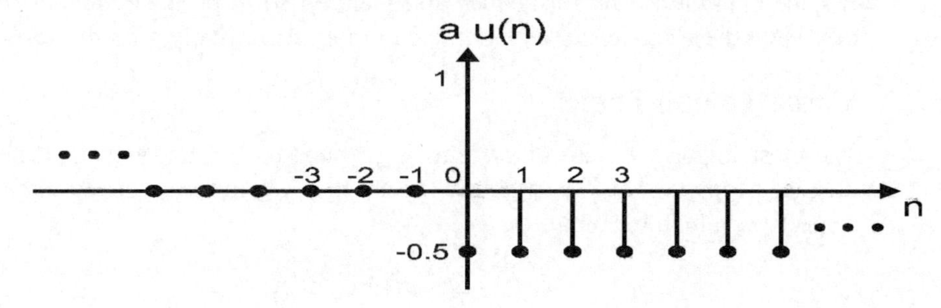

Figure 3.5
Multiplication of the unit step sequence by a = –0.5

Any arbitrary sequence can be represented by a summation of scaled and shifted unit impulses. For instance,

$$
\begin{aligned}
x(n) &= \{3,\ 0.5,\ -1\ -2.5\ 0\ 1\ 2\} \\
&= 3\delta(n-3)+0.5\delta(n-2)-1\delta(n-1)-2.5\delta(n)+0\delta(n+1)+1\delta(n+2)+2\delta(n+3)
\end{aligned}
$$

3.3.3 Vector addition and multiplication

Vector addition is the element-by-element summation of two sequences. Some useful sequences can be produced from the elementary impulse and step sequences. For example, a (finite-duration) rectangular signal sequence of unit amplitude can be produced from the (vector) summation of two delayed unit step sequences:

$$y(n) = u(n-n_1) - u(n-n_2)$$

In this case, the first and last non-zero elements are at indices n_1 and (n_2-1) respectively, with $n_2 > n_1$.

3.3.4 Block diagram representation

The above basic operations on discrete-time signals and sequences can be represented in block diagram form. They are shown in Figure 3.6.

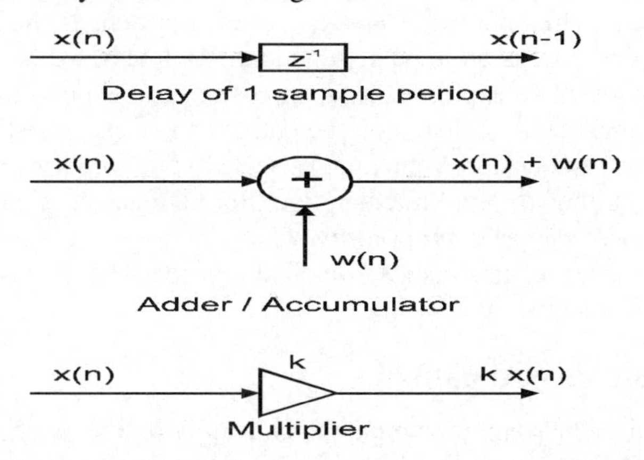

Figure 3.6
Block diagrams of the basic operations on a discrete-time sequence

3.4 Classification of systems

If some operation is performed on a sequence $x(n)$ to produce another sequence $y(n)$, we may view these sequences as input and output, respectively, of a discrete-time system.

3.4.1 Linear vs non-linear

The most important class of systems is perhaps the linear systems. Linear systems have the nice property that if $y_1(n)$ and $y_2(n)$ are the system responses to inputs $x_1(n)$ and $x_2(n)$ respectively, then for input

$$x(n) = a_1 x_1(n) + a_2 x_2(n)$$

where a_1 and a_2 are some arbitrary constants, the output of the linear system will be a similar summation of the individual responses:

$$y(n) = a_1 y_1(n) + a_2 y_2(n)$$

This property is sometimes called the superposition principle. This property implies that if we know the system's responses to some typical inputs such as impulse and step functions, and an arbitrary input can be expressed as a linear combination of these elementary functions, then the system's response to this arbitrary input is known.

A non-linear system is one that is not linear. In other words, the superposition principle does not hold. In real life, all systems are non-linear. However, a non-linear system can usually be approximated as linear within some constraints.

3.4.2 Time-variant vs time-invariant

A time-invariant (or shift-invariant) system has the property that if $x(n)$ produces $y(n)$, then $x(n-n_d)$ produces $y(n-n_d)$ for all n and any n_d. That is, a delay of input samples implies a corresponding delay in the output.

Linear time-invariant (LTI) systems also have the commutative property. This means that if subsystems are arranged in series (or cascade), then the order in which they are arranged can be changed without affecting the final output. This re-arrangement of subsystems sometimes offers advantages such as reducing complexity of the implementation.

A linear time-invariant system can be completely characterized by its impulse response. The impulse response $h(n)$ of a system is the output of the system obtained when the input is an impulse function. Previously we have shown as an example how a sequence can be expressed as a summation of the delayed and scaled impulse functions. For a linear system, therefore, the output can be expressed as a summation of the system's impulse response correspondingly scaled. The impulse response of a system can be of finite duration or of infinite duration. But in practice, impulse responses that are very long are truncated in an appropriate way.

In the rest of this course, we shall consider only linear time-invariant systems unless otherwise stated.

3.4.3 Causal vs non-causal

In a causal system, the output depends only on the present and/or previous values of the input. This seems to be an obvious property since we simply cannot anticipate the future. However, if the data is recorded and processed offline, the algorithm operating on this data set need not be causal since 'future' data are available too.

3.4.4 Stable vs unstable

A stable system produces a finite, or bounded, output in response to a bounded input. This implies that if the system in its equilibrium state is disturbed by a finite amplitude signal, the output will not diverge or grow without limit. In practice, the output of an unstable system will eventually be limited through numerical overflow or saturation of the electronic devices concerned. Systems with feedback have the potential to become unstable.

A stable linear time-invariant system is invertible. For an invertible system, knowledge of the output allows us to find the input uniquely. That means if an input $x(n)$ to a system S produces an output $y(n)$, then the inverse system S^{-1} will produce output $x(n)$ if the input is $y(n)$.

3.5 The concept of convolution

Convolution is the process by which an input interacts with a linear system to produce an output. The input and output of a linear time-invariant system can be easily related through the impulse response of the system. The process is best illustrated by an example.

The impulse response $h(n)$ of a system is shown in Figure 3.7. For the sake of simplicity, it is a triangular function with duration of 5 samples. An input $x(n)$ expressed as a weighted combination of impulse functions is given by

$$x(n) = \sum_{k=0}^{N} x(k)\delta(n-k)$$

The sequence indices are from 0 to N.

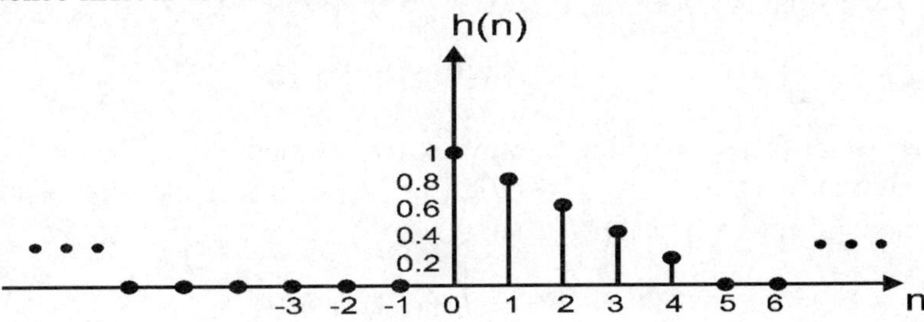

Figure 3.7
Impulse response, h(n), of a system

Since the system is LTI, by time-invariance, the input $\delta(n-k)$ will produce output $h(n-k)$ as shown in Figure 3.8.

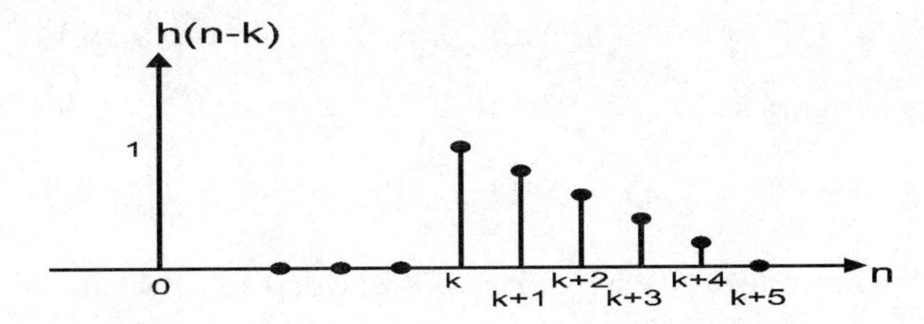

Figure 3.8
Response of the system to δ (n–k)

By linearity, the output corresponding to the weighted sum is the combination of the impulse responses. This is shown graphically in Figure 3.9.

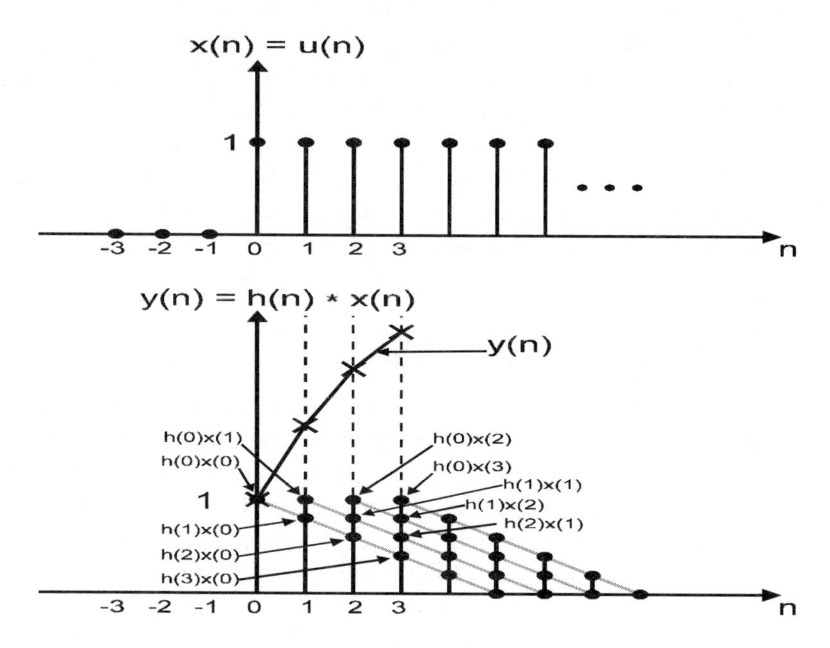

Figure 3.9
Response of the system to x(n)

The output at time instant $n = 0$ is given by

$$y(0) = h(0)x(0)$$

At instant $n = 1$, the output has two components. The first one is the effect of the current input $x(1)$, given by $h(0)x(1)$. The second is the delayed effect of the impulse $x(0)$ at $n = 1$, given by $h(1)x(0)$. Thus,

$$y(1) = h(1)x(0) + h(0)x(1)$$

Similarly, subsequent outputs are

$$\begin{aligned} y(2) &= h(2)x(0) + h(1)x(1) + h(0)x(2) \\ y(3) &= h(3)x(0) + h(2)x(1) + h(1)x(2) + h(0)x(3) \\ &\vdots \\ y(n) &= h(n)x(0) + h(n-1)x(1) + \cdots + h(0)x(n) \end{aligned}$$

In general,

$$y(n) = \sum_{k=0}^{N} x(k)h(n-k)$$

Alternatively, it can be written as (by a change of variables)

$$y(n) = \sum_{k=0}^{N} h(k)x(n-k)$$

This is known as the convolution sum relating the input and output of a discrete-time system. The value of N is usually chosen to be the length of the impulse response sequence. Notice that each output sample is computed from a product of terms involving a sample of the impulse response sequence and a previous input sample.

Convolution will be denoted by the symbol $*$ and the above two equations then become

$$\begin{aligned} y(n) &= x(n) * h(n) \\ &= h(n) * x(n) \end{aligned}$$

respectively. This implies that the convolution operation is commutative. It is also distributive

$$[w(n)+x(n)] * h(n) = [w(n) * h(n)] + [x(n) * h(n)]$$

and associative

$$[w(n) * x(n)] * h(n) = w(n) * [x(n) * h(n)]$$

$$y(n) = \sum_{k=-\infty}^{\infty} x(k)h(n-k) = \sum_{k=-\infty}^{\infty} h(k)x(n-k)$$

These equations can be generalized to data sequences of infinite duration:
This convolution operation is also called linear convolution.

3.6 Autocorrelation and cross-correlation of sequences

In a lot of DSP applications we need to compare the similarity between one set of data with another. These data sets are typically sampled values of two signals. In other words the correlation between these data sets need to be established.

$$r_{12}(k) = \frac{1}{N} \sum_{n=0}^{N-1} x_1(n)x_2(n+k)$$

A formula for computing the cross-correlation function for two discrete-time signals is

Note that correlation is evaluated with one of the signals shifted by k samples. In this definition x_2 is shifted to the left by k samples and products of the corresponding pair of points are summed. This is necessary because two signals may be completely correlated but are out of phase with one another. An example is given in Figure 3.10 where one signal is a delayed version of another. The correlation when $k = 0$ is zero, indicating no correlation between them.

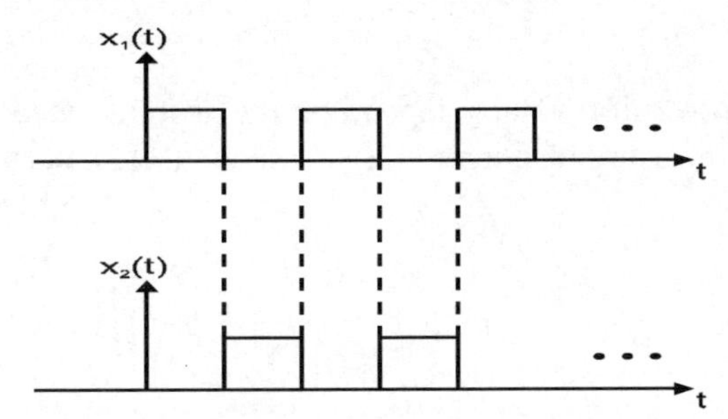

Figure 3.10
Two completely correlated but out of phase signals

It is important to note that correlation coefficients are computed using block-processing techniques. This means that the accuracy is dependent on the size of the block chosen.

The finite amount of data that are used to evaluate the correlation gives rise to another problem. As x_2 is shifted further and further to the left the ends of the signals no longer overlap. This means that the actual number of product pairs will decrease as k increases, leading to a corresponding decrease in $r_{12}(k)$. This is known as the end effect.

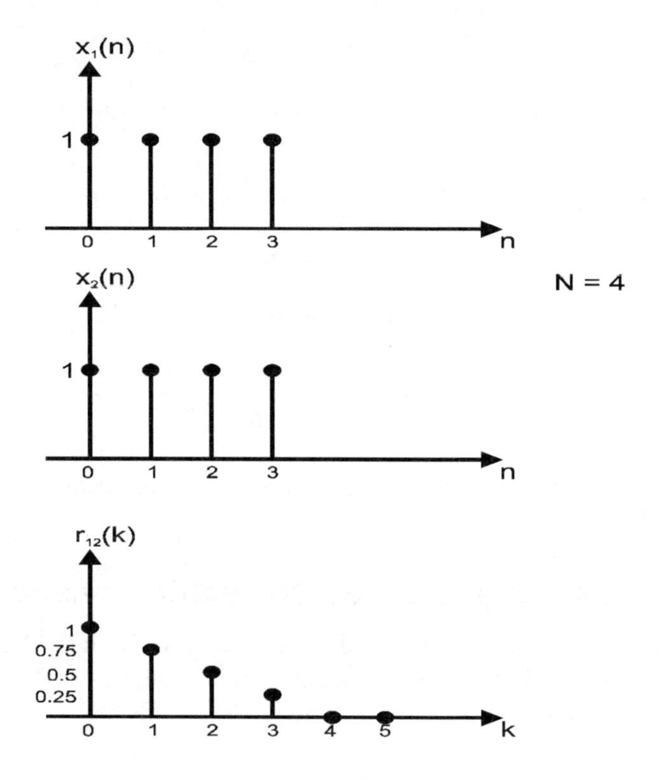

Figure 3.11
The end effect on the cross-correlation of two constant amplitude signals

To overcome the end effect, a correction has to be made to the values computed by equation 9. Assuming that we have two constant valued sequences of finite duration. The decrease in $r_{12}(k)$ purely as consequence of the end effect can be observed. It is shown in Figure 3.11. Thus the corrected values are given by

$$r_{12}(k)_{\text{corrected}} = r_{12}(k) + \frac{k}{N} r_{12}(0)$$

These cross-correlation values depend on the absolute values of the data considered. In order to obtain a measure of likeness of two sequences, a normalized definition is needed.

$$\rho_{12}(k) = \frac{r_{12}(k)}{\frac{1}{N}\left[\sum_{n=0}^{N-1} x_1^2(n) \sum_{n=0}^{N-1} x_2^2(n)\right]^{1/2}}$$

These values are known as the cross-correlation coefficients.

A signal sequence can also be correlated with itself. In this case, the coefficients are called autocorrelation coefficients. Note that the unnormalized autocorrelation coefficient at zero shift (or lag) is the energy of the signal.

$$r_{11}(0) = \frac{1}{N}\sum_{n=0}^{N-1} x_1^2(n)$$

It is also true that

$$r_{11}(0) \geq r_{11}(k) \quad \text{for } k > 0$$

It should also be noted that cross-correlation and autocorrelation functions are not unique. Therefore we cannot deduce the waveform from these functions. However, these functions highlight some of the properties of these signals, which are sometimes not obvious.

3.6.1 Periodic sequences

In the above discussions on correlation and convolution, we considered general procedures that apply to all sequences that are of finite length. If the sequences we are dealing with are periodic in nature, more care has to be taken, especially if the two sequences are of unequal length.

Suppose we have two sequences:

$$x = \{2,3,1,4\}$$
$$y = \{1,2,3\}$$

that are periodic with periods of four and three samples respectively. The cross-correlation function (unnormalized) is shown below:

Lag (k)	$r_{xy}(k)$
0	3.75
1	5.5
2	5.75
3	3.75
4	5.5
5	5.75

The cross-correlation sequence is computing using equation 9 with N=4, the longer period of the two sequences. It is obvious that $r_{xy}(k)$ is periodic, with a period of three samples. Since the result does not reflect the full periodicity of the longer sequence, it must be incorrect.

If we use N=6, and append 2 zeros to sequence x and 3 zeros to sequence y, the cross-correlation function becomes:

Lag (k)	$R_{xy}(k)$
0	11/6
1	7/6
2	3/6 = 0.5
3	4/6
4	9/6
5	17/6
6	11/6

The linear cross-correlation of the two sequences has a period of 6.

As a general rule, if the periods of two sequences are different, say N_1 and N_2, then N_1–1 augmenting zeros must be appended to the sequence with period N_2 and N_2–1 augmenting zeros appended to the one with period N_1. Both sequences are now of length N_1+N_2-1 and the correct linear cross-correlation result will be obtained. This technique is called zero padding.

Convolution can be viewed as the cross-correlation of one sequence with the reverse of a second sequence, the above is also valid for computing linear convolution. Zero padding should be applied if the two periodic sequences are of unequal periods.

3.6.2 Implementation

The correlation and convolution procedures require N multiplications and N–1 additions per output sample. For large values of N, it is very computationally expensive. They can generally be much more efficiently implemented by transforming the time sequence into the frequency domain. This will be considered in the next chapter.

4

Frequency-domain representation of discrete-time signals

So far we have been looking at signals as a function of time or an index in time. Just like continuous-time signals, we can view a time signal as one that consists of a range of frequencies. A signal can be observed as a trace on an oscilloscope or we can observe it through a spectrum analyzer, which displays the strength of the frequency components that make up that signal. Analytically, Fourier analysis provides us with the connection between the time-domain and frequency-domain view of the signal. It tells us that provided some conditions are satisfied; the two views are equivalent. Thus sometimes it is more convenient to describe a signal in the time-domain whereas the frequency-domain description will be more effective in other circumstances.

As we shall discover in this chapter, the frequency-domain (Fourier) representation not only gives us an alternative view of discrete-time signals, it also provides us with a way to compute certain time-domain operations like convolution and correlation more efficiently. Tools for Fourier analysis consist of the discrete Fourier series (DFS) for periodic signals and the discrete Fourier transform (DFT) for aperiodic signals. Transforming signals from the time to the frequency domain through the DFT is computationally expensive. In the early 1960s, Cooley and Tukey discovered an efficient algorithm for the computation of DFTs, called the fast Fourier transform (FFT). This discovery made real-time computation of convolution and filtering a reality. Some would say that this is when digital signal processing as a discipline was established. Since then a large variety of similar algorithms were proposed and studied. Some of them improve on the original algorithm while others tackle situations that the FFT is not designed for. Today, all DSP processors are able to compute the FFT in software sufficiently fast for most but the most demanding applications. In those situations, specific VLSI devices are available commercially.

4.1 Discrete Fourier series for discrete-time periodic signals

Any periodic discrete-time signal, with a period of N, can be expressed as a linear combination of N complex exponential functions.

$$x(n) = \sum_{k=0}^{N-1} c_k e^{j2\pi kn/N} \qquad (1)$$

where

$$j = \sqrt{-1}$$

and

$$e^{j\theta} = \cos\theta + j\sin\theta$$

Equation 1 is called the discrete-time Fourier series (DTFS).
Given the signal $x(n)$, the Fourier coefficients can be calculated by

$$c_k = \frac{1}{N}\sum_{n=0}^{N-1} x(n) e^{-j2\pi kn/N} \qquad (2)$$

Note that these Fourier coefficients are generally complex-valued. They provide a description of the signal in the frequency domain. The coefficient c_k has a magnitude and phase associated with a (normalized) frequency given by

$$\omega_k = 2\pi k/N$$

These Fourier coefficients form the (discrete) frequency spectrum of the signal. This normalized frequency can be de-normalized if we know the sampling frequency (F_s), or

$$\omega_s = 2\pi/T$$

the time lapse (T) between two samples which are related by
Since

$$0 \leq \omega_k \leq 2\pi$$

The denormalized frequency ω takes on values in the range

$$0 \leq \omega_k \leq \omega_s$$

It is easy to verify that the sequence of coefficients given by equation (2) is periodic with a period of N. This means that the frequency spectrum of a periodic signal is also periodic.

Example: 4.1
Determine the discrete spectrum of a periodic sequence $x(n)$ with a period N=4 given by

$$x(n) = \{0,\ 1,\ 1,\ 0\} \qquad n = 0,1,2,3$$

Solution:
From equation (2), we have

$$c_k = \frac{1}{4}\sum_{n=0}^{3} x(n) e^{-j2\pi kn/4} \qquad k = 0,\ldots,3$$

$$= \frac{1}{4}\left[x(1)e^{-j\pi k/2} + x(2)e^{-j\pi k}\right]$$

and

$$c_0 = \frac{1}{4}[1+1] = \frac{1}{2}$$

$$c_1 = \frac{1}{4}\left[e^{-j\pi/2} + e^{-j\pi}\right] = \frac{1}{4}(-1-j)$$

$$c_2 = \frac{1}{4}\left[e^{-j\pi} + e^{-j2\pi}\right] = 0$$

$$c_3 = \frac{1}{4}\left[e^{-j3\pi/2} + e^{-j3\pi}\right] = \frac{1}{4}(-1+j)$$

The magnitude of this discrete spectrum is shown in Figure 4.1.

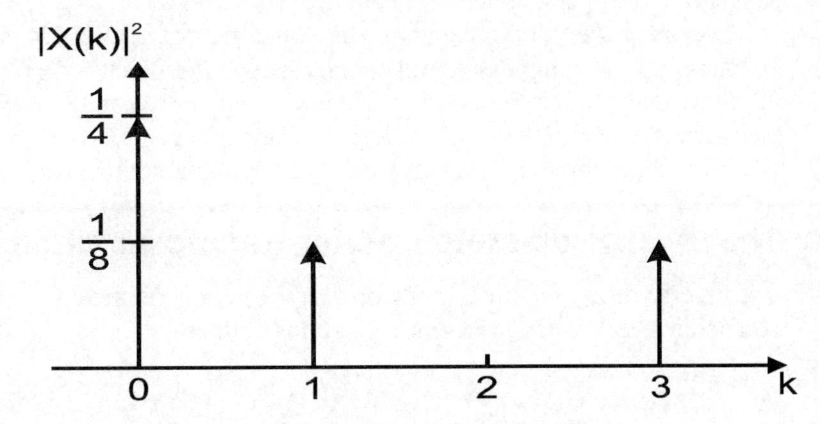

Figure 4.1
The discrete magnitude squared spectrum of the signal in Example 4.1.

4.2 Discrete Fourier transform for discrete-time aperiodic signals

When a discrete-time signal or sequence is non-periodic (or aperiodic), we cannot use the discrete Fourier series to represent it. Instead, the discrete Fourier transform (DFT) has to be used for representing the signal in the frequency domain. The DFT is the discrete-time equivalent of the (continuous-time) Fourier transforms. As with the discrete Fourier series, the DFT produces a set of coefficients, which are sampled values of the frequency spectrum at regular intervals. The number of samples obtained depends on the number of samples in the time sequence.

A time sequence $x(n)$ is transformed into a sequence $X(\omega)$ by the discrete Fourier transform.

$$X(k) = \sum_{n=0}^{N-1} x(n)e^{-j2\pi kn/N} \qquad k = 0,1,\ldots,N-1 \tag{3}$$

This formula defines an N-point DFT. The sequence $X(k)$ are sampled values of the continuous frequency spectrum of $x(n)$. For the sake of convenience, equation 3 is usually written in the form

$$X(k) = \sum_{n=0}^{N-1} x(n) W_N^{kn} \qquad k = 0, 1, \ldots, N-1 \tag{4}$$

where

$$W = e^{-j2\pi/N}$$

Note that, in general, the computation of each coefficient $X(k)$ requires a complex summation of N complex multiplications.

Since there are N coefficients to be computed for each DFT, a total of N^2 complex additions and N^2 complex multiplications are needed. Even for moderate values of N, say 32; the computational burden is still very heavy. Fortunately, more efficient algorithms than direct computation are available. They are generally classified, as fast Fourier transform algorithms and some typical ones will be described later in the chapter.

4.3 The inverse discrete Fourier transform and its computation

The inverse discrete Fourier transform (IDFT) converts the sequence of discrete Fourier coefficients back to the time sequence and is defined as

$$x(n) = \frac{1}{N} \sum_{k=0}^{N-1} X(k) e^{j2\pi kn/N} = \frac{1}{N} \sum_{k=0}^{N-1} X(k) W_N^{-kn} \qquad k = 0, 1, \ldots, N-1 \tag{5}$$

The formulas for the forward and inverse transforms are identical except for the scaling factor of $1/N$ and the negation of the power in the exponential term (W_N^{-kn} instead of W_N^{kn}). So any fast algorithm that exists for the forward transform can easily be applied to the inverse transform. In the same way, specific hardware designed for the forward transform can also perform the inverse transform.

4.4 Properties of the DFT

In this section, some of the important properties of the DFT are summarized. Let

$$x(n) \leftrightarrow X(k)$$

denote the DFT between $x(n)$ and $X(k)$ in the discussions below.

4.4.1 Periodicity of the DFT

Consider $X(k+N)$ which is given by

$$\begin{aligned} X(k+N) &= \sum_{n=0}^{N-1} x(n) e^{-j2\pi kn/N} e^{-j2\pi Nn/N} \\ &= \sum_{n=0}^{N-1} x(n) e^{-j2\pi kn/N} \\ &= X(k) \end{aligned}$$

This implies that $X(k)$ is periodic with a period of N, even though $x(n)$ is aperiodic. Thus for a discrete-time signal, we cannot obtain the full spectrum (frequencies from negative infinity to infinity).

4.4.2 Linearity

If both $x_1(n)$ and $x_2(n)$ are of the same length and

$$x_1(n) \leftrightarrow X_1(k) \quad \text{and} \quad x_2(n) \leftrightarrow X_2(k)$$

then

$$x(n) = ax_1(n) + bx_2(n) \leftrightarrow X(k) = aX_1(k) + bX_2(k)$$

where a and b are arbitrary constants.

4.4.3 Parseval's relation

This relationship states that the energy of the signal can be calculated from the time sequence or from the Fourier coefficients.

$$\sum_{n=0}^{N-1} |x(n)|^2 = \frac{1}{N} \sum_{k=0}^{N-1} |X(k)|^2$$

4.4.4 Real sequences

If $x(n)$ is real, then

$$x(n) = x^*(n) \quad \text{and}$$
$$X(k) = X^*(-k)$$

i.e. the real part of $X(k)$ is an even function or is symmetrical about $k = 0$ and the imaginary part is odd function. This kind of symmetry in $X(k)$ is also known as hermitian symmetry.

4.4.5 Even and odd functions

If $x(n)$ is an even function or has even symmetry about $n = 0$, i.e. $x(n) = x(-n)$, then $X(k)$ will also be an even function. If $x(n)$ is an odd function, i.e. $x(n) = x(n)$, then $X(k)$ will also be an odd function.

Furthermore, if $x(n)$ is real and even, then $X(k)$ is real and even. If $x(n)$ is real and odd, then $X(k)$ is imaginary and odd.

4.4.6 Convolution

$$x(n) = x_1(n) * x_2(n) \leftrightarrow X(k) = X_1(k)X_2(k)$$

Convolution in the time domain becomes a point-by-point multiplication in the frequency domain. Thus a convolution operation can be performed by first performing the DFT of each time sequence, obtain the product of the DFTs, and then inverse transform the result back to a time sequence. This is called circular convolution, which is somewhat different from the linear convolution discussed in the previous chapter. The computation of linear convolution using DFT is explored in more detail in a later section.

$$r_{12}(n) = \sum_{k=-\infty}^{\infty} x_1(k)x_2(k-n) \leftrightarrow S_{12}(k) = X_1(k)X_2(-k)$$

4.4.7 Correlation

$S_{12}(k)$ is called the cross-energy density spectrum. In the case where $x_1(n) = x_2(n)$, it is called the power spectrum. In other words, the power spectrum of a signal is the DFT of its autocorrelation sequence.

4.4.7 Time delay

$$x(n-m) \leftrightarrow W^{km} X(k)$$

A delay in the time domain is equivalent to a multiplication by a complex exponential function in the frequency domain.

4.4.9 Frequency shifting

$$W^{-h} x(n) \leftrightarrow X(k-1)$$

This is the frequency domain equivalent of the previous property. A shift (or delay) in the frequency sequence is equivalent to the multiplication of the time sequence by a complex exponential function. Alternatively, we can say that if the time sequence is multiplied by a complex exponential, then it is equivalent to a shift in the frequency spectrum.

$$x(n)\cos(2\pi \ln / N) \leftrightarrow \frac{1}{2} X(k+l) + \frac{1}{2} X(k-l)$$

4.4.10 Modulation

In the above frequency shifting property, if the complex exponential function is replaced by a real sinusoidal function, then the multiplication in the time domain is equivalent to shifting the half the power of the spectrum up and half of it down by the same amount. Multiplication by a sinusoid is the modulation operation performed in communication systems. This is illustrated in Figure 4.2.

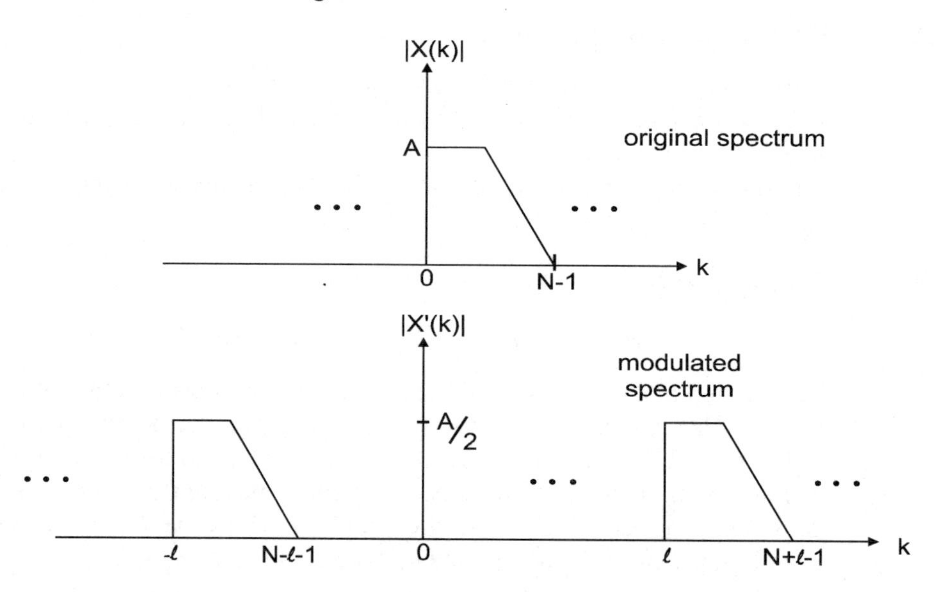

Figure 4.2
The effect of modulation on the discrete frequency spectrum

4.4.11 Differentiation in the frequency domain

$$nx(n) \leftrightarrow j\frac{dX(\omega)}{d(\omega)}$$

Differentiation in the frequency domain is related to the multiplication of the time signal by a ramp. This property is useful in the computation of the group delay of digital filters. Details can be found in the chapter on FIR filters.

4.5 The fast Fourier transform

The DFT is computationally expensive. In general, an N-point DFT requires N complex multiplications and N–1 complex additions. If N=2^r, where r is a positive integer, then an efficient algorithm called the fast Fourier transform (FFT) can be used to compute the DFT. Cooley and Tukey first developed this algorithm in the 1960s.

The basic idea of the FFT is to rewrite the DFT equation into two parts:

$$\begin{aligned} X(k) &= \sum_{n-0}^{N/2-1} x(2n)W_N^{2nk} + W^k \sum_{n=0}^{N/2-1} x(2n+1)W_N^{2nk} \qquad k = 0,1,\ldots,N-1 \\ &= X_1(k) + W_N^k X_2(k) \end{aligned} \tag{6}$$

The first part,

$$\begin{aligned} X_1(k) &= \sum_{n=0}^{(N/2)-1} x(2n)W_N^{2kn} \\ &= \sum_{n=0}^{(N/2)-1} x(2n)W_{N/2}^{kn} \end{aligned}$$

is the DFT of the even sequence and the second part, $X_2(k)$, is the DFT of the odd sequence. Notice that the factor W^{2nk} appears in both DFTs and need only be computed once. The FFT coefficients are obtained by combining the DFTs of the odd and even sequences using the formulas:

$$X(k) = X_1(k) + W_N^k X_2(k) \qquad \text{for } k = 0,1,\ldots,\frac{N}{2}-1$$

$$X\left(k+\frac{N}{2}\right) = X_1(k) - W_N^k X_2(k) \qquad \text{for } k = 0,1,\ldots,\frac{N}{2}-1$$

The complex factor W_N^k is known as the twiddle factor.

These subsequences can be further broken down into even and odd sequences until only 2-point DFTs are left. So each N/2-point DFT is obtained by combining two N/4-point DFTs, each of that is obtained by combining two N/8-point DFTs, etc. There are a total of r stages since N=2^r.

Computation of two-point DFTs is trivial. The basic operation is illustrated in Figure 4.3.

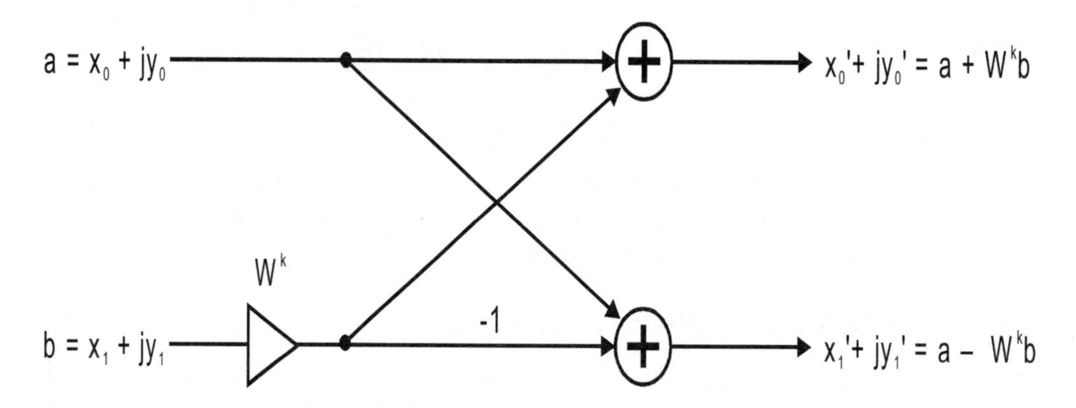

Figure 4.3
The butterfly operation of the decimation-in-time FFT

It is usually known as a butterfly operation. Figure 4.4 illustrates the three stages required in the computation of an 8-point FFT. The twiddle factors are usually pre-computed and stored in memory.

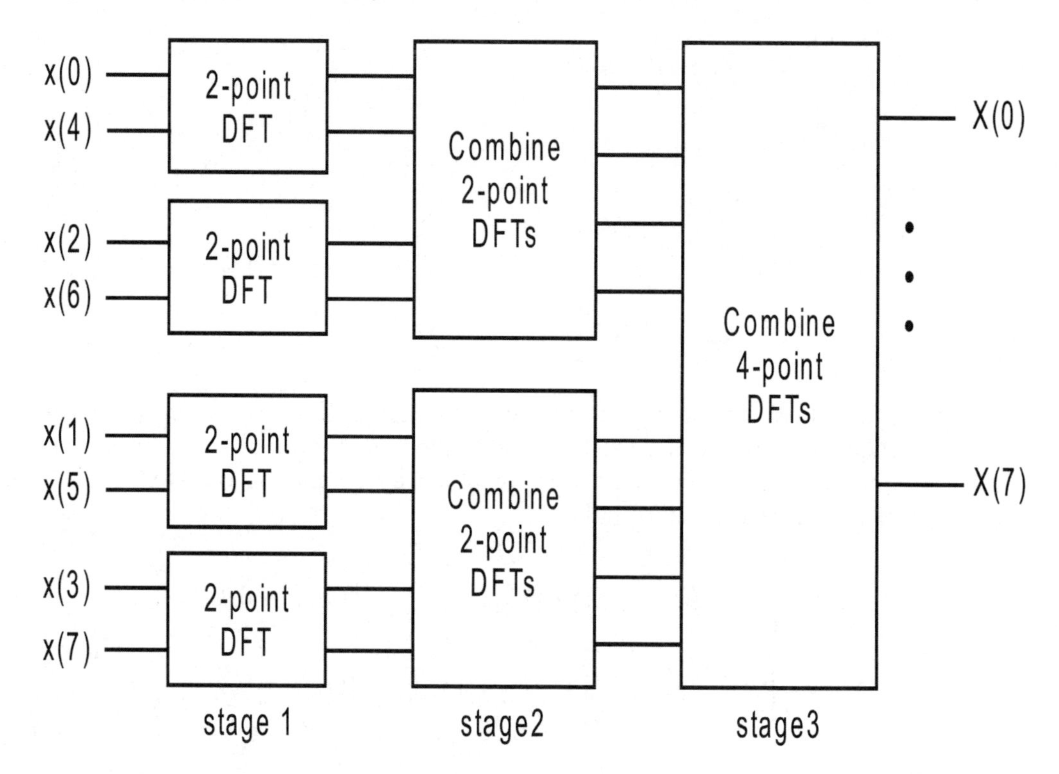

Figure 4.4
The three stages in an 8-point decimation-in-time FFT

Note that if we want the DFT coefficients to come out at the natural order, the input sequence has to be rearranged. This reordering is known as bit reversal. We can see why it is called bit reversal if we represent the index sequence in binary form. The following table illustrates this.

Natural order	Binary form	Bit reversed	Reordered index
0	000	000	0
1	001	100	4
2	010	010	2
3	011	110	6
4	100	001	1
5	101	101	5
6	110	011	3
7	111	111	7

Compare the bit-reversed indices with that of Figure 4.4.

An efficient algorithm for bit reversal will be discussed later.

This FFT algorithm is also referred to as the radix-2 decimation-in-time FFT. Radix 2 refers to the fact that 2-point DFTs are the basic computational block in this algorithm. Decimation-in-time refers to the breaking up (decimation) of the data sequence into even and odd sequences. This is in contrast to decimation-in-frequency a little later.

4.5.1 Computational savings

An N-point FFT consists of $N/2$ butterflies per stage with $\log_2 N$ stages. Each butterfly has one complex multiplication and two complex additions. Thus there are a total of $(N/2)\log_2 N$ complex multiplications compared with N^2 for DFT, and $N \log_2 N$ complex additions compared with $N(N-1)$ for the DFT. A substantial saving when N is large.

4.5.2 Decimation-in-frequency algorithm

Partitioning the data sequence into two halves, instead of odd and even sequences can derive another radix 2 FFT algorithm. Thus,

$$
\begin{aligned}
X(k) &= \sum_{n=0}^{(N/2)-1} x(n)W_{\mathrm{N}}^{\mathrm{kn}} + \sum_{n=N/2}^{N-1} x(n)W_{\mathrm{N}}^{\mathrm{kn}} \\
&= \sum_{n=0}^{(N/2)-1} x(n)W_{\mathrm{N}}^{\mathrm{kn}} + W_{\mathrm{N}}^{\mathrm{Nk/2}} \sum_{n=0}^{N/2-1} x(n+\frac{N}{2})W_{\mathrm{N}}^{\mathrm{kn}} \\
&= \sum_{n=0}^{(N/2)-1} \left[x(n) + (-1)^k x\left(n+\frac{N}{2}\right)\right] W_{\mathrm{N}}^{\mathrm{kn}}
\end{aligned}
$$

The FFT coefficient sequence can be broken up into even and odd sequences and they have the form

$$
X(2k) = \sum_{n=0}^{(N/2)-1} g_1(n)W_{\mathrm{N/2}}^{\mathrm{kn}}
$$

$$
X(2k+1) = \sum_{n=0}^{(N/2)-1} g_2(n)W_{\mathrm{N/2}}^{\mathrm{kn}}
$$

where

$$g_1(n) = x(n) + x\left(n + \frac{N}{2}\right)$$

$$g_2(n) = \left[x(n) - x\left(n + \frac{N}{2}\right)\right] W_N^n \qquad n = 0, 1, \ldots, (N/2) - 1$$

The computation of the $g_1(n)$ and $g_2(n)$ sequences involves the butterfly operation as shown in Figure 4.5. It is similar to the butterfly for the decimation-in-time FFT except for the position of the twiddle factor.

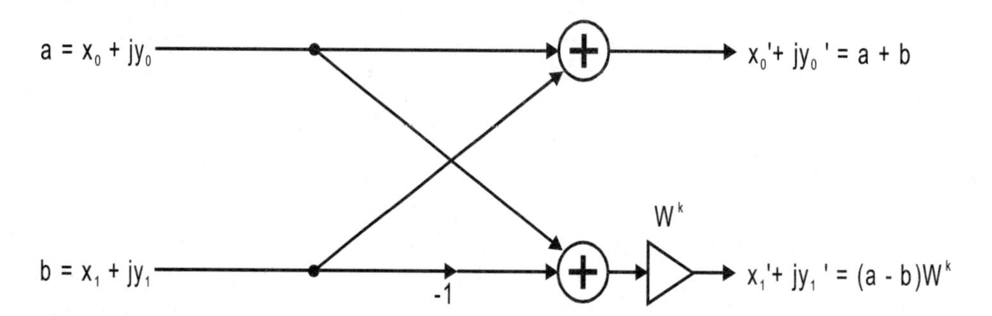

Figure 4.5
Butterfly computation in the decimation-in-frequency FFT

The even and odd coefficient sequences can each be divided into the first and second halves and the same procedure repeated until only 2-point DFTs are required. Figure 4.6 shows the three stages of an 8-point radix-2 decimation-in-frequency FFT. Note that the data sequence appears in its natural order whereas the FFT output occurs in bit reversed order.

The computational complexity is the same as the decimation-in-time algorithm.

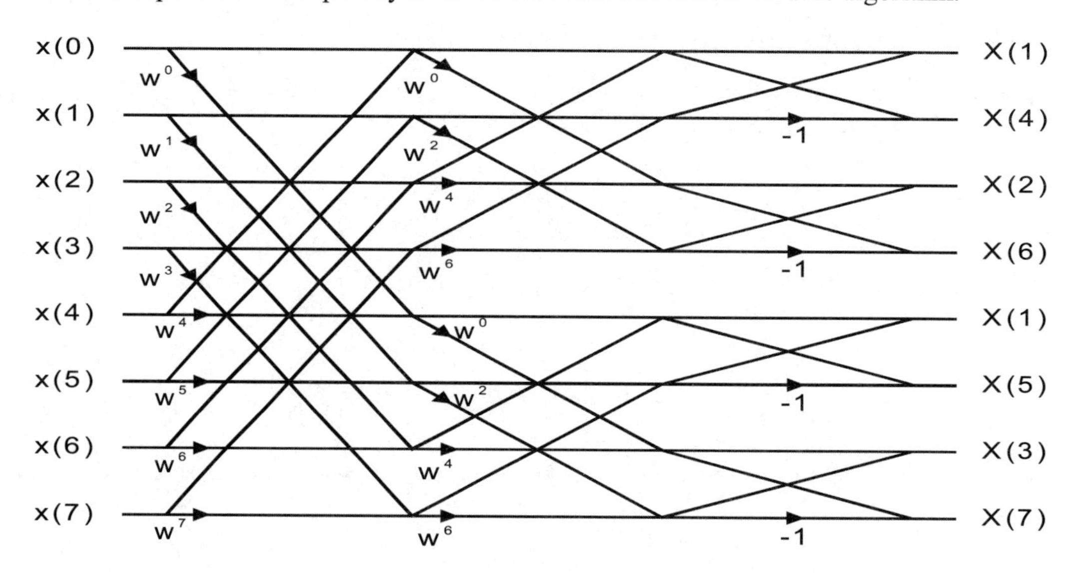

Figure 4.6
The three stages in an 8-point decimation-in-frequency FFT

4.5.3 Other fast algorithms

There are literally tens of different variations of the basic radix-2 FFT algorithms. Some further computational saving can be obtained by using radix-4 or mixed-radix (split-radix) variants. Other algorithms deal with cases where the number of data points is not a power of two, typically when N is a prime number. One of the most well known in this category is the Winograd Fourier transform (WFT). It does not make use of the butterfly and the idea is completely different from the Cooley and Tukey algorithm. It is very fast in the sense that it requires fewer multiplications and is particularly useful for short length transforms. But both the mathematics involved and the implementation are considerably more complex. Thus they are outside of the scope for this introductory course.

4.6 Practical implementation issues

4.6.1 Bit reversal

As shown above, bit reversal is required for both the decimation-in-time and decimation-in-frequency FFT algorithms. A simple algorithm for generating a list of bit reversed numbers is attributed to Bruneman. It goes as follows:

- Start with {0,1}. Multiply by 2 to get {0,2}.
- Add 1 to the list of numbers obtained above. In this case, it is {1,3}.
- Append the list in step 2 to that in step 1 to get {0,2,1,3}.
- The list obtained in step 3 now becomes the starting list in step 1. The steps are repeated until the desired length of the list is obtained.

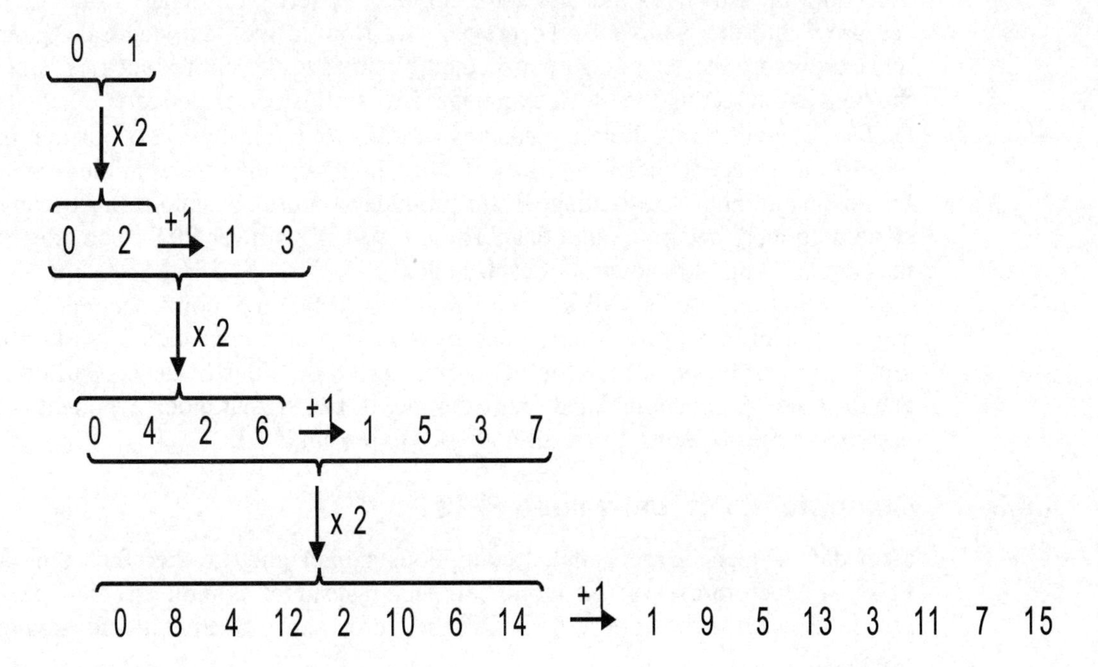

Figure 4.7
Bit reversal using Bruneman's algorithm

Figure 4.7 illustrates the above steps in diagram form to obtain a 16-point bit reversed list. It should be pointed out that more efficient algorithms have been developed for bit reversal. But we shall not cover them here.

4.6.2 Fixed point implementations

Fixed point implementations of the FFT algorithms are quite common as a number of commercially available DSP processors do not have hardware floating point multipliers. The ones that do have floating point hardware integrated on chip are usually several times more expensive than fixed point ones.

Consider first the butterfly calculations. It involves a complex multiplication, a complex addition and a complex subtraction. These operations can potentially cause the data to grow by 2 bits from input to output. For example, in Figure 4.3, if x_0 is 07FFH where *xxxx*H represents a hexadecimal number, then x_0' could be 100FH. Precautions must be taken to avoid data overflow because of this bit growth. We shall describe three techniques to prevent overflow in an FFT.

One way to ensure that overflow does not occur is to include enough extra sign bits, called guard bits, in the FFT input data to ensure that bit growth does not result in overflow. Data can grow by 2 bits in a butterfly but a data value cannot grow by this maximum amount in two consecutive stages. The number of guard bits necessary for an N-point FFT is $\log_2 N+1$. For example, each of the input samples of a 32-point FFT must contain 6 guard bits. In a 16-bit implementation, the remaining 10 bits will be available for data (one sign bit, nine magnitude bits). This method requires no data shifting and is therefore very fast. The disadvantage is that the number of bits available for data will be severely limited for large N.

Another way to avoid data overflow is to scale the outputs down by a factor of two unconditionally after each stage. This approach is called unconditional block floating point scaling. Initially, two guard bits are included in the input data to accommodate the maximum bit growth in the first stage. In each butterfly of a stage calculation, the data can grow into the guard bits. To prevent overflow in the next stage, the guard bits are replaced before the next stage is executed by shifting the entire block of data one bit to the right and updating the block exponent. This shift is performed after every stage except the last. The input data therefore can have 14 bits. In total, $(\log_2 N)-1$ bits are lost because of shifting. Thus unconditional block floating point-scaling results in the same number of bits lost as in input data scaling. But it produces a more accurate result because the FFT starts with more accurate input data. The tradeoff is a slower FFT calculation because of the extra shifting of the output of each stage.

The third method is called conditional block floating point scaling. As the name suggests, it differs from the previous method in that output data is shifted only if bit growth occurs. If one or more output grows, the entire block of data is shifted to the right and the block exponent updated. For example, if the original block exponent is 0 and data is shifted three positions, the resulting block exponent is +3.

4.6.3 Computation of real-valued FFTs

Most data sequences are sampled continuous signals and are therefore real-valued. The FFTs are therefore performed on real data instead of complex-valued data assumed earlier. Since the imaginary part will be zero, some computational savings can be achieved.

One way to achieve computational saving is by computing two real FFTs simultaneously with one complex FFT. Suppose $x_1(n)$ and $x_2(n)$ are two real data sequences of length N. We can form a complex sequence by using $x_1(n)$ as the real part and $x_2(n)$ as the imaginary part of the data.

$$x(n) = x_1(n) + jx_2(n)$$

Alternatively, the original sequences can be expressed in terms of $x(n)$ by

$$x_1(n) = \frac{1}{2}\left[x(n) + x^*(n)\right]$$

$$x_2(n) = \frac{1}{2j}\left[x(n) - x^*(n)\right]$$

Since the DFT is a linear operation, the DFT of $x(n)$ can be expressed as

$$X(k) = X_1(k) + jX_2(k)$$

where

$$X_1(k) = \frac{1}{2}\left\{DFT\left[x(n)\right] + DFT\left[x^*(n)\right]\right\}$$

$$X_2(k) = \frac{1}{2j}\left\{DFT\left[x(n)\right] + DFT\left[x^*(n)\right]\right\}$$

Since the DFT of $x^*(n)$ is $X^*(N-k)$,

$$X_1(k) = \frac{1}{2}\left[X(k) + X^*(N-k)\right]$$
$$X_2(k) = \frac{1}{2j}\left[X(k) - X^*(N-k)\right] \quad (7)$$

In this way, we have computed the DFT of two real sequences with one FFT. Apart from the small amount of additional computation to obtain $X_1(k)$ and $X_2(k)$, the computational requirement is halved.

Another way of achieving computational savings on real sequences is to compute $2N$-point DFT of real data with one N-point complex FFT. First, subdivide the original sequence $g(n)$ into two sequences

$$x_1(n) = g(2n)$$

$$x_2(n) = g(2n+1)$$

with $x_1(n)$ the even and $x_2(n)$ the odd data elements of the sequence. Then form a complex sequence as before and perform the FFT on it. The DFT of $x_1(n)$ and $x_2(n)$ are given by equation 7. Notice that separating a sequence into odd and even sequences is basically what is done with the decimation-in-time FFT algorithm. Hence the $2N$-point DFT can be assembled from the two N point DFTs using the formulas given by the decimation-in-time algorithm. That is,

$$G(k) = X_1(k) + W_{2N}^{k} X_2(k) \quad k = 0, 1, ..., N-1$$

$$G(k+N) = X_1(k) + W_{2N}^{k} X_2(k) \quad k = 0, 1, ..., N-1$$

4.6.4 Computational complexity

The measure of computational complexity is given in terms of the number of complex multiplications and additions. In some older books, only the number of complex multiplications is compared. This is because conventionally, a multiplication requires substantially more CPU cycles than additions. This is no longer true with the advent of digital signal processors that are optimized for this kind of DSP computations. The amount of time required for a multiplication is roughly the same as that for additions. Therefore we should compare the total amount of computation needed. Also, the overhead computations become more significant as the basic operations are optimized.

We shall discuss the architectures of DSP devices in more detail in a later chapter. But note that most DSP operations we have encountered so far involve the basic operation called multiply-and-accumulate (MAC). Typically, there are loops involving the multiplication of two numbers, which means that the results of the multiplications are added together. The looping overhead is also optimized.

Some processors have bit reversal-addressing capability built in. So other overhead operations like those involved in the computation of real data discussed above become more significant. These factors should be noted when comparing computational complexity.

4.7 Computation of convolution using DFT

It has been pointed out in the previous section that convolution can be performed using DFT. Each convolution involves two forward DFTs and one IDFT. Convolution is important because filtering of discrete-time signals using finite impulse response (FIR) digital filters is basically a linear convolution of the signal sequence with the impulse response sequence of the filter. It is also a very fundamental operation in DSP and therefore deserves more detailed study.

4.7.1 Circular convolution

When we compute and multiply the N-point DFTs of two N-point sequences, we obtain N DFT coefficients. Inverse transforming these N coefficients using IDFT gives us an N-point sequence. From the previous chapter we understand that, in general, the linear convolution of two N-point sequences will result in a sequence longer than N. Obviously, convolution via DFT is not exactly the same as linear convolution. It is called circular convolution.

The convolution is circular because of the periodic nature of the DFT sequence. Recall that an N-point DFT of an aperiodic sequence is periodic with a period of N. Also recall that the IDFT is essentially a DFT with a small difference. Therefore, the N-point IDFT operation will also produce a periodic sequence with period N. Thus the resulting time domain sequence is periodic or circular.

Mathematically, we can define circular convolution as follows:

$$y(n) = x(n) \otimes h(n) = \sum_{m=0}^{N-1} x(m) h\left[(n-m) \bmod N\right] \tag{8}$$

where (m mod N) is the remainder of m/N and is called 'm modulo N'.

Notice that the operations involved in circular convolution are similar to linear convolution. The only differences are in the limits of the summation (from 0 to $N-1$) and in the 'modulo N' of the index of one of the sequences.

Linear convolution, as computed using the equation given in Chapter 3, is essentially a sample-by-sampling processing method. However, circular convolution, computed using DFT and IDFT is a block processing method.

Example: 4.2
Perform the circular and linear convolution of the following sequences:

$$X_1(n) = \{1, \quad 2, \quad 1, \quad 2\}$$
$$X_2(n) = \{1, \quad 2, \quad 3, \quad 4\}$$

Solution:
Linear convolution of the two sequences gives:

$$\begin{aligned} y(n) &= x_1(n) * x_2(n) \\ &= \{1, \quad 4, \quad 8, \quad 14, \quad 15, \quad 10, \quad 8\} \end{aligned}$$

Circular convolution using DFT:

$$DFT\{x_1(n)\} = \{6, \quad 0, \quad -2, \quad 0\}$$
$$DFT\{x_2(n)\} = \{10, \quad -2 + j2, \quad -2, \quad -2 - 2j\}$$

The product is

$$\begin{aligned} X(k) &= DFT\{x_1(n)\} \cdot DFT\{x_2(n)\} \\ &= \{60, \quad 0, \quad 4, \quad 0\} \end{aligned}$$

and inverse transform gives

$$\begin{aligned} x(n) &= IDFT\{X(k)\} \\ &= \{16, \quad 14, \quad 16, \quad 14\} \end{aligned}$$

In the above example, linear convolution produces a sequence of length 7. Recall from the previous chapter that, in general, two sequences with lengths L and M respectively, will produce a convoluted sequence of length $L+M-1$. So a DFT of length

$$N \geq L + M - 1$$

will be needed to represent the correct linear convolution result. In order to use N point DFTs, we need to increase the original sequences to length N. The simplest way to do it is to pad these sequences with zeros. This will not affect the spectra of these sequences since they are aperiodic. But by sampling the frequency spectra at N equally spaced points we have sufficient points to represent the result in the time domain after IDFT. So a circular convolution is equivalent to linear convolution of two finite length aperiodic sequences provided the number of points N is sufficiently long.

Example: 4.3
Find the linear convolution of the two sequences given in Example 4.2 by using DFT techniques.
Solution:
We need N=7 point DFTs. Therefore

$$x_1'(n) = \{1, \quad 2, \quad 1, \quad 2, \quad 0, \quad 0, \quad 0\}$$
$$x_2'(n) = \{1, \quad 2, \quad 3, \quad 4, \quad 0, \quad 0, \quad 0\}$$

The forward DFT of these sequences are

$$DFT\{x_1'(n)\} = \begin{bmatrix} 6 \\ 0.2225 - j3.4064 \\ 0.9010 + j0.0477 \\ -0.6235 - j2.0358 \\ -0.6235 + j2.0358 \\ 0.9010 - j0.0477 \\ 0.2225 + j3.4064 \end{bmatrix}$$

$$DFT\{x_2'(n)\} = \begin{bmatrix} 10 \\ -2.0245 - j6.2240 \\ 0.3460 + j2.4791 \\ 0.1784 - j2.4220 \\ 0.1784 + j2.4220 \\ 0.3460 - j2.4791 \\ -2.0245 + j6.2240 \end{bmatrix}$$

Inverse transform the point-by-point product of these two DFT sequences gives

$$x'(n) = IDFT\left\{DFT\left(x_1'(n)\right) \cdot DFT\left(x_2'(n)\right)\right\}$$
$$= \{1, \quad 4, \quad 8, \quad 14, \quad 15, \quad 10, \quad 8\}$$

which is the linear convolution result obtained in Example 4.2.

4.7.2 Convolution of long data sequences

If the data sequence is long, it is computationally very expensive to perform convolution or filtering using a single DFT. Even if this is possible, it may not be desirable as the delay between the first input sample and the full result is generated becomes prohibitively long. In this case, it is better if the data sequence is divided up into subsequences and shorter DFTs are performed on each subsequence. The question is how to combine the results from these subsequences so that the correct result is obtained.

Two methods will be described: overlap-add and overlap-save.

4.7.2.1 Overlap-add method

Assume that the length of the impulse response sequence is of length M. The data sequence is divided into non-overlapping subsequences of length L. The size of the DFTs and IDFT is therefore $N=L+M-1$. Hence $M-1$ zeros will need to be appended to the data subsequence. The impulse response sequence is appended with $L-1$ zeros.

If $x_m(n)$ and $X_m(k)$ represents the mth (zero-padded) subsequence and its DFT respectively, and $H(k)$ is the DFT of the zero-padded impulse response sequence, then the output of the mth segment is

$$Y_m(k) = H(k)X_m(k)$$

The IDFT of $Y_m(k)$ gives $y_m(n)$ which is of length N. Since the original subsequences are of length L (<N), the last M–1 points of $y_m(n)$ must be overlapped and added to the first M–1 points of the next subsequence $y_{m+1}(n)$. An illustration of the overlap save method for three subsequences (or blocks) of data is shown in Figure 4.8.

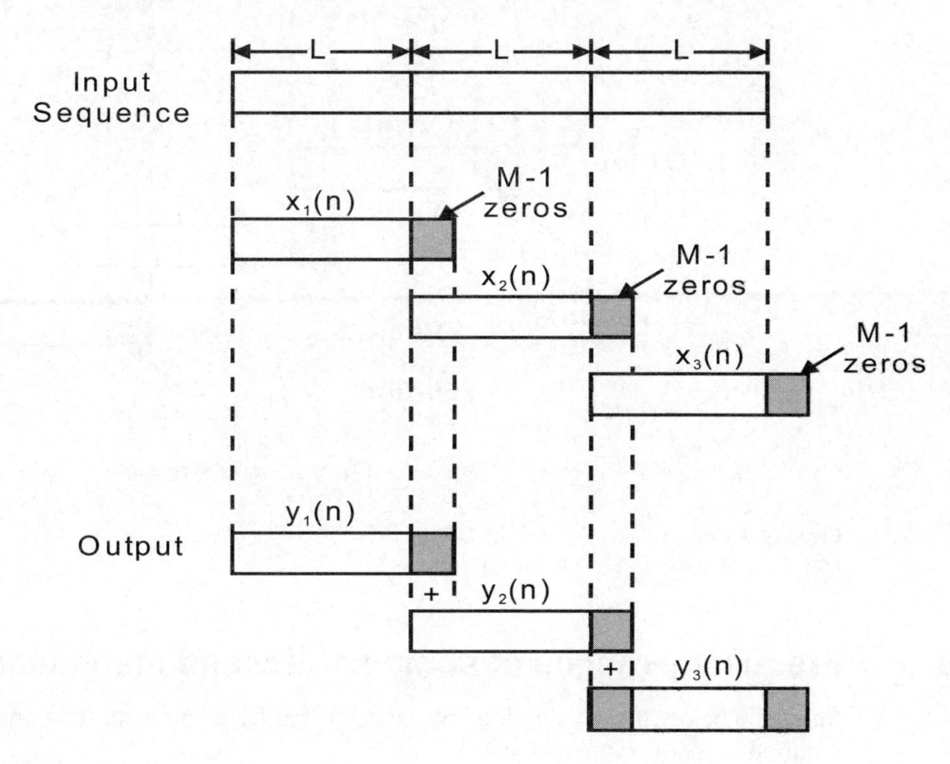

Figure 4.8
Illustration of the overlap-add method

4.7.2.2 Overlap-save method

With this method the size of each subsequence is $N=L+M-1$. The first M–1 points in each subsequence comes from the previous subsequence. For the first subsequence, M–1 zeros will be added in its place. The size of DFTs and IDFT is N. The output of each processed subsequence $y_m(n)$ will have N points. The first M–1 points of $y_m(n)$ are discarded, leaving it with L points.

This method is called overlap-save because the last M–1 points of each subsequence are saved for the processing in the next block. Figure 4.9 illustrates this method graphically.

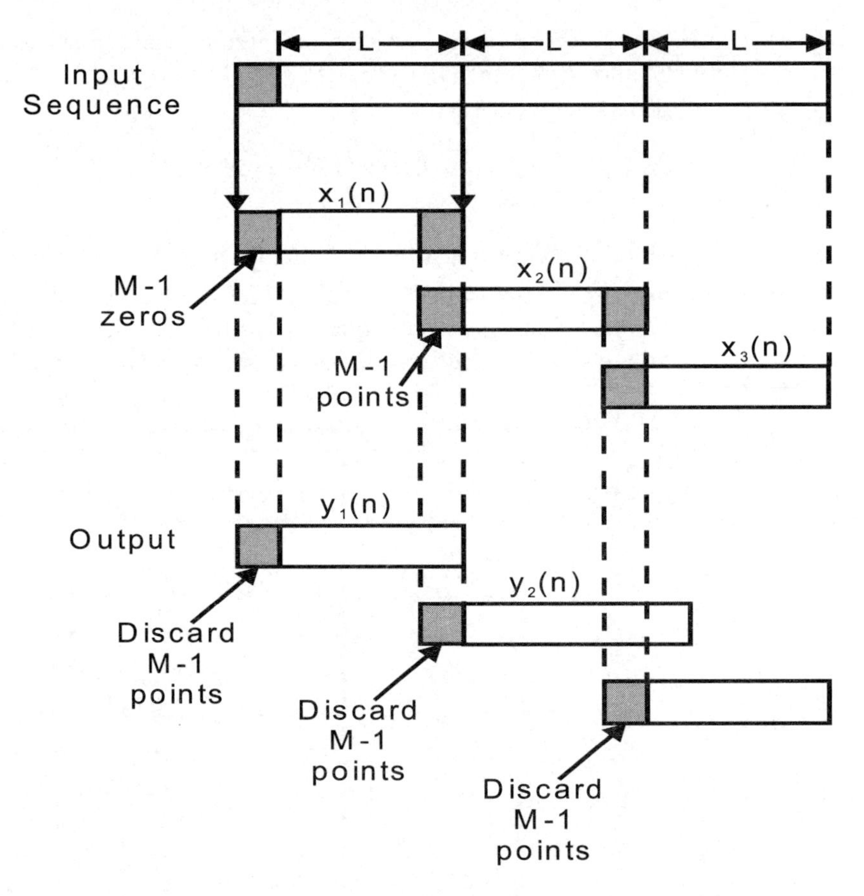

Figure 4.9
Illustration of the overlap-save method

4.8 Frequency ranges of some natural and man-made signals

The table below gives us an idea of the sampling frequency and the computational speed required for some typical signals.

Type of Signal	Approx. Frequency Range (Hz)
Voiced speech (e.g. vowels)	100–4000
Speech (fricatives)	100–8000
Electrocardiogram (ECG)	0–100
Electroencephalogram (EEG)	0–100
Earthquake and seismic signals	0.01–10
Radio broadcast	3×10^4–3×10^6
Microwave	3×10^8–3×10^{10}
Infrared	3×10^{11}–3×10^{14}
Gamma and X-rays	3×10^{17}–3×10^{18}

5

DSP application examples

It may be useful at this point to discuss in considerable detail some common DSP applications. Some of the contents of this chapter will be better appreciated after learning about digital filter design and implementations. However, these examples will help us recapture the broad picture of the usefulness of DSP in a variety of engineering areas.

The first application is on periodic signal generation using wave tables. Wave table generation techniques have become more popular with some of the new audio cards available for PCs. It is a very flexible way to generate periodic waveforms like sinusoidal waves. Wave table synthesis lies at the heart of many computer music application programs. This example also serves to illustrate the important concepts learnt in previous chapters on sampling.

The second application is in the area of communication systems. We shall describe the implementation of a certain wireless transmitter using DSP techniques. The specific advantages of this approach will be discussed. This example will illustrate the choice of sampling frequency or ADC resolution and the effects of these choices on the spectral characteristics of the generated signal.

The third application is in the area of speech synthesis. Speech synthesis is a very broad area that requires understanding of the characteristics of speech signals, phonetics, and at a higher level, linguistics. A simple way to model and synthesize speech signals by a model of the human vocal tract will be described. This application illustrates the filtering of some simple signals to create signals with certain desired spectral characteristics. The filtering required is in fact time varying.

We shall also describe some applications of DSP in image processing. Image (and video) processing is one of the major applications of DSP. Some photographic outlets offer instant photo enlargements and cropping, enhancement of old photographs, and other imaging services. None of these would be possible without the aid of DSP.

Finally, the application of active noise control will be discussed. Active noise control (ANC) is based on the simple physics of destructive interference of propagating acoustic waves. DSP systems are now powerful to enable real-time ANC systems to be developed with applications in air conditioning ducts, aircraft, cars and magnetic resonance imaging (MRI) systems.

5.1 Periodic signal generation using wave tables

Many DSP applications require the generation of periodic waveforms such as sinusoids, periodic square waves, sawtooth signals, etc. An example application is the generation of dual-tone multi-frequency (DTMF) signals for touch-tone telephone handsets.

Figure 5.1 shows the two frequency groups of a DTMF keypad.

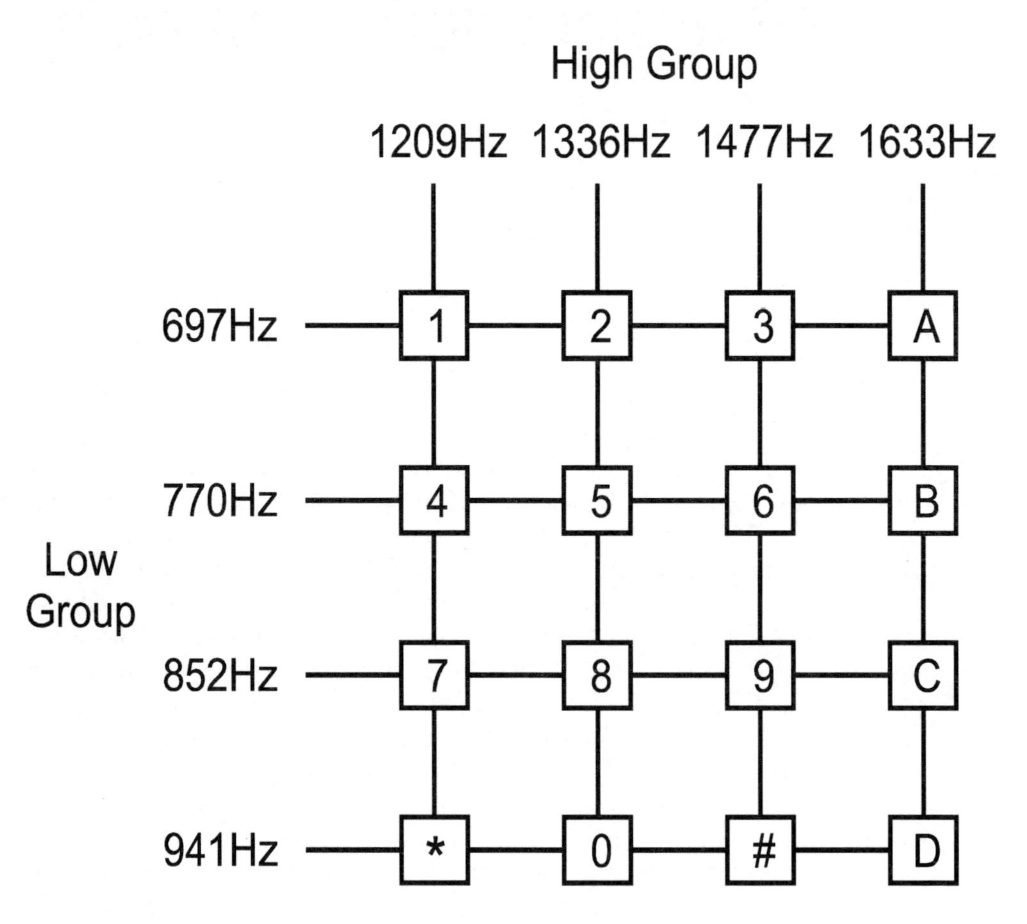

Figure 5.1
Frequencies for a DTMF keypad

When a key on the keypad is pressed, a signal, which is the sum of two audible sinusoidal tones, is generated. Each key on the keypad is uniquely defined by a frequency pair $\{f_L, f_H\}$, one from the low and one from the high frequency group. The digitally generated signal is mathematically given by

$$y(n) = \cos(\omega_L n) + \cos(\omega_H n)$$

where

$$\omega_L = 2\pi f_L / f_s$$
$$\omega_H = 2\pi f_H / f_s$$

f_s is the sampling frequency, and n is the integer time index.

5.1.1 Digital waveform generation using wave tables

One approach to generate such periodic waveforms is to design a digital filter with an impulse response $h(n)$ corresponding to one period of the waveform, one wishes to generate. The periodicity is generated by exciting this digital filter with a train of impulses separated by the fundamental period of the waveform. This process is shown in Figure 5.2.

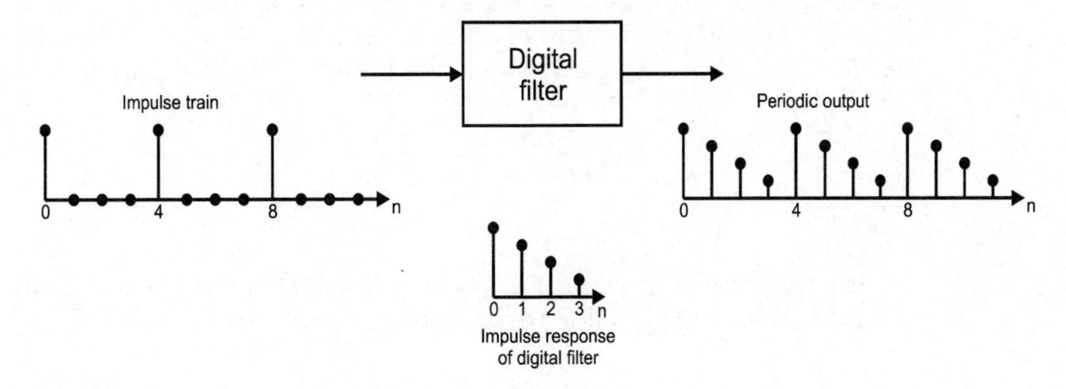

Figure 5.2
Periodic digital waveform generation using digital filters

A more efficient approach is to pre-compute the samples of the waveform and store them in the system's memory (RAM or ROM). The data are arranged as a circular buffer and accessed when needed. The period of the waveform can be controlled by either varying the speed of cycling around the table or by accessing a subset of the table at a fixed speed. This approach is called wave table synthesis, which is used very successfully in computer music. Many audio cards available for PCs now use wave table generators.

5.1.2 Sampling frequency

A sinusoidal signal does not necessarily remain periodic when sampled at a given frequency. In order that $y(n)$ remains periodic in the time index n with a period of, say, D samples, it is necessary that one whole period of the sinusoid fit within the D samples. This requires that the sampling frequency is an integral multiple of the analog frequency f. That is,

$$f_s = Df$$

Due to the periodicity, only the samples for one period of the signal need be calculated and stored. f is now the fundamental frequency of the wave table. A typical sampling frequency for DTMF generation is 8 kHz.

There are sinusoids of 8 different frequencies that need to be generated for DTMF signals. One way to do it is to have 8 wave tables. Since there are only a few entries per table, this approach is not impractical. Another way is to change the fundamental frequency using a single table. This is the approach that we shall examine in more detail.

5.1.3 Generating integer multiples of the fundamental frequency

The frequency f can be changed either by changing the sampling frequency f_s or by changing the effective length D of the basic period. The first approach is not practical in our application since we need to deal with two different frequencies each time. Thus the second approach will be used.

The fundamental frequency generated from a wave table with D entries and a sampling frequency is given by

$$f = \frac{f_s}{D}$$

Replacing D by a smaller value, d will increase the frequency of the sinusoid. For instance, if $d=D/2$, the frequency is doubled. This also means that only every other entry in the wave table will be accessed for each period. So the new generated frequency will be

$$f' = \frac{f_s}{d} = \frac{f_s}{D/2} = 2\frac{f_s}{D} = 2f$$

Given the desired frequency f_d and a table length D, we will be using only c regularly spaced samples from the full wave table and c is given by

$$c = D\frac{f_d}{f_s}$$

Now we have assumed that d and c are integers. This clearly restricts the choices of frequencies we can generate. We shall now consider how we can generate other frequencies where both d and c are real numbers.

5.1.4 Generating arbitrary frequencies

Any frequency f that we want to generate must be within the nyquist interval

$$|f| \le \frac{f_s}{2}$$

This requires that c satisfies the condition

$$|c| \le \frac{D}{2}$$

Negative values of c correspond to negative frequencies. This is useful for introducing 180° phase shifts in waveforms.

Since c is no longer an integer, some truncation or rounding will have to be done in order to get values from the wave table.

Alternatively, we can interpolate between sample values from the wave table for more accurate synthesis. Linear interpolation is usually sufficient. Let $w(i)$ and $w(j)$ be the i and j entries in the wave table and we need a sample with a real-valued index q which is between i and j. The linearly interpolated value is given by

$$y = w(i) + (q - i)[w(j) - w(i)]$$

This interpolation is shown in Figure 5.3.

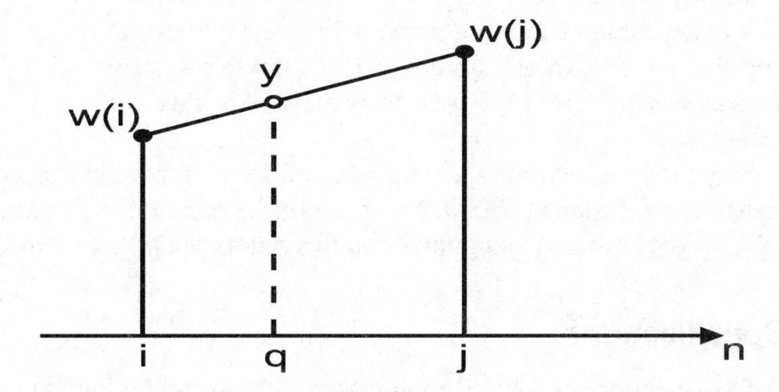

Figure 5.3
Linear interpolation between two samples

Interpolation naturally produces the most accurate results with rounding coming next. Truncation is the most inaccurate of the three methods. The inaccuracies become smaller as the length of the wave table D increases. In computer music applications, wave table sizes vary between 512 and 32 768.

5.1.5 DTMF example

Coming back to the generation of the DTMF frequencies. Assume that the sampling frequency is 8 kHz and the size of the wave table is 200. The fundamental frequency is therefore 400 Hz.

For the low frequency group, the values of c are given by

$$c_1 = \frac{(200)(697)}{8000} = 17.425$$

$$c_2 = \frac{(200)(770)}{8000} = 19.25$$

$$c_3 = \frac{(200)(852)}{8000} = 21.3$$

$$c_4 = \frac{(200)(941)}{8000} = 23.525$$

Similarly, the values of c for the high frequency groups are found to be 30.225, 33.4, 36.925 and 40.825.

5.2 Wireless transmitter implementation

A paging systems standard has been created in Europe in 1989 by the ETSI (European Telecommunications Standards Institute). It is called the ERMES (European radio message system). The main objective is to allow roaming throughout Europe and to guarantee receiver compatibility.

There is an older system called POCSAG, which transmits at 512 or 1200 bits per second. With new service needs foreseen in the most populated areas of Europe, ERMES was designed to transmit at 6250 bits per second. The modulation format that has been chosen is called 4-PAM/FM modulation. Since each symbol is encoded by two bits, the actual symbol rate is 3125 symbols (bauds) per second.

It is desirable to design a transmitter that can transmit using the older POCSAG system and the newer ERMES system. Such a flexible system will have the benefit of reduced system design cost. One way to achieve this flexibility is to implement it using DSP techniques.

Since radio spectrum is scarce, specifications for radio transmitters are very tight. The transmitters designed should be very precise and stable. Again, this is one of the major advantages of digital techniques and this points to DSP implementation.

5.2.1 Specifications

All the specifications of the modulation method can be found in

- 'European radio message system – part 4: air interface specification', ETSI DE/PS 2 01-4, version 0.2.1, November 1990.
- 'European radio message system – part 6: base station conformance specification', ETSI DE/PS 2 01-4, version 0.2.1, November 1990.

The modulated signal generation process is illustrated in block diagram form in Figure 5.4.

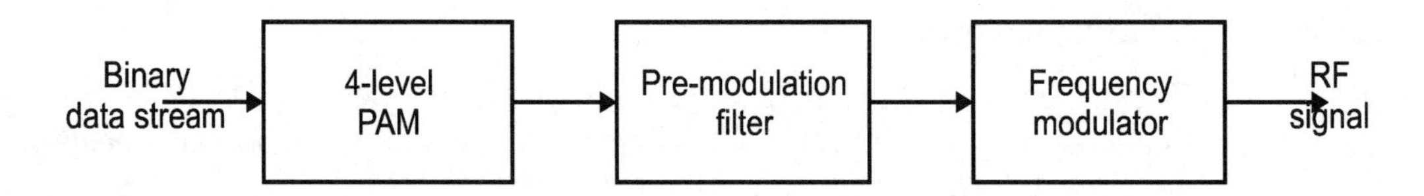

Figure 5.4
Generation of 4-PAM/FM signals

Using this technique, the communication of two data bits is achieved by the transmission of one of four signaling frequencies. The modulated signal is required to have a continuous phase (no sudden jumps in phase). This and the pre-modulation pulse shaping (or filtering) of the data stream constrain the transmitted radio frequency (RF) spectrum. The four signal frequencies are as shown below.

Nominal frequency	Symbol
$f0 + 4687.5$ Hz	10
$f0 + 1562.5$ Hz	11
$f0 - 1562.5$ Hz	01
$f0 - 4687.5$ Hz	00

Here $f0$ is the intended operating frequency. In a data stream, the most significant bit shall be transmitted first. The transmission rates are

Data rate: 6.25 kbits per second

Symbol rate: 3.125 kbauds

The binary signal is filtered by the pre-modulation filter to create a smooth signal rather than one with sudden jumps in levels. The specifications for this filter are shown in Figure 5.5.

The specifications are given in terms of frequency characteristics with upper and lower limits. Apart from the amplitude spectrum, the phase spectrum is also specified as group delay characteristics. Group delay is defined as the negative rate of change of phase. So a constant group delay implies a linear phase function. In most cases for data transmission, this is the ideal. These specifications are derived from a 10th order low-pass Bessel filter with a 3 dB bandwidth of 3.9 kHz.

The rise or fall time for the frequency transition between two successive symbols shall be 88 microseconds with a tolerance of 2 microseconds.

The RF spectrum of the output of the transmitter shall conform to the mask shown in Figure 5.6.

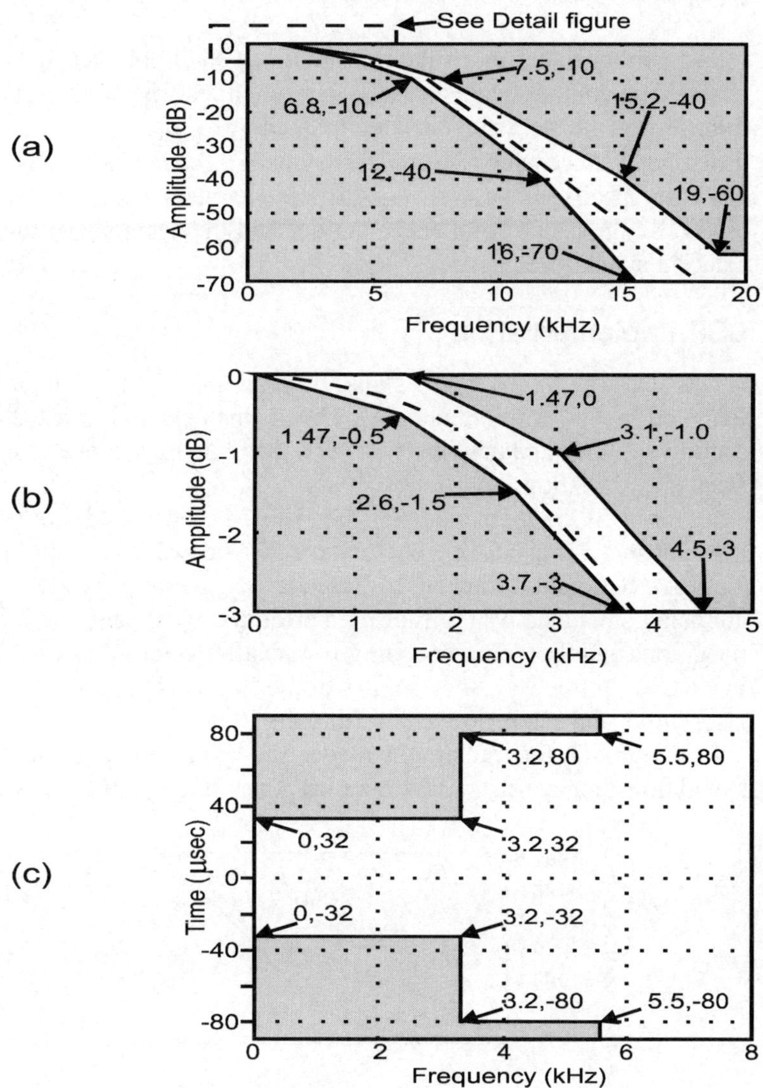

Figure 5.5
Specifications of the premodulation filter in ERMES

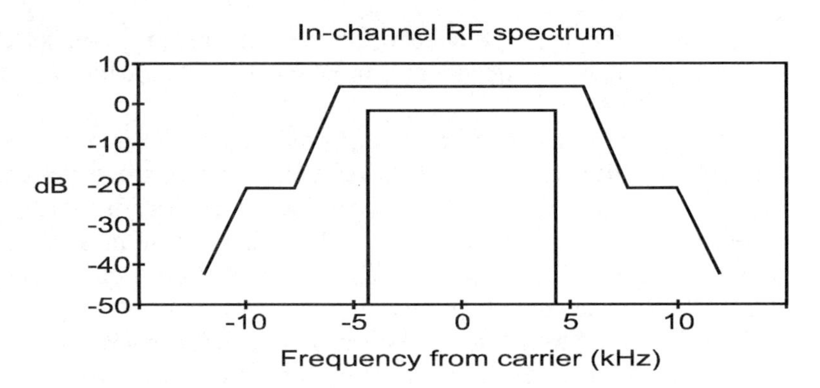

Figure 5.6
Specifications of the RF signal spectrum for ERMES

The center frequency of the transmissions shall not exceed 15 Hz either way from the intended operating frequency (f_0). The intended operating frequency may be forced to differ from the nominal channel frequency by up to 185 Hz (frequency offset). The difference between any two adjacent symbol frequencies shall be 3125 ± 15 Hz.

These specifications are very difficult to meet using conventional analog circuitry, especially the center frequency stability and the accuracy of the difference between two adjacent symbol frequencies.

5.2.2 DSP implementation

A classical way to generate a signal of this kind is to use an analog pre-modulation filter followed by an FM modulator. However, this approach presents problems regarding the stability of the parameters. It is very difficult to achieve a precision of 0.1% in the frequency deviation.

In the DSP implementation, the 4-PAM coder and the pre-modulation filter are implemented in the following manner. Since we have four possible symbols in the alphabet, there are a total of 16 possible transitions between symbols. These transitions are being smoothed by the filtering performed by the pre-modulation filter. All these 16 transitions and their corresponding results after filtering can be pre-calculated and stored in a ROM. In this way, the samples of the filtered signal are a function of the current and last symbol and can be looked up from the ROM.

In Figure 5.7, the transition between the symbol 00 and the symbol 10 is shown. The stored transition consists of 64 samples per symbol, 16 bits per sample.

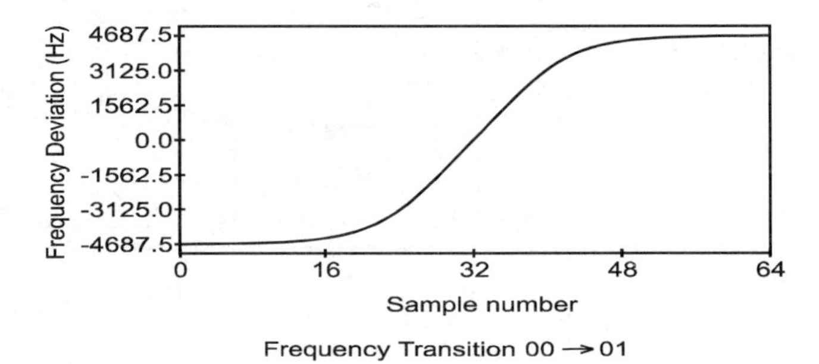

Figure 5.7
Frequency transition between two symbols in ERMES

The FM modulator block diagram is shown in Figure 5.8. The frequency samples obtained from the previous process are passed to an integrator, which is simply an accumulator of all past values.

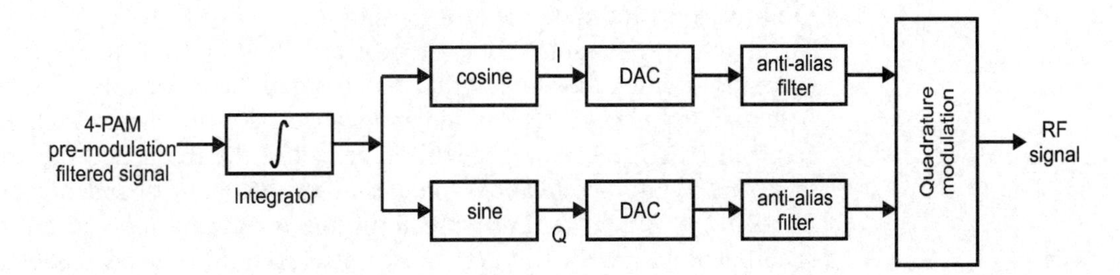

Figure 5.8
Block diagram of frequency modulator using DSP

This generates the phase samples corresponding to the intended frequencies. The output from the integrator becomes the arguments of the sine and cosine tables. The cosine table gives the in-phase (I) component of the modulated signal and the sine table gives the quadrature (Q) component. The tables are quantized with 8-bit resolution. These samples are converted to analog signals using two DACs and a quadrature modulator, which can be obtained commercially as a chip, is used to generate the RF signal.

The specification allows frequency offsets that can be introduced intentionally. A very precise offset can be introduced by adding or subtracting a certain constant from the filter output samples.

POCSAG can also be implemented using the same scheme by including a different transition table corresponding to the POCSAG specification. In this way, the transmitter can switch between the two systems by simply getting its samples from a different table.

The symbol rate is low enough to allow us to choose the very high sampling rate of 64 samples per symbol. This high oversampling rate implies that very simple anti-aliasing filters can be used after the DAC. High sampling rate also reduces the time uncertainty of the transitions.

Figure 5.9 shows the analog I and Q signals generated with input symbols (10, 00, 10, 00).

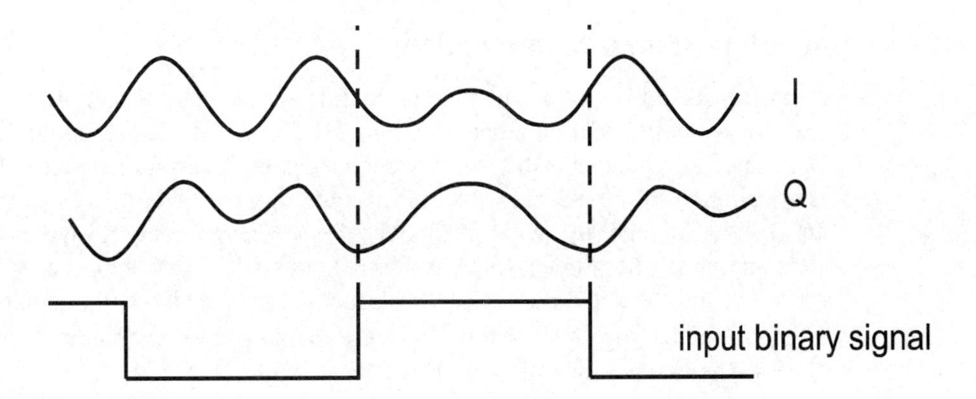

Figure 5.9
The in-phase and quadrature signals

5.2.3 Other advantages

There are some other advantages in using the DSP approach that may not be obvious initially. Two major ones are briefly discussed below:

- **Quadrature modulator compensation**
  The quadrature modulator can process the I and Q signals to produce the RF output is an integrated circuit that consists of analog circuitry. Hence there are potential frequency offsets, and imbalance between the two channels. The result of these defects is generally a spread in the frequency spectrum. Since there are tight specifications on the allowable out-of-band power (at least 72 dB for ERMES), this spread in the spectrum may mean that these specifications are violated. If these imperfections can be modeled and the parameters of the model can be measured on-line or off-line, then compensation can be applied to the discrete-time signal before modulation so that the resulting modulated signal is near perfect.

- **Power amplifier linearization**
  The most efficient power amplifiers have non-linear transfer characteristics. If the modulation format depends on the amplitude of the signal, then this non-linearity introduced by the power amplifier will severely distort the original signal. This is the reason why constant envelope type of modulations such as FM, is used in radio systems.

On the other hand, the most efficient modulation schemes do not produce constant envelope signals. Hence they require linear amplifiers which are much less efficient. By using DSP implementations, we can pre-compensate for the amplifier non-linearity by pre-distorting the discrete-time signal. In this way, the most efficient modulation schemes can be used with the most efficient power amplifiers. This is obviously not possible with analog implementations.

5.3 Speech synthesis

Digital speech processing has been one of the most important areas of DSP. It is the application of digital speech and image (including video) processing that leads to the explosion of multimedia communication that we are experiencing at the moment.

5.3.1 Speech production mechanism

Speech signals consist of a sequence of sounds. These sounds and the transition between them carry the information that needs to be conveyed. These sound sequences obey certain rules. Linguistics is the study of such rules for a certain language. The study of the classification of the basic sounds is in the realm of phonetics.

In order to come up with a model of speech production, we need to have an understanding of the human vocal system. It consists of two main parts: the vocal cords (or glottis), and the vocal tract. The vocal tract in turn consists of three main parts:

- The pharynx – connection from the esophagus to the mouth.
- The oral cavity – the mouth.
- The nasal tract – begins at the velum and ends at the nostrils.

The source of energy comes from the air pressure exerted by the lungs, bronchi and trachea. Speech is produced when an acoustic wave is radiated from this vocal system when air is expelled from the lungs and the air flow is perturbed by constrictions somewhere in the vocal tract. When the velum is lowered, the nasal tract is acoustically coupled to the vocal tract to produce nasal sounds.

5.3.2 Classification of sounds

The basic units of speech sounds in the English language are called phonemes. There are two main types of phonemes: vowels and consonants. More detailed classifications are also available. But for our discussions, we shall assume only these two types of sounds.

Vowels are produced when the vocal tract is excited by pulses of air caused by the vibration of the vocal cords. The vibration is periodic in nature and the period is the pitch of that sound. The shape of the vocal tract determines the resonant frequencies of the tract, called formants. For vowels, there are typically three formants between the frequencies 200 Hz and 3 kHz. The exact frequencies of the formants vary from person to person. Figure 5.10 shows a typical frequency spectrum of a vowel.

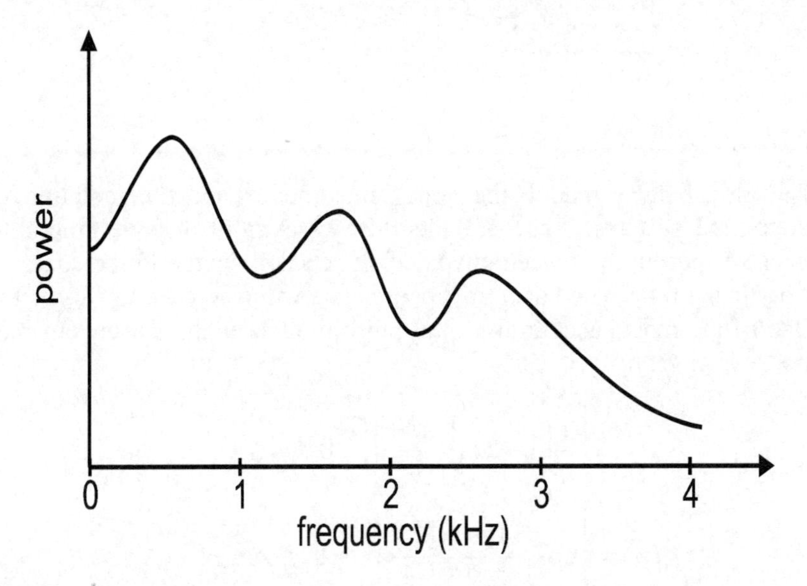

Figure 5.10
Typical spectrum of a vowel

In the production of consonants, the vocal cord is totally relaxed in general, although there are exceptions. In this way, air flows into the vocal tract without the periodic excitation generated by the vocal cord. Consonants can be broadly classified into:

- **Nasals**
  Nasals are produced when the vocal tract is totally constricted at some point along the oral cavity. The velum is lowered and the air flows through the nasal tract, radiating through the nostrils.
- **Fricatives**
  Fricatives are produced when the steady air flow becomes turbulent in the region of a constriction in the vocal tract.
- **Stops**
  Stops are transient sounds produced by building up pressure behind a total constriction somewhere in the oral tract, and suddenly releasing the pressure.

5.3.3 Speech production model

In order to synthesize speech sounds artificially; we need a model of the speech production system described above. We have looked at one briefly in Chapter 1. Figure 5.11 shows a more detailed model.

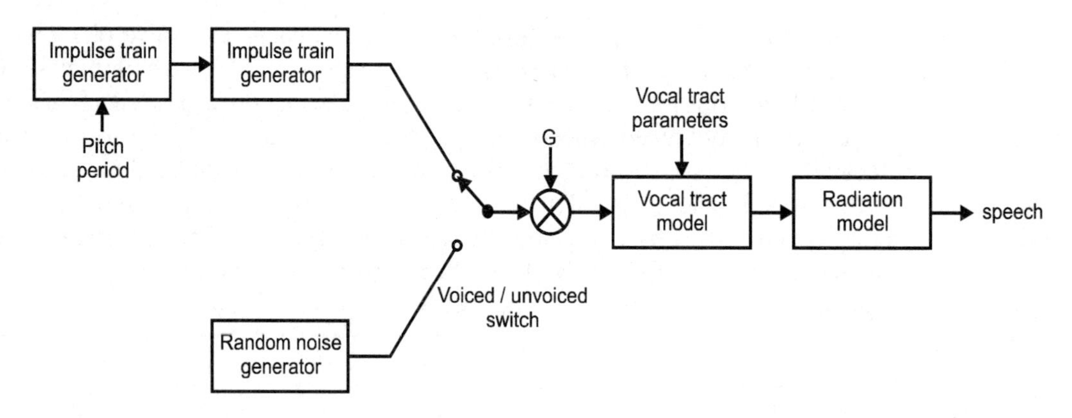

Figure 5.11
A speech production model

The glottal pulse model, the vocal tract model, and the radiation model are linear discrete-time systems. They are therefore essentially discrete-time filters. In order to synthesize speech, the voiced/unvoiced switch will switch to the source for the sound at that particular time. The vocal tract parameters will also need to vary with time.

One of the most successful glottal pulse models is the Rosenberg model. Its impulse response is given by

$$g(n)=\begin{cases}\frac{1}{2}\left[1-\cos\left(\frac{\pi N}{N_1}\right)\right] & 0\le n\le N_1\\ \cos\left[\frac{\pi(n-N_1)}{2N_2}\right] & N_1\le n\le N_1+N_2\\ 0 & \textit{otherwise}\end{cases}$$

The vocal tract model is usually a linear predictive model. It is so called because the current speech sample is generated from a number of past samples plus the current

$$s(n)=\sum_{k=1}^{y}a_k s(n-k)+u(n)$$

excitation. This can be described in equation form as

Here a_k is the coefficient for the model and it changes from one phoneme to another, and $u(n)$ is the input sample to the vocal tract model. The prediction order, p, is typically 10 to 12.

In most cases, the radiation model is ignored.

In the practical sessions, there will be an opportunity to experiment with the model.

5.4 Image enhancement

In this course, we only deal with the processing of one-dimensional signals and images are inherently two-dimensional. However, image processing is a very important DSP application area. We shall consider briefly the application of DSP techniques to the enhancement of images. This will give us some insights into what this area is about. Some operations are also non-linear as opposed to linear operations we have discussed so far.

Image enhancement is the processing of images to improve their appearance. There are a variety of methods, which are suitable for different objectives. Some objectives are to improve the image quality and visual appearance to a human viewer. Other ones include the sharpening of an image to aid in the automatic machine recognition of objects. But the overall objective is to make the processed image better in some sense than the unprocessed one.

We shall consider two types of enhancement: contrast and dynamic range enhancement, and noise reduction. For simplicity, we shall only use gray-scale images.

5.4.1 Contrast enhancement

A simple way to improve the contrast or the dynamic range of image pixel intensities is by a technique called gray-scale modification. It applies a transformation T to the original image to produce the enhanced image. This transformation is often represented by a table.

Consider an image of 5×5 pixels represented by 3 bits. There are a total of eight levels with 0 being the darkest to 7 being the brightest. The pixel values are shown in Figure 5.12. We can see that the pixel values are between 2 and 5. So only 4 out of 7 possible levels are used. Making use of the full range of values can produce a better contrast.

2	3	3	4	5
2	3	3	4	5
2	3	4	4	5
2	3	4	4	5
2	3	4	4	5

Figure 5.12
Pixel values of a 5 × 5 pixel region

The problem is how we can find a suitable transformation that will do a good job. Computing the histogram of the image and studying its characteristics can identify a suitable transformation. The histogram is just a tabulation or a graph of the number of pixels that have specific intensities. The histogram of Figure 5.12 is shown in Figure 5.13.

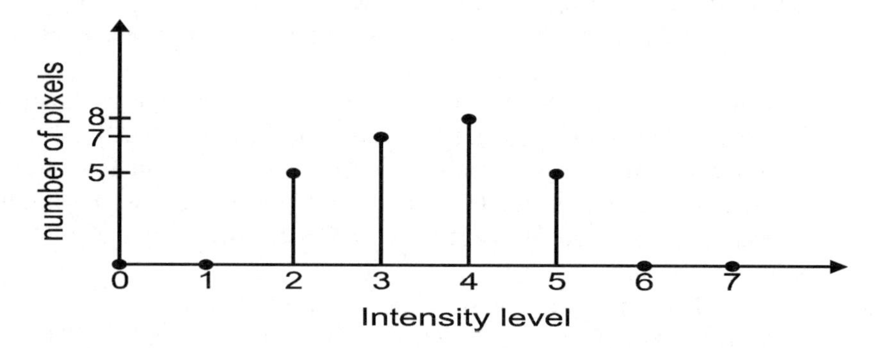

Figure 5.13
Histogram of the image in Figure 5.12

This histogram showed us that the dynamic range is not well utilized as discussed above. A transformation that will improve the contrast is shown in Figure 5.14 and the resulting output image after the transformation is in Figure 5.15.

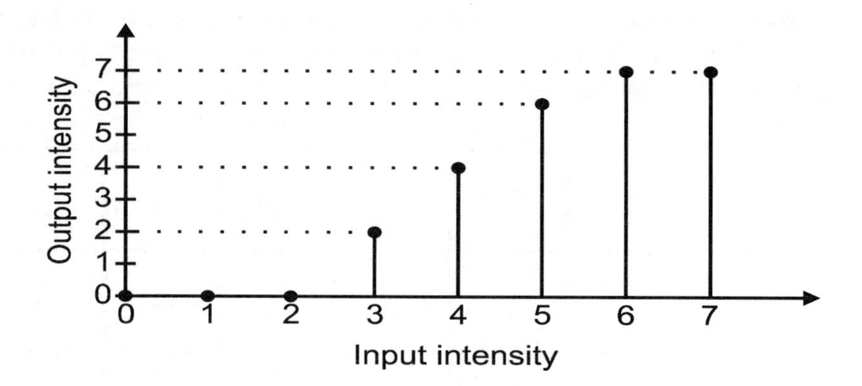

Figure 5.14
A contrast enhancement transformation

0	2	2	4	6
0	2	2	4	6
0	2	4	4	6
0	2	4	4	6
0	2	4	4	6

Figure 5.15
Image of Figure 5.12 after contrast enhancement transformation

The transformation can be obtained automatically by defining a desired histogram. In most cases, the desired histogram is usually a uniform distribution of gray level values within the image. This will make the number of pixels at any one gray level about the same as another. The transformation T must be monotonically non-decreasing like the one in Figure 5.14. It is given by

$$T_i = \frac{M-1}{N} \sum_{j=0}^{i} n_i$$

where N is the total number of pixels in the image, n_i is the number of pixels at gray level i, and M is the total number of gray levels possible.

This simple procedure often produces significant improvements in image quality or intelligibility to the viewer.

5.4.2 Noise reduction

There are two main types of noise in images. One is the uniform random noise similar to those for one-dimensional images. Another type one is known as impulse noise or salt-and-pepper noise. They appear as isolated bright or dark pixels in the image. They can occur due to random bit error during transmission.

The energy of a typical image is primarily in the low frequency region. Therefore, (two-dimensional) low-pass filtering will be quite effective in removing a substantial amount of uniform random noise. This is done at the expense of removing some details of the image. It should be noted that edges that exist in the image produce high frequency components. If these components are removed or reduced in energy, then the edges will become fuzzier.

Median filters are very effective in removing impulse noise while preserving edges. They are non-linear filters however, and therefore the process cannot be reversed. In median filtering, a window or mask slides along the image. This window defines a local area around the pixel being processed. The median intensity value of the pixels within that window becomes the new intensity value of the pixel being processed.

Figure 5.16(a) shows a 3×3 window. The pixel being processed in the middle of this window. The numbers within the window are the intensity levels of the pixels in that window. Figure 5.16(b) indicates the processed output. Note that the intensity of the pixel in the middle is now replaced by the median value.

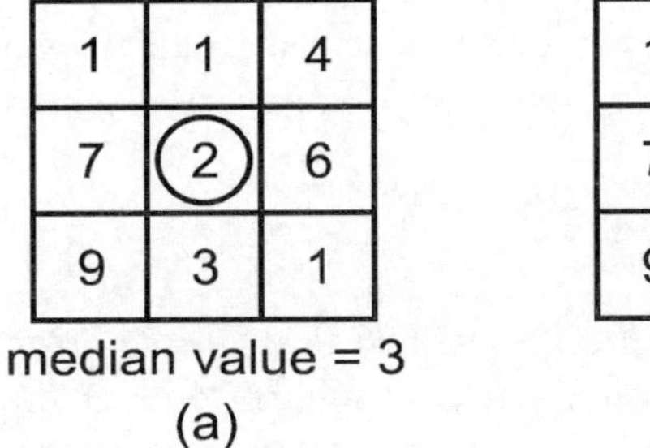

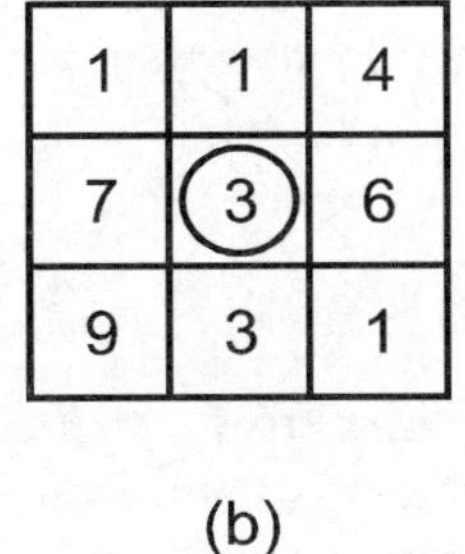

Figure 5.16
Median filtering using a 3 × 3 *window*

An important parameter in median filtering is the size of the window. Different results can often be obtained by using different window sizes. The choice also depends on the characteristics of the image and the noise. As a rule of thumb, images with lots of variations require the use of smaller windows while larger windows can be applied to images that have more uniform intensity areas.

5.5 Active noise control

Active noise control (ANC) is based on the simple physics of destructive interference of propagating acoustic waves. The concept that acoustic wave interference can be controlled to produce zones of quietness has been known since sounds waves were first modeled by linear equations. In fact, a US patent was granted in the 1930s for an analog ANC system. DSP devices are now powerful enough to allow us to design and implement digital ANC systems that operate in real-time.

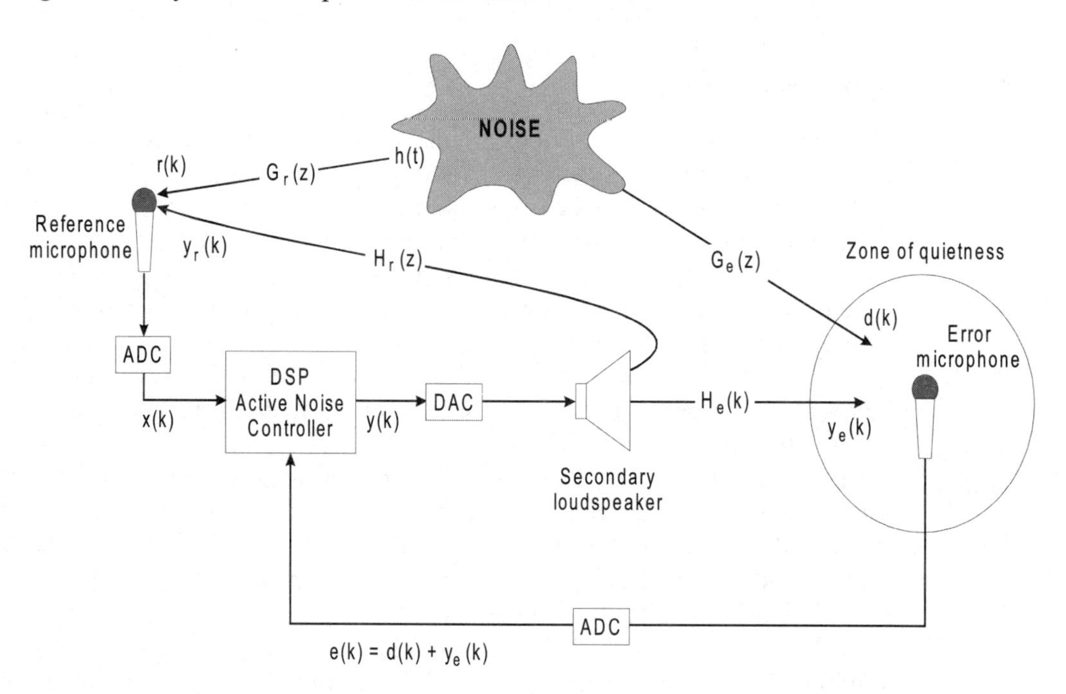

Figure 5.17
A single channel ANC system

Figure 5.17 illustrates the subsystems involved in a single channel digital ANC system. The secondary loudspeaker produces an acoustic signal, which on arrival at the error microphone, is a 180° phase-shifted version of the original signal *d*(*k*). If

$$y_e(k) = -d(k)$$

then the resulting error signal *e*(*k*) obtained through the error microphone is zero and a zone of quietness is setup around the error microphone. Owing to the complex nature of even the simplest acoustic environment, in practice zero error is unlikely to be achieved.

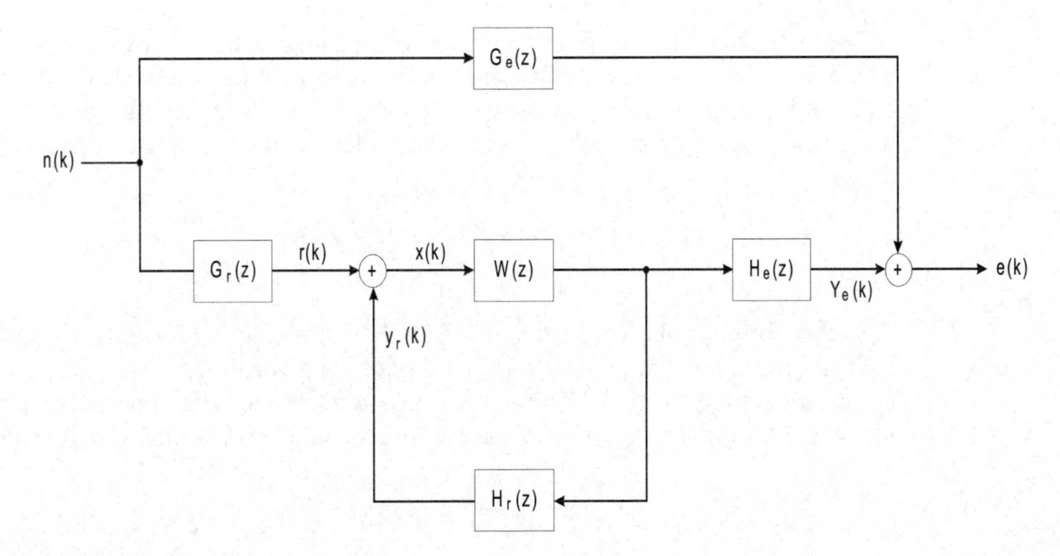

Figure 5.18
Model block diagram of ANC system

Figure 5.18 shows a system block diagram, which models the system in Figure 5.17. Here $H_e(z)$ models the acoustic path from the loudspeaker to the error microphone, and $H_r(z)$ models that from the loudspeaker to the reference microphone. The reference microphone provides the noise controller with an input signal whose spectral content is similar to that of $d(k)$. The reference microphone signal is linearly filtered to produce an appropriate loudspeaker output $y(k)$. The active noise controller is modeled by $W(z)$.

If we want $e(k) = 0$, without going into the details, the transfer function of the ANC filter is required to be as follows:

$$W_{\text{req}}(z) = \frac{-G_e(z)}{G_r(z)H_e(z) - G_e(z)H_r(z)}$$

This is an analytical solution to the problem provided we have full knowledge of the characteristics of the acoustic paths. This is generally not the case. Furthermore, acoustic path characteristics may change with time since they are affected by events such as movements of objects within the location concerned. Most practical ANC systems therefore use adaptive filtering techniques, which allows the system to adaptively model the acoustic paths. It also has the advantage that the adaptation algorithm can devote its efforts to solve for $W_{\text{req}}(z)$ only for those frequencies, which are actually contained in the noise signal. In this way, more efficient use of the filter coefficients can be obtained.

It should be noted that many adaptation algorithms require prior knowledge of the acoustic path $H_e(z)$ in order to solve for $W_{\text{req}}(z)$. Fortunately, the impulse response is usually well defined and easily measurable.

The inherent filter inside the active noise controller can either be an FIR or an IIR filter. As we shall see in the next two chapters, there are advantages and disadvantages for each type of filter in this application. FIR filters are stable and the equations for solving for the filter coefficients are easier to handle compared with IIR filters. But the order of FIR filter required is much higher compared to an IIR filter with similar spectral characteristics.

The filter weights are usually adapted or updated using a least-mean-squared (LMS) type of algorithm. This type of algorithm basically attempted to minimize the mean of the error signal squared, i.e. $e^2(k)$. At each iteration k, the filter coefficient $w(k)$ are updated using the currently available information. The simplest updating formula is given below:

$$w(k+1) = w(k) - 2\mu e(k)u(k)$$

In this formula, the parameter μ affects how fast the coefficients are changing. μ is typically much less than 1 to ensure stability of the algorithm.

Block diagram of an LMS adaptive algorithm for an IIR filter system is shown in Figure 5.19. $A(z)$ and $B(z)$ are the transfer functions related to filter coefficients.

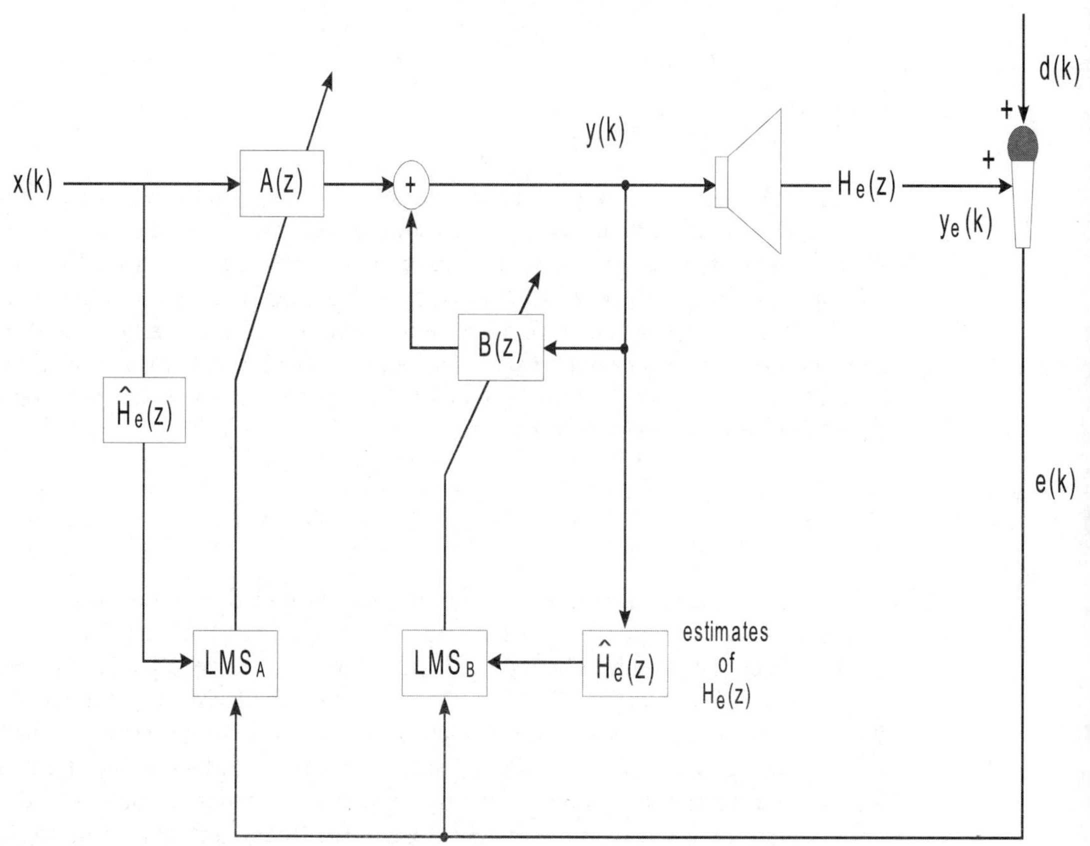

Figure 5.19
Adaptive IIR system for ANC

The order of the filter used is dependent on the complexity of the acoustic path $H_e(z)$ that is reflected by the impulse response of this path. Filter orders between 100 and 200 are not uncommon. Naturally, the filter order is also dependent on the sampling frequency since it affects the total delay that the filter can model. For instance, a sampling frequency of 2 kHz has a sampling period of 5 ms, so a filter order of 100 can model an acoustic path delay of 500 ms or half a second.

To create a larger or multiple zones of quietness, we can use a multi-channel noise controller similar to the one illustrated in Figure 5.20.

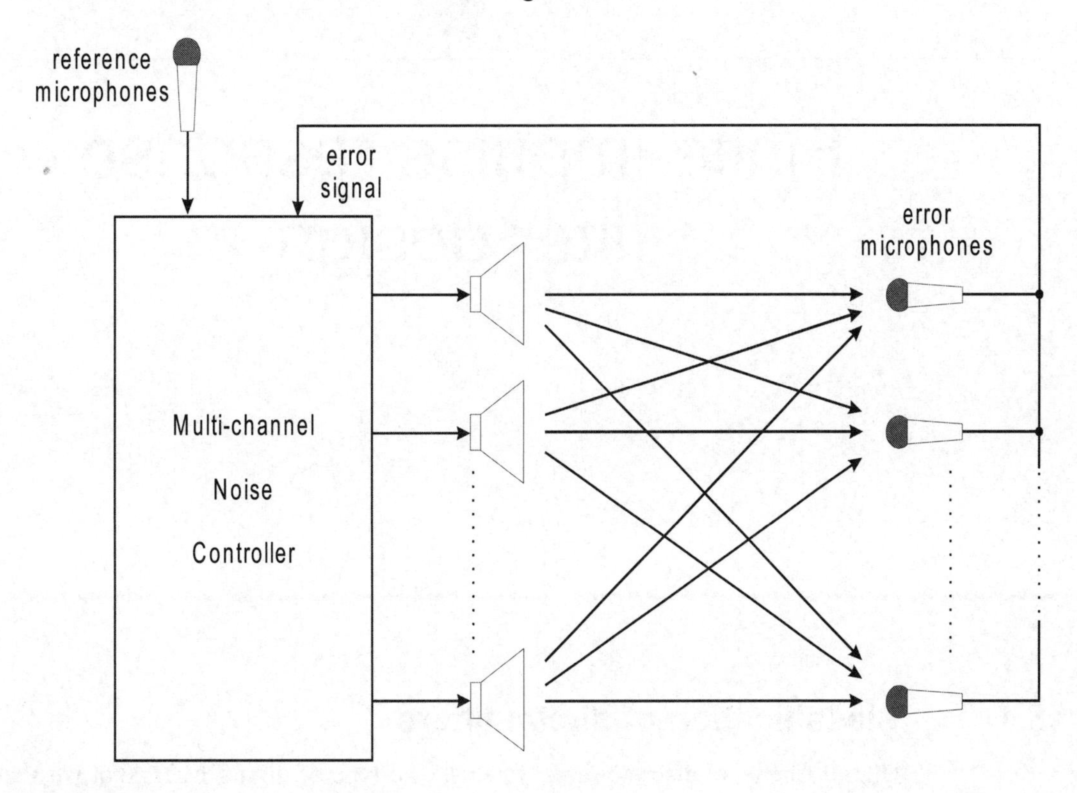

Figure 5.20
A multi-channel ANC system

To get a feel of the values of the parameters used, let's consider a MRI system. MRI scanners produce very high levels of low frequency acoustic noise, which can make the scanning procedure very traumatic and also interfere with patient-operation communication. A two-channel ANC system could be used with a 120th order FIR filter operating at a sampling frequency of 2 kHz. Frequencies below 350 Hz are reduced by around 10-20 dB but those frequencies above 350 Hz are not reduced so that voice communication between patient and operator is not affected.

5.6 To probe further

We can only touch on a few applications in several areas that may be of interest to our readers. As indicated in chapter 1, DSP application areas are very broad. A good source of review articles can be found in 'IEEE signal processing magazine'

Other popular electronics magazines also feature practical projects using DSP techniques.

6

Finite impulse response filter design

6.1 Classification of digital filters

Digital filters are discrete-time systems. The type of digital filters that we shall design in this course is linear. Therefore, they possess all the properties of linear discrete-time systems discussed in Chapter 3. All linear discrete-time operations on an input sequence can be viewed as a filtering of the sequence to produce an output sequence. This is the reason why digital filters are so important in DSP.

Non-linear filters are also commonly used, especially in areas such as image processing. The median filter discussed in section 5.4 for image enhancement is a typical non-linear digital filter.

Linear systems are characterized by their impulse responses. An impulse response can either have a finite or an infinite duration. A finite impulse response $h(n)$ has its non-zero values extending over a finite time interval and is zero beyond that interval. The following finite impulse response

$$h(n) = \{h_0, h_1, h_2, ..., h_N, 0, 0, 0, ...\}$$

has non-zero values in the interval

$$0 \leq n \leq N$$

and is referred to as a finite impulse response (FIR) filter or system of order N. So an Nth order FIR digital filter has an impulse response with a length of $(N+1)$ samples. The samples of the impulse response function (h_0, h_1, etc) are usually called filter coefficients, filter weights, and filter tap coefficients/weights.

If the impulse response function has an infinite duration, we have an infinite impulse response (IIR) filter. It is obvious that IIR filters cause computational problems since we cannot compute an infinite number of terms. But the type of IIR filters that are designed have their input and output samples interrelated through a linear difference equation. The output sequence can then be computed recursively. This is the reason why IIR filters are also known as recursive filters and FIR filters as non-recursive filters.

In this chapter, we shall concentrate on FIR filters. IIR filters will be discussed in detail in the next chapter.

6.2 Filter design process

The general digital filter design process can be broken up into four main steps:

- Approximation
- Synthesis and realization
- Performance analysis
- Implementation

These steps are illustrated in diagram form in Figure 6.1.

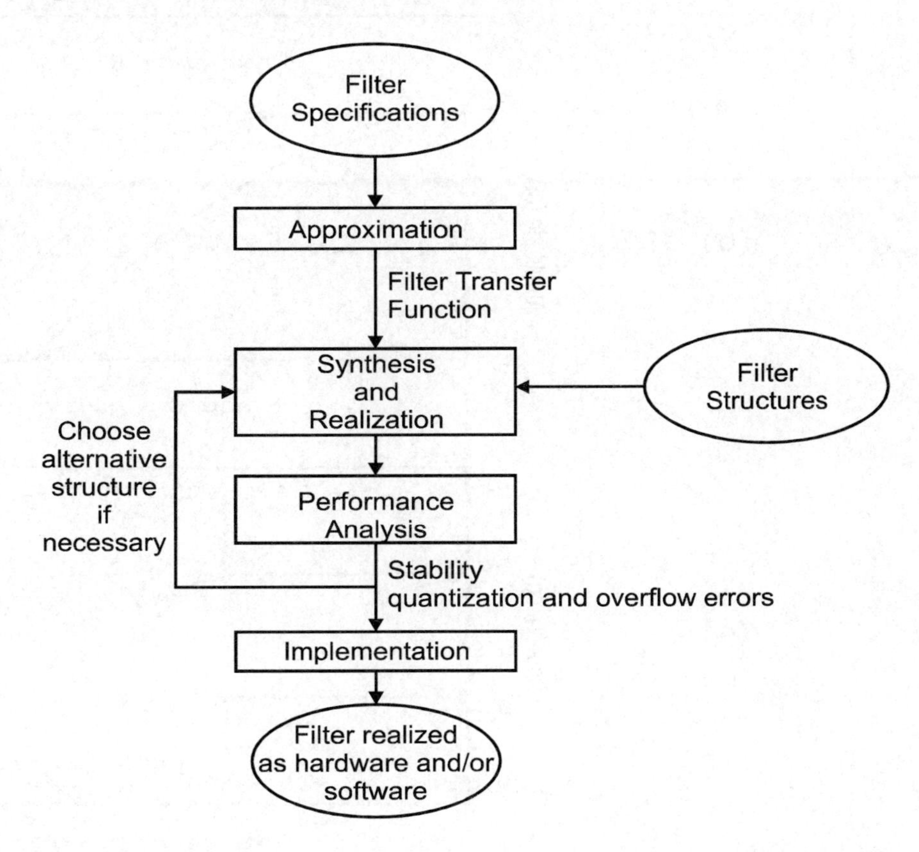

Figure 6.1
The filter design process

6.2.1 Approximation

The design process normally starts with the specifications and requirements of the filter, which are intimately related to the application at hand. These specifications may include frequency domain characteristics such as magnitude and phase responses. There may also be some time domain requirements such as maximum delay.

Most specifications define the upper and lower limits to each of these characteristics. Typical examples can be found in many communication system standards documents. The pre-modulation filter of the ERMES standard for paging systems is shown in Figure 6.2. Alternatively, a desired or ideal response may be given with the maximum amount of deviations from the ideal specified.

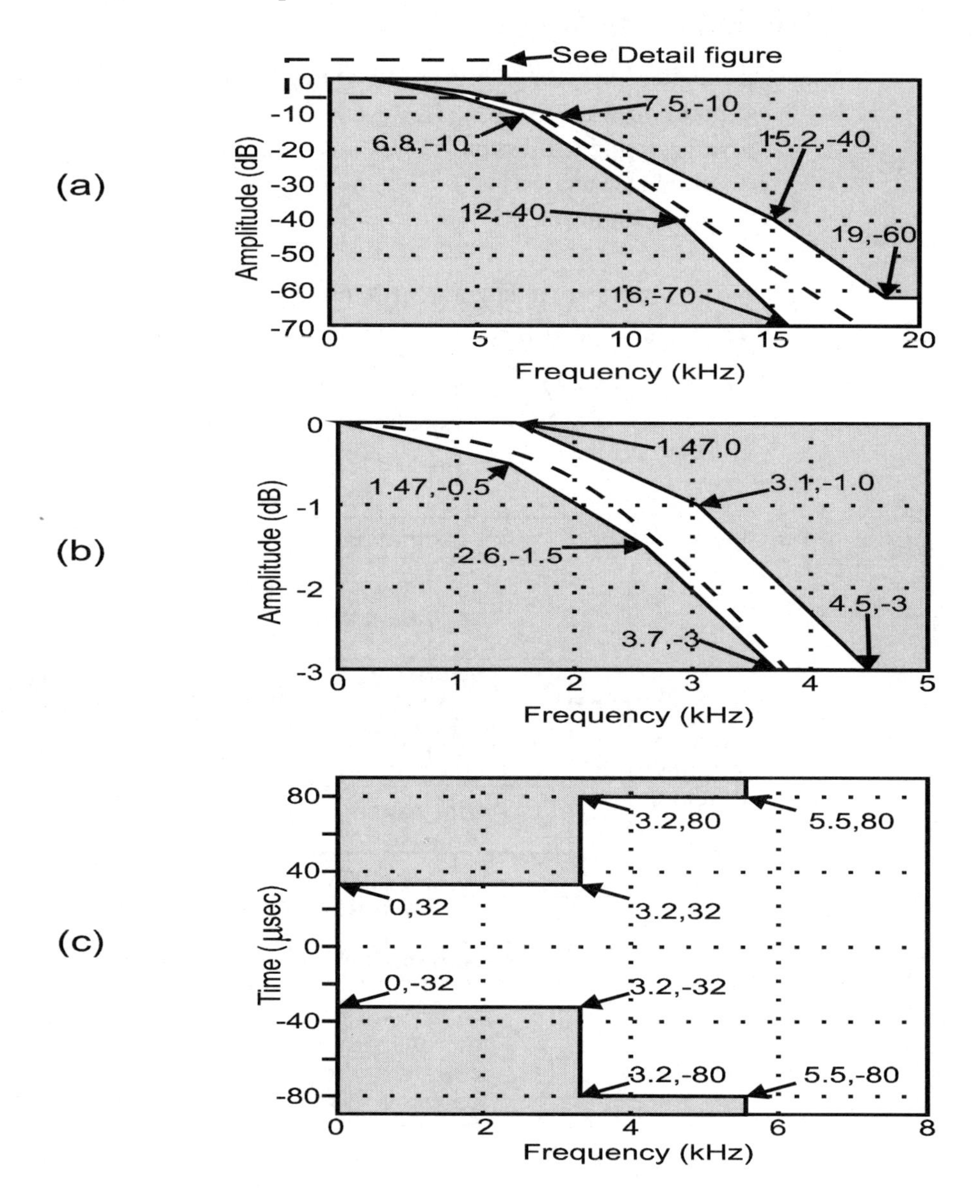

Figure 6.2
Pre-modulation filter specifications in the ERMES system

Given the filter specifications, the first step of the design process is to find a filter transfer function that will satisfy these specifications. This process is called approximation. It is so called because what we are doing is in fact finding a transfer function that approximates the ideal response that is specified.

The methods for solving the approximation problem for digital filters can be classified as direct or indirect. With direct methods, the problem is solved in the discrete-time (and hence discrete-frequency) domain. For indirect methods, a continuous-time transfer function is first obtained using well-established methods in analog filter design. This transfer function is then transformed into a discrete-time transfer function. Indirect methods are more commonly used for IIR filters, whereas FIR filter design methods are mostly direct ones.

These solution methods can also be classified as closed-form or iterative. Closed form methods make use of closed-form formulas and are usually completed in a definite number of steps. Iterative methods make use of optimization techniques that start with an initial solution and the solution is refined progressively until some pre-determined performance criteria are satisfied. The number of iterations is unknown and depends on the initial solution and the effectiveness of the optimization techniques employed.

6.2.2 Synthesis and realization

Once the transfer function has been determined, it has to be realized into a discrete-time linear network. This procedure is analogous to the filter realization procedure for analog filters where suitable circuit topology and circuit element values are chosen to realize a certain filter transfer function. A number of realization methods has been proposed and studied in the past. The best realization of a given transfer function depends very much on the application. General considerations include the number of adders and multipliers required, and the sensitivity of the network to finite precision arithmetic effects.

Digital filter realization will be discussed in detail in Chapter 8.

6.2.3 Performance analysis

Even though the filter coefficients are determined to a high degree of precision in the approximation step, digital hardware has a finite precision. The accuracy of the output will depend on the type of arithmetic used: fixed-point or floating-point. This is particularly so for fixed-point arithmetic. The designer must ensure that the error introduced by finite precision will not cause violations of the filter specifications. Furthermore, arithmetic overflow and underflow effects must be examined.

It cannot be over-emphasized how important this design step is, especially for IIR filters. While FIR filters are guaranteed to be stable, IIR filters can exhibit instability due to quantization errors introduced in the computational process.

Finite precision effects in digital filters will be discussed in detail in Chapter 8.

6.2.4 Implementation

Digital filters can be implemented either in software or hardware or a combination of both. Software implementations require a decision to be made on the type of computer or microprocessor the software will eventually run on. DSP chips, which are designed specifically for DSP type of operations, are very effective. In Chapter 9 we shall outline the architectures and characteristics of some of the more commonly used and commercially available devices on the market.

Note that the ease of software development depends very much on the quality of the development tools. While the performance of some DSP chips may be similar, the quality of tools available may be very different. The DSP designer should be aware of this fact. Some software tools are developed by the DSP chip manufacturers while others are third party. Some of these tools are described in Chapter 10.

In very demanding applications, the filter may need to be hard-wired or implemented as an application specific integrated circuit (ASIC) in order to obtain the speed required. It may also be necessary that some of the other functions such as analog-to-digital conversion and digital-to-analog conversion be integrated on the same device. However, development time will generally be longer and the cost is much higher.

6.3 Characteristics of FIR filters

Since FIR filters are linear discrete-time systems, the output sequence is related to the input and the impulse response of the filter by the convolution sum:

$$y(n) = \sum_{m=0}^{M} x(m)h(n-m)$$

This equation indicates that any particular output sample is only dependent on N input samples for an Nth order filter. Therefore FIR filters are also known *as* non-recursive filters. Also note that the summation on the right-hand side is a convolution between $x(n)$, the input sequence and $h(n)$, the impulse response of the filter. Hence they are also called convolution filters. From the statistical viewpoint, the output sample is a weighted average of the N input sample values. Thus the name moving-average (MA) filter is also used. But the name 'FIR' is most commonly seen in publications.

One of the major advantages of FIR filters is the ease with which exact linear phase filters can be designed. A filter with linear phase characteristics will not distort the input signal and is desirable in a number of applications such as digital communications. Design methods for FIR filters are generally linear and efficient. Another important property of FIR filters is that they are guaranteed to be stable. Furthermore, they can be efficiently realized on general and special purpose hardware. For instance, most DSP chips have special instructions to facilitate the implementation of an FIR filter.

6.3.1 Frequency response

The frequency response of an Nth order FIR filter is given by

$$H(\omega) = \sum_{n=0}^{N-1} h(n)e^{-j\omega n}$$

where ω is in radians per second. Strictly speaking, the exponent should be $(-j\omega Tn)$ where T is the sampling period. But we shall assume that $T = 1$ for simplicity, unless otherwise stated.

Notice that even though the filter is a discrete-time system, the frequency variable is continuous and is periodic with period 2π. This is an important point to remember especially if we are evaluating the frequency response using DFT. For a length-N impulse response, the DFT equation will give us N frequency points. If N is small, we may not get an accurate picture of the response. In these cases, the original impulse response $h(n)$ may need to be padded with an appropriate number of zeros in order to provide us with a more accurate frequency response curve.

Recall that the frequency response of a digital system is generally complex valued and consists of a magnitude and a phase. The function can be written as

$$H(\omega) = A(\omega)e^{j\theta(\omega)}$$

where $A(\omega)$ is the amplitude function/response and $\theta(\omega)$ is the phase function/response. The magnitude response is therefore given by

$$M(\omega) = |H(\omega)| = |A(\omega)|$$

The DFT of a length-N impulse response $h(n)$ is defined as

$$C(k) = \sum_{n=0}^{N-1} h(n)e^{-j2\pi kn/N} \qquad k = 0,1,...,N-1$$

Example 6.1
An FIR filter has impulse response

$$h(n) = \{1,3,5,3,1\}$$

The magnitude and phase responses are shown in Figure 6.3.
The five DFT coefficients are given by

$$C(k) = \begin{Bmatrix} 13.00, \\ -4.2361 - j3.0777, \\ 0.2361 + j0.7256, \\ 0.2361 - j0.7256, \\ -4.2361 + j3.0777 \end{Bmatrix}$$

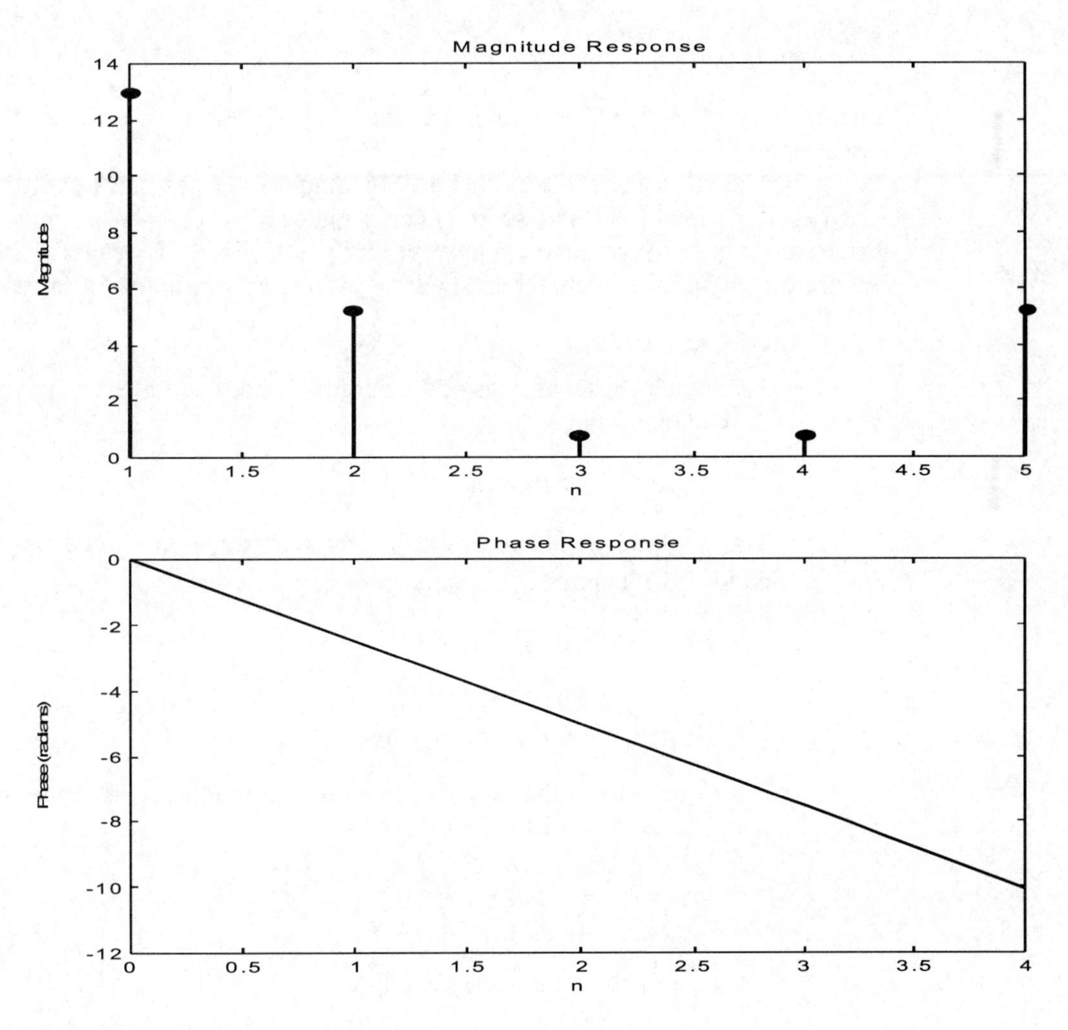

Figure 6.3
Magnitude and phase responses in Example 6.1

6.3.2 Linear phase filters

Linear phase refers to the phase response being a linear function of frequency. The FIR filter in the example given in the previous section has linear phase characteristics as shown in the phase response in Figure 6.4.

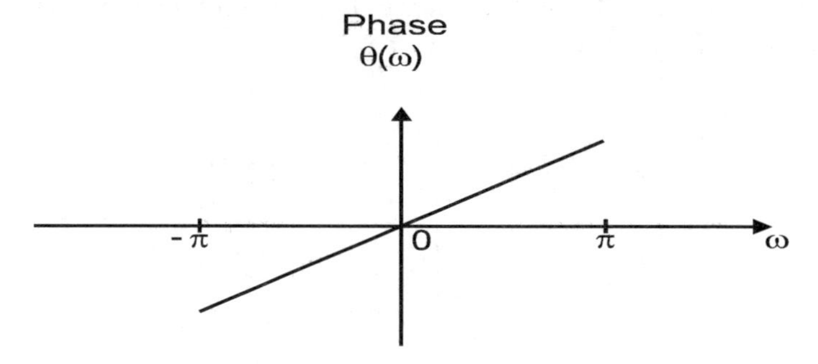

Figure 6.4
A linear phase response

Another way of saying that a filter has linear phase response is to say that it has a constant group delay response.

It can be shown mathematically that an FIR digital filter possesses exact linear phase properties if its impulse response is either symmetric (with even symmetry) or anti-symmetric (with odd symmetry) about the midpoint. Since the length of the impulse response of a digital filter can either be odd or even, there are in total four types of linear phase FIR filters.

- **Type 1:**
  The impulse response has odd length (N is odd) and is even symmetric about its midpoint. Thus

$$h(n) = h(N - n - 1)$$

The amplitude response has even symmetry about $\omega = 0$ and $\omega = \pi$. It is also periodic with period 2π. That is,

$$A(\omega) = A(-\omega)$$
$$A(\pi + \omega) = A(\pi - \omega)$$
$$A(\omega + 2\pi) = A(\omega)$$

Figure 6.5 shows the impulse response and amplitude spectrum of a typical type 1 linear phase FIR filter.

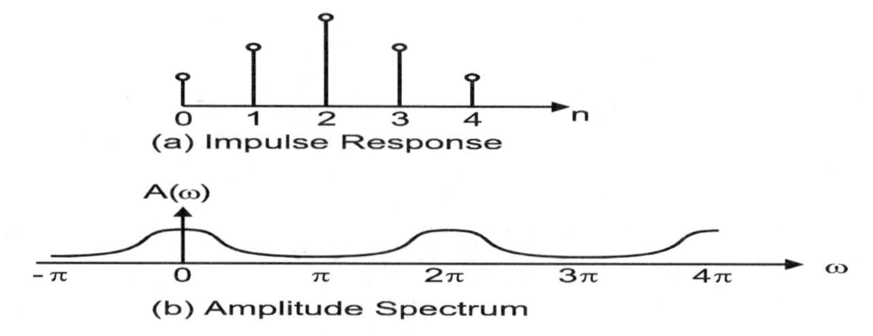

Figure 6.5
Type 1 linear phase responses

- **Type 2:**
  The impulse response has even length and is even symmetric about its midpoint M. Note that in this case M is not an integer. The amplitude spectrum is even about $\omega = 0$ and odd about $\omega = \pi$. The spectrum is also periodic with a period of 4π, instead of 2π.

$$A(\omega) = A(-\omega)$$
$$A(\pi + \omega) = A(\pi - \omega)$$
$$A(\omega + 2\pi) = A(\omega)$$

An example is shown in Figure 6.6.

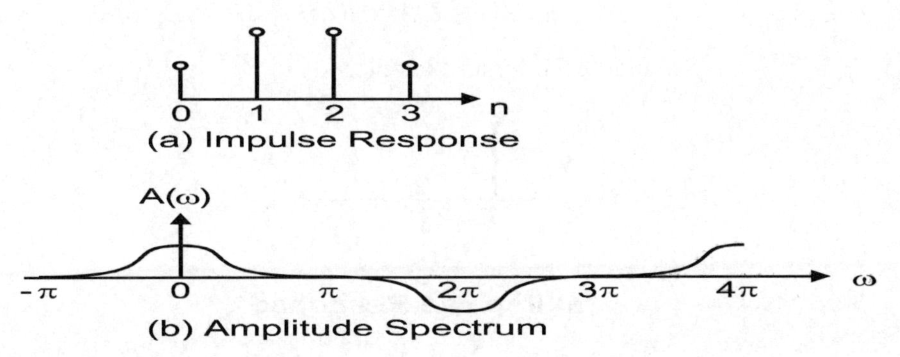

Figure 6.6
An example of type 2 linear phase responses

The frequency response of this type of filter must be zero at $\omega = \pi$. They will make good low-pass filters but are unsuitable for high-pass designs.

- **Type 3:**
  The impulse response has odd length and odd symmetry about the midpoint. The amplitude spectrum is odd about $\omega = 0$ and $\omega = \pi$. It has a period of 2π.

$$A(\omega) = A(-\omega)$$
$$A(\pi + \omega) = -A(\pi - \omega)$$
$$A(\omega + 4\pi) = A(\omega)$$

Figure 6.7 shows an example.

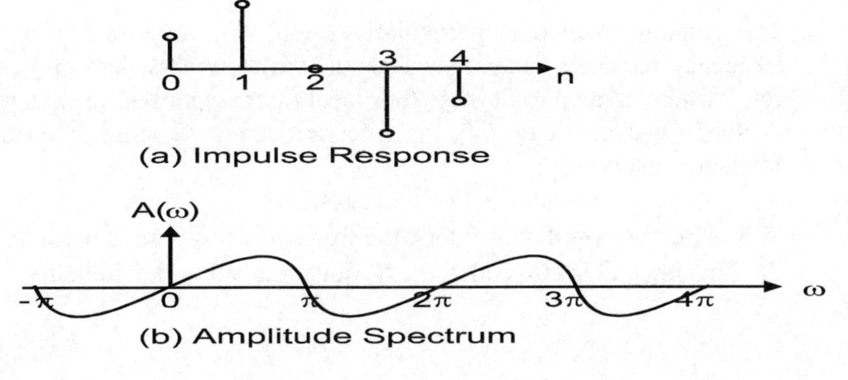

Figure 6.7
An example of type 3 linear phase responses

Note that this type of filter has frequency responses which must be zero at zero frequency ($\omega = 0$) and at $\omega = \pi$. They are therefore not suitable for low-pass and high-pass designs. Furthermore, they introduce a phase shift of 90°.

- **Type 4:**
  The impulse response has even length and odd symmetry about the midpoint. The amplitude spectrum is odd about $\omega = 0$ and even about $\omega = \pi$. It has a period of 4π.

$$A(\omega) = -A(-\omega)$$
$$A(\pi + \omega) = -A(\pi - \omega)$$
$$A(\omega + 2\pi) = A(\omega)$$

See Figure 6.8 for an example.

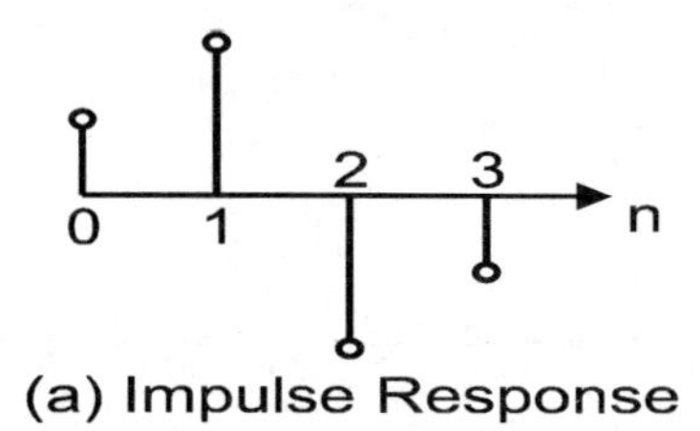

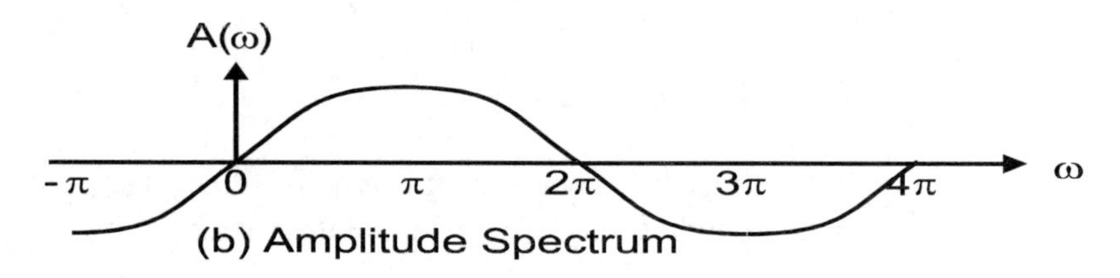

Figure 6.8
Example of type 4 linear phase responses

The frequency response of this type of filters must be zero at zero frequency but not necessarily so at $\omega = \pi$. Therefore they should not be used for low-pass designs but they can make good high-pass filters. Like type 3 filters, they also introduce a phase shift of 90°.

6.4 Window method

The window method is particularly useful for designing filters with simple desired frequency response curves, such as an ideal low-pass, high-pass, band-pass and band-reject filters. Examples of these four ideal filter responses are shown in Figure 6.9.

Notice that in Figure 6.9, the filter frequency responses are only specified over the frequency interval

$$-\pi \leq \omega \leq \pi$$

This is because for digital filters, the frequency response is periodic in ω with a period of 2π. This interval is also called the Nyquist interval in the literature.

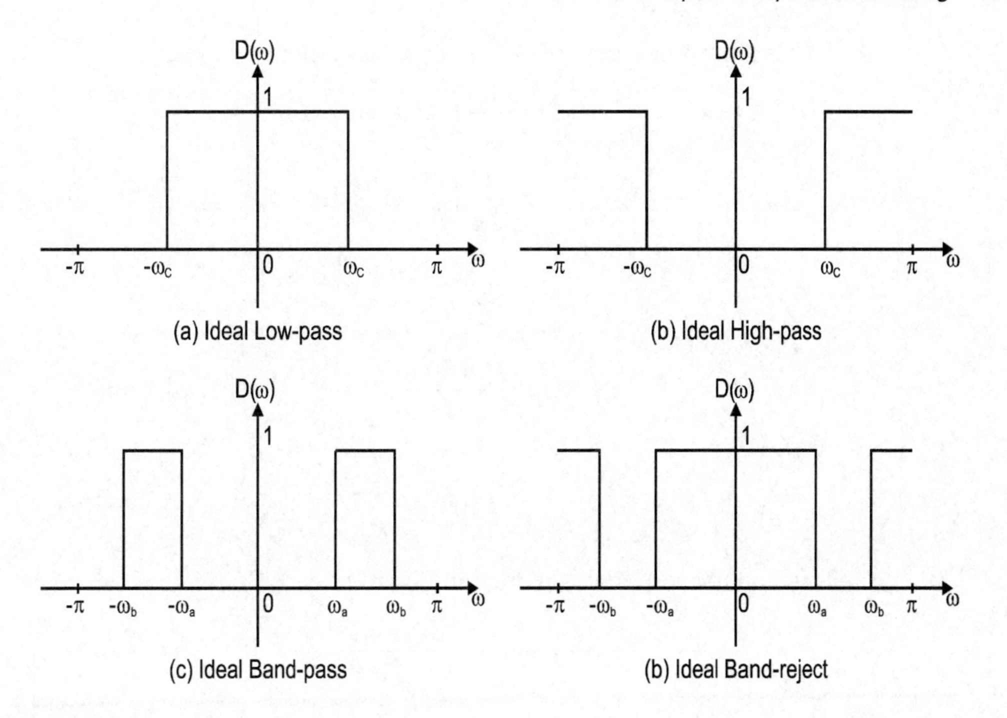

Figure 6.9
Four ideal filter frequency responses

The (continuous) frequency response and the (discrete-time) impulse response are related by the discrete-time Fourier transform (DTFT) relationships:

$$D(\omega) = \sum_{k=-\infty}^{\infty} d(k)e^{-j\omega k}$$

$$d(k) = \int_{-\pi}^{\pi} D(\omega)e^{j\omega k} \frac{d\omega}{2\pi}$$

So, given the desired frequency response, the filter impulse response can be obtained by using the inverse DTFT equation. The filter coefficients will simply be the impulse response samples. However, the impulse response obtained by using the inverse DTFT will in general have two undesirable properties: non-causal and infinite duration. For instance, consider a desired low-pass filter response given by

$$D(\omega) = \begin{cases} 1, & \text{if } |\omega| \le \omega_c \\ 0, & \text{if } \omega_c \le |\omega| \le \pi \end{cases}$$

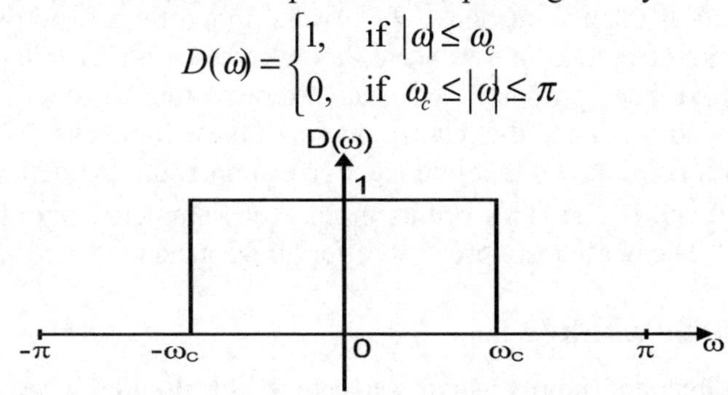

Figure 6.10
An ideal low-pass filter response

This is shown in Figure 6.10. The impulse response will be

$$d(k) = \frac{1}{2\pi}\int_{-\pi}^{\pi} D(\omega)e^{j\omega k}\,d\omega$$

$$= \frac{1}{2\pi}\int_{-\omega_c}^{\omega_c} 1\cdot e^{j\omega k}\,d\omega$$

$$= \frac{1}{2\pi jk}\left[e^{j\omega_c k} - e^{-j\omega_c k}\right]$$

$$\frac{\sin(\omega_c k)}{\pi k} \qquad -\infty < k < \infty$$

with

$$d(0) = \frac{\omega_c}{\pi}$$

This impulse response is plotted in Figure 6.11.

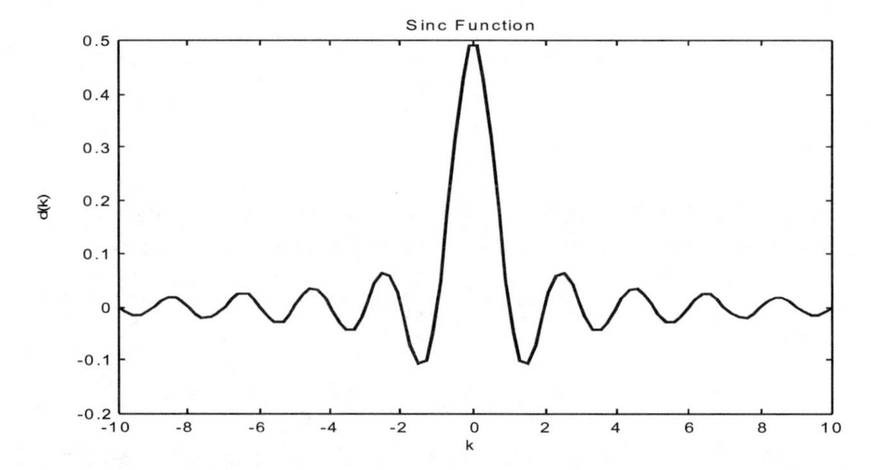

Figure 6.11
Impulse response of the ideal low-pass filter

In order to truncate the impulse response to a finite duration, a window function can be used. The ideal impulse response $d(k)$ is multiplied by a window function, which has a finite duration, resulting in a truncated impulse response. A number of different window functions have been proposed. We shall examine four of them in order to illustrate the effects of windowing and the relative merits of these functions.

It is worth pointing out that in the above example, the frequency response is symmetric about $\omega = 0$ and is real. This results in an even symmetric impulse response that is also real-valued. The phase response is zero for all frequencies.

6.4.1 Rectangular window

The most direct and simple way to truncate the ideal impulse response $d(k)$ is to keep the values of $d(k)$ within a certain interval, say, $-M$ to M. This is equivalent to multiplying $d(k)$ by a rectangular function given by

$$w(n) = \begin{cases} 1, & |n| \le M \\ 0, & \text{otherwise} \end{cases}$$

as shown in Figure 6.12.

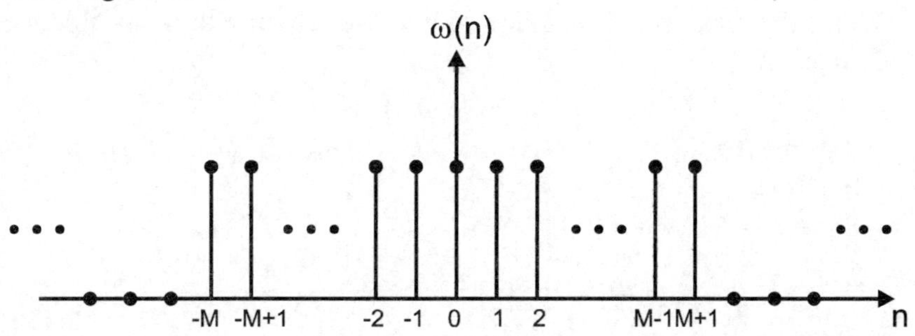

Figure 6.12
A rectangular window

The resulting impulse response $h_w(n)$ has either $N = 2M$ or $N = 2M+1$ non-zero values. In our following discussions, we shall assume that N is odd. The arguments can easily be extended to the case where N is even.

$$h_w(n) = [d_{-M}, d_{M+1}, ..., d_{-1}, d_0, d_1, ..., d_{M-1}, d_M]$$

The windowed impulse response $h_w(n)$ is still non-causal, i.e. it has non-zero values before the time origin n = 0. To make it causal we can simply shift the time origin to the first non-zero sample and re-index the entries. The impulse response (and hence the filter coefficients) of the FIR filter is therefore

$$h(n) = d_w(n - M) \qquad n = 0,1,..., N-1$$

This process is illustrated in Figure 6.13.

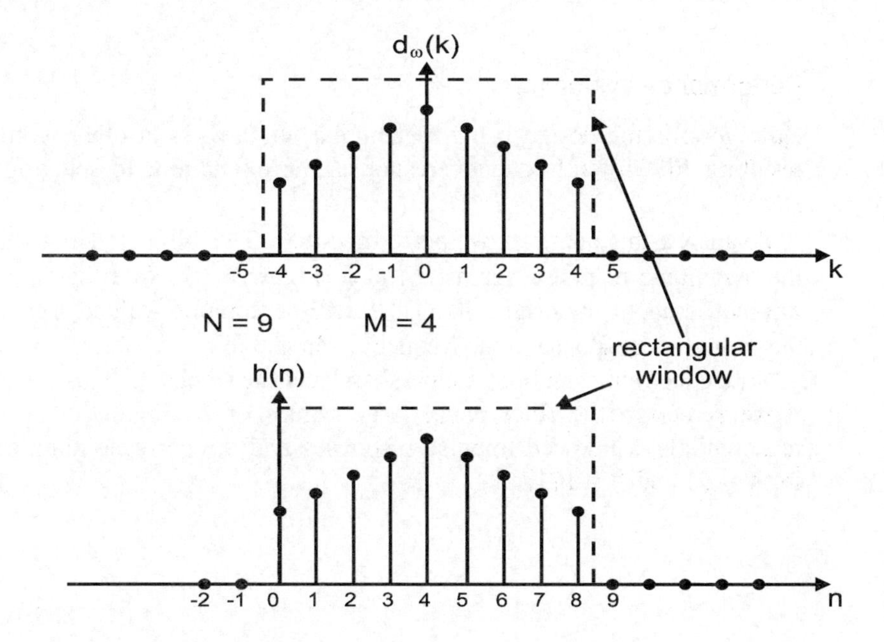

Figure 6.13
Rectangular windowed impulse response

Example 6.2

Find the rectangularly windowed impulse response of an ideal low-pass filter with cut-off frequency

$$\omega_C = \pi/4$$

Assume $N = 11$.

Solution:

Since $N = 11$, $M = (N–1)/2 = 5$.

$$d_w(n) = \frac{\sin(\pi n/4)}{\pi n} \qquad -5 \leq n \leq 5$$

$$= \left[\frac{\sqrt{2}}{10\pi}, 0, \frac{\sqrt{2}}{6\pi}, \frac{1}{2\pi}, \frac{\sqrt{2}}{2\pi}, \frac{1}{4}, \frac{\sqrt{2}}{2\pi}, \frac{1}{2\pi}, \frac{\sqrt{2}}{6\pi}, 0, \frac{\sqrt{2}}{10\pi}\right]$$

The filter impulse response $h(n)$ is plotted in Figure 6.14.

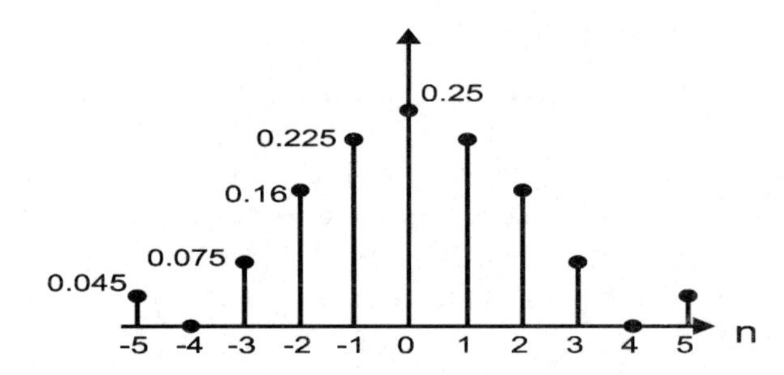

Figure 6.14
Filter impulse response of Example 6.2

6.4.1.1 Performance evaluation

How good is the design using rectangular windows? In other words, how close is the resulting FIR filter frequency response approximation, to the original ideal response $D(\omega)$?

To answer this question, we performed the DTFT of $h(n)$, denoted by $H(\omega)$ and plotted the magnitude response against $D(\omega)$ in Figure 6.15. Note that since $h(n)$ is no longer symmetric about the origin, its DTFT will be complex-valued. Hence we only compare the magnitude response in the frequency range 0 to π.

Notice the ripples in both the passband and the stopband. This can be more clearly seen if we re-design the filter using large values of N. Figures 6.16 and 6.17 show the rectangularly windowed impulse responses and the corresponding magnitude responses for $N = 51$ and $N = 101$.

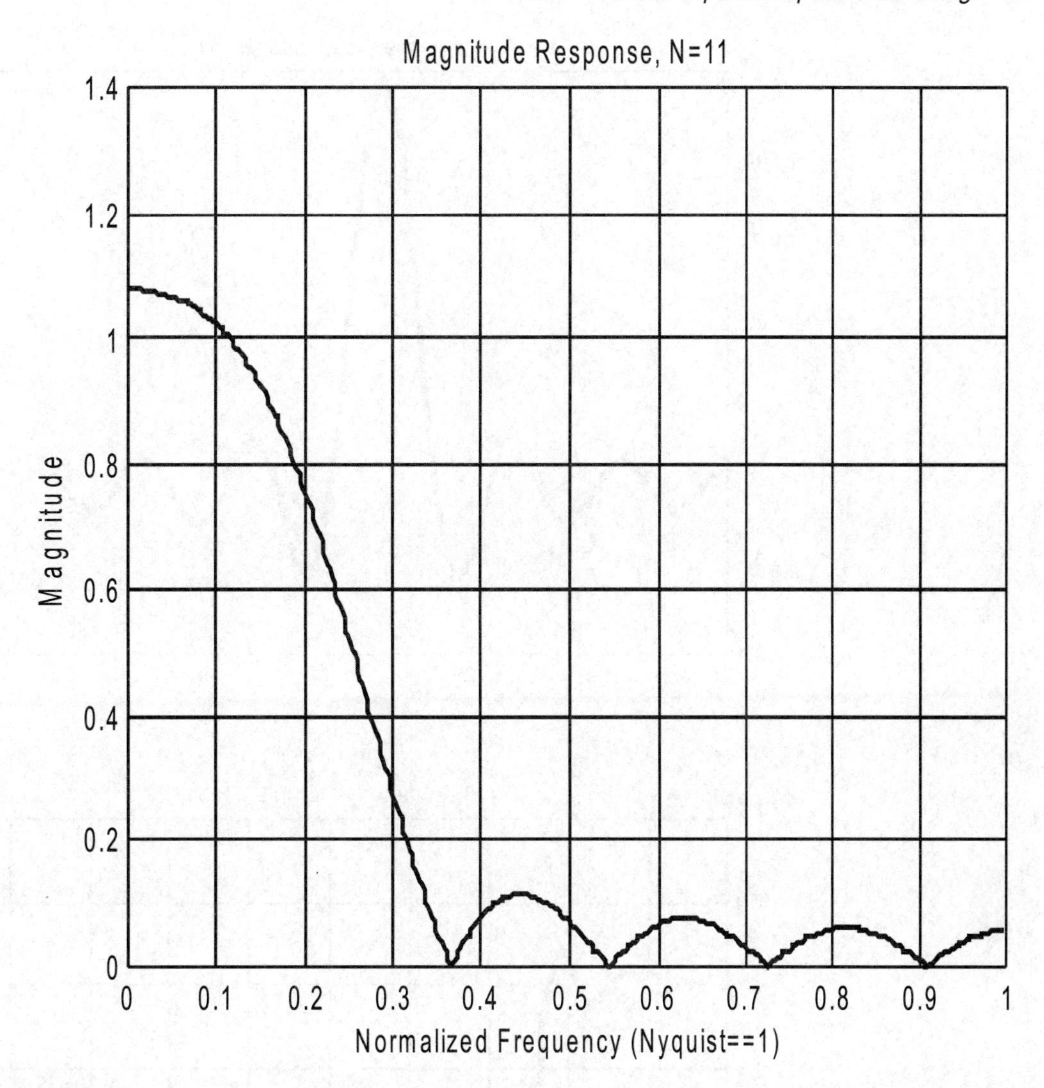

Figure 6.15
Magnitude response of rectangular windowed filter with N=11

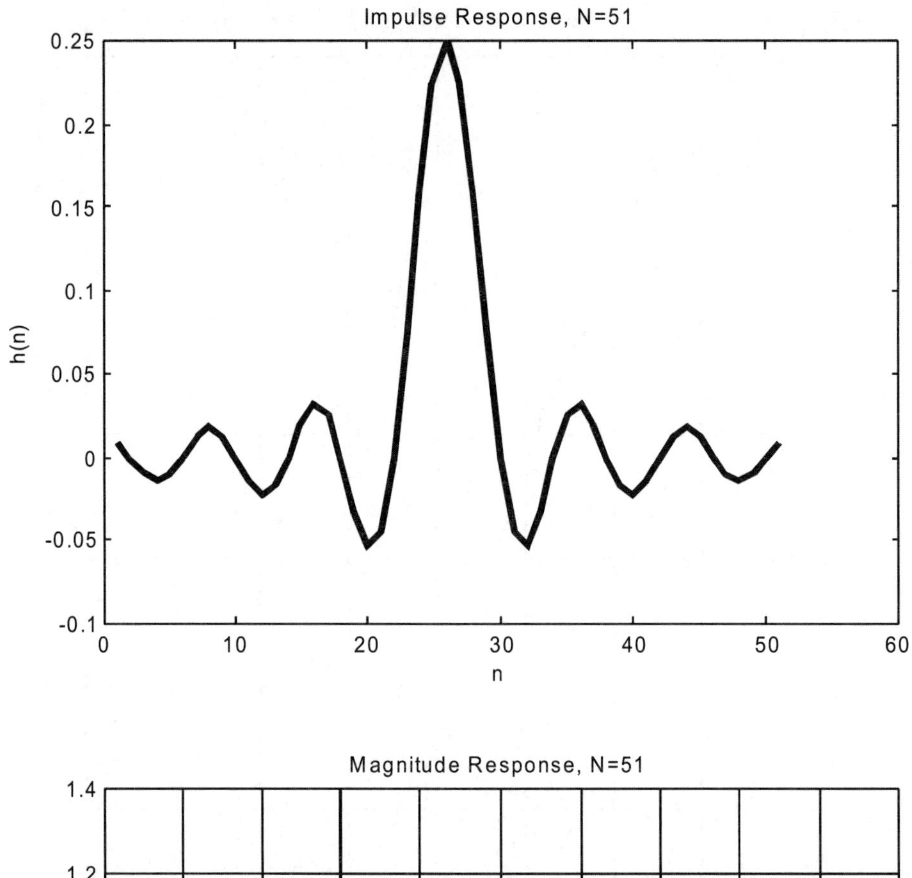

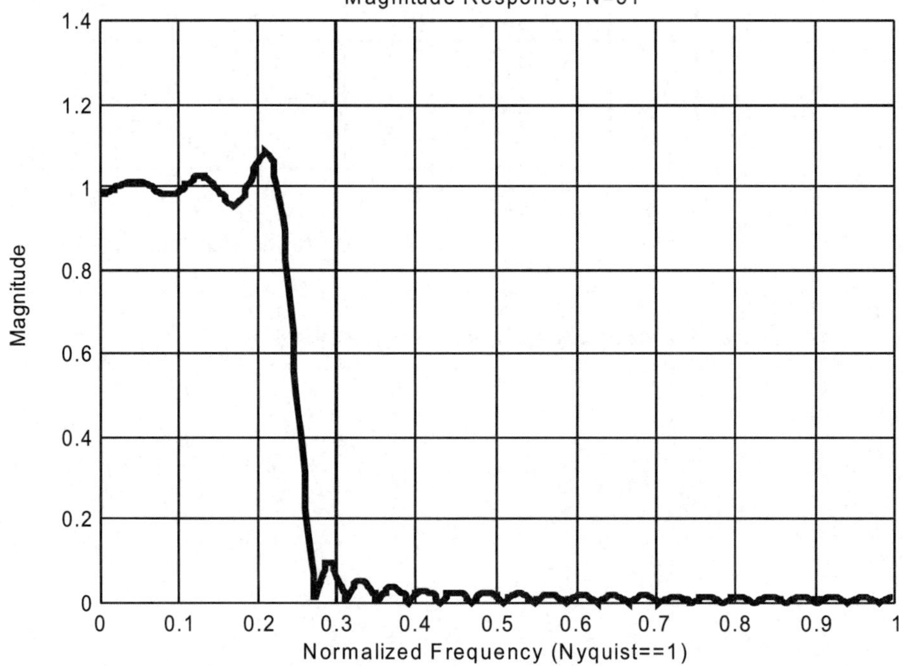

Figure 6.16
Impulse and magnitude responses of truncated ideal LPF with N=51

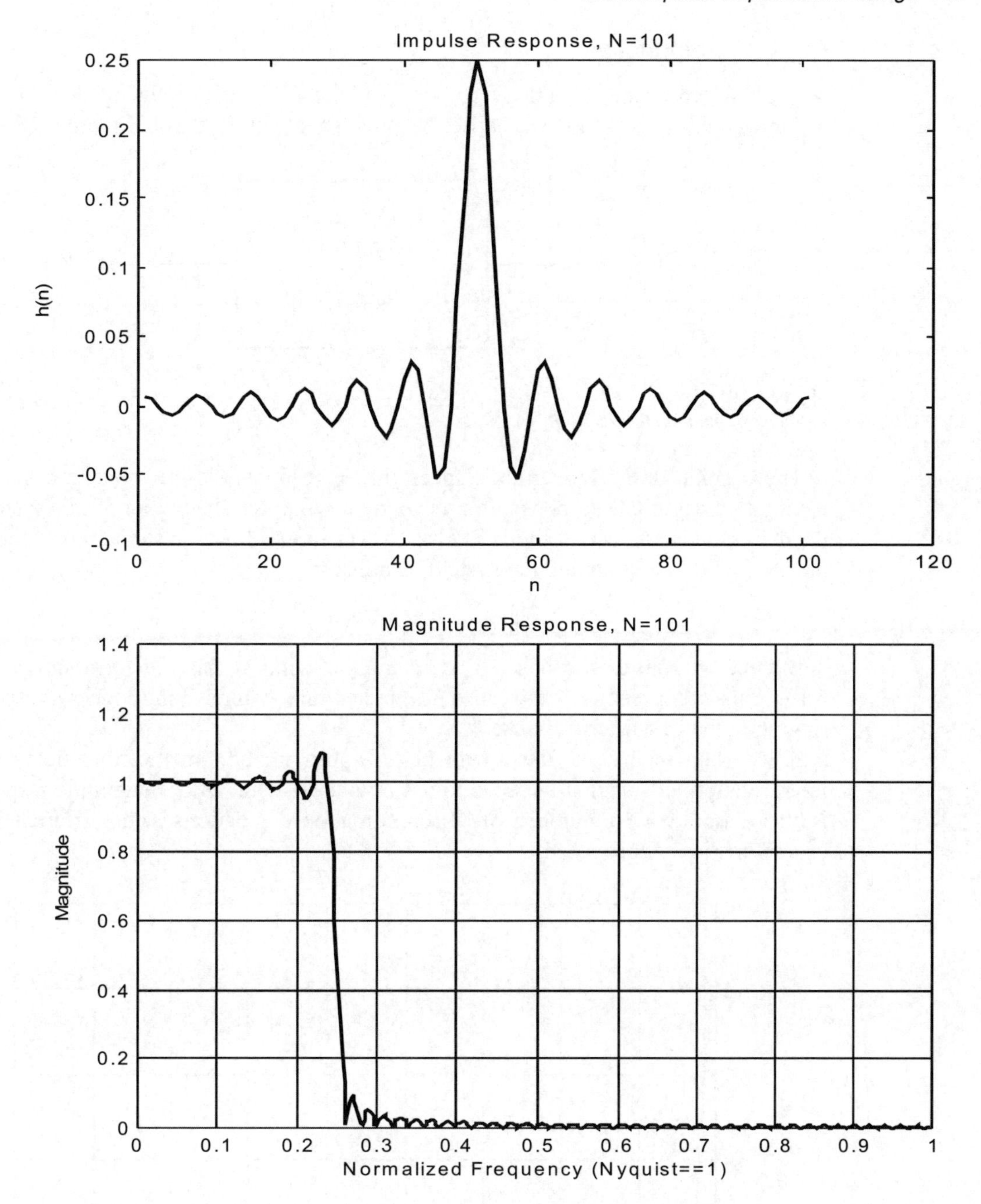

Figure 6.17
Impulse and magnitude responses of truncated ideal LPF with N=101

One would expect that as N increases, the approximation would become better. This is indeed the case except for the region close to the transition between passband and stopband. This area corresponds to a discontinuity in the ideal desired frequency response. The truncation of the Fourier series introduces ripples in the frequency response due to the non-uniform convergence of the Fourier series at a discontinuity. This phenomenon is known as the Gibb's phenomenon. For this reason, the approximation at the band edge will always be poor for the rectangular window design regardless of how large N is.

6.4.1.2 Another interpretation

We can interpret the magnitude response of the FIR filter by using our knowledge of linear systems discussed in the earlier chapter. This is illustrated in Figure 6.18.

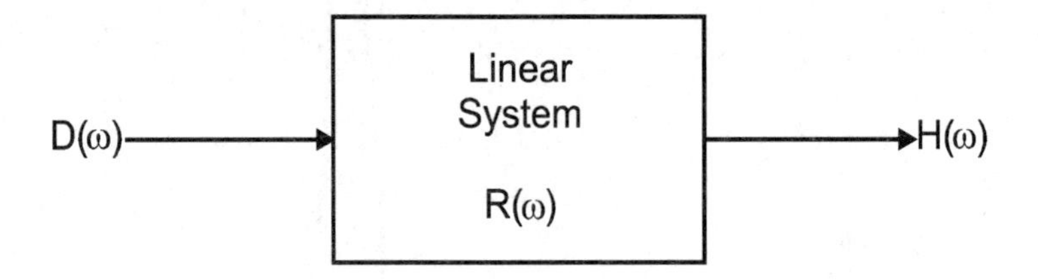

Figure 6.18
Frequency domain interpretation

The original desired magnitude spectrum (or response) is the input to a linear system having the magnitude response of a rectangular window. The output of this system is then the magnitude response of the FIR filter. According to linear system theory, the output is the product of the input and the system responses:

$$H(\omega) = R(\omega) = D(\omega)$$

But the magnitude response of a rectangular window has the form shown in Figure 6.19. This interpretation shows that the rectangular window introduces the ripples and ringing in the FIR filter response.

If we want to design filters with better approximation in the transition region, this interpretation tells us that we need to use a window with better magnitude response. This is the reason why a number of different window functions with different frequency characteristics are proposed.

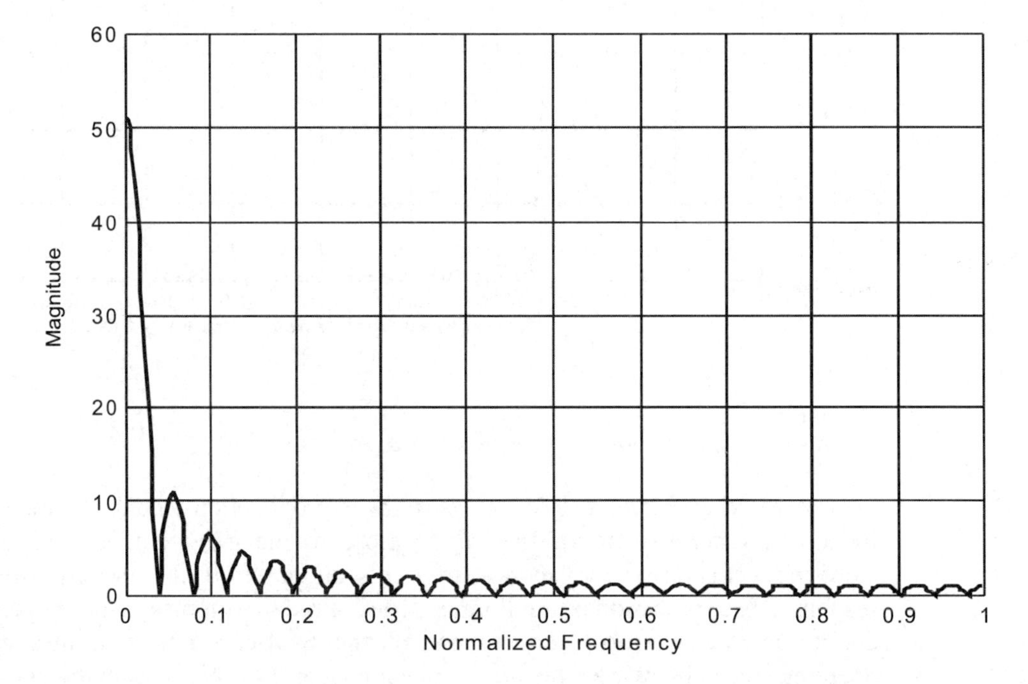

Figure 6.19
Magnitude response of a rectangular window

6.4.1.3 Summary of filter characteristics

- The ripple size decreases with increasing filter order N. Approximation well within the passband and stopband becomes better as the filter order increases.
- The transition width decreases with increasing filter order. For any order N, the filter response is always equal to 0.5 at the cut-off frequency.
- Ripple size near the passband to stopband transition remains roughly the same as N increases. The maximum ripple size is about 9%. This is known as the Gibb's phenomenon.

6.4.2 Hamming window

Since discontinuities in the time function give rise to ringing in the frequency response, we can replace the rectangular window with a window function that tapers off smoothly at both ends. This will reduce the ripple effect. The Hamming window is a popular one in this class of window functions.

The Hamming window is defined mathematically as

$$w(n) = 0.54 - 0.46\cos\left(\frac{2\pi n}{N-1}\right) \qquad n = 0,1,\ldots,N-1$$

Figure 6.20 plots this window function.

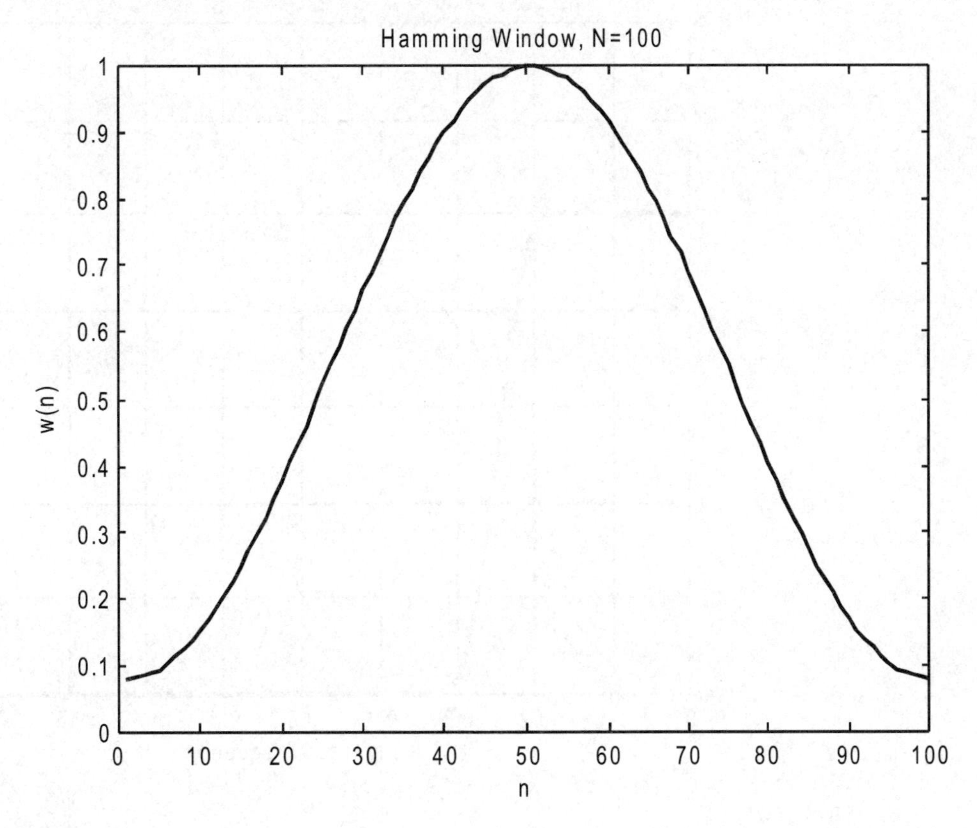

Figure 6.20
The Hamming window

Notice that this equation defines the window samples as already shifted (indices from 0 to N–1). So the impulse response of the FIR low-pass filter designed using the Hamming window is

$$h(n) = w(n)d(n-M)$$
$$= \left[0.54 - 0.46\cos\left(\frac{2\pi n}{N-1}\right)\right]\cdot\frac{\sin[(n-M)\omega_c]}{(n-M)\pi}$$

Figure 6.21 shows a length-51 low-pass filter with cut-off at $\pi/4$ as in the example in the previous section.

Comparing this response with that shown in Figure 6.16, which was designed using the rectangular window, it is obvious that the Hamming window design is better. The ripples in the both the passband and the stopband are virtually eliminated. The cost involved is a wider transition width.

The Hamming window function has the same form as the raised cosine function familiar to digital communication engineers. The only differences are in the scalar value in the constant and cosine terms. The Hamming window does not taper to end values of zero. Instead it goes to a value of 0.08. The maximum stopband ripple is about 53 dB below the passband gain.

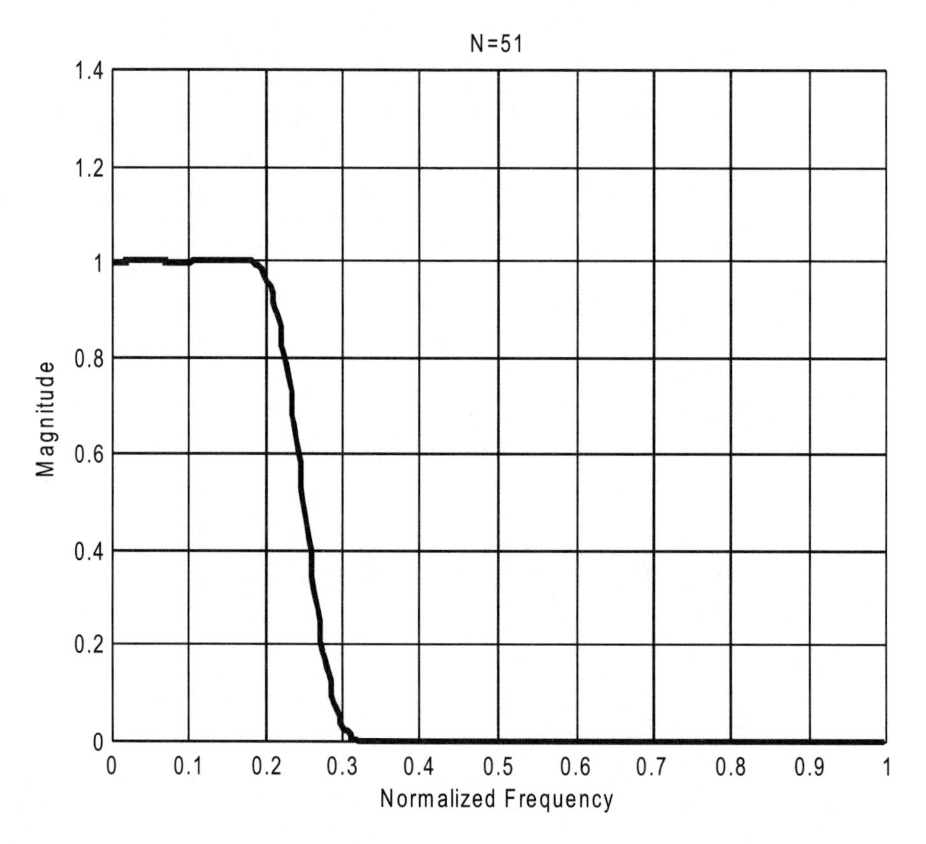

Figure 6.21
Hamming windowed magnitude response

6.4.3 Blackman window

The Blackman window exhibits an even lower maximum stopband ripple (about 74 dB down) in the resulting FIR filter than the Hamming window. It is defined mathematically as

$$w(n) = 0.42 - 0.5\cos\left(\frac{2\pi n}{N-1}\right) + 0.08\left(\frac{4\pi n}{N-1}\right) \qquad n = 0, 1, \ldots, N-1$$

Its magnitude and impulse responses are plotted in Figure 6.22. Note that the width of the main lobe in the magnitude response is about 50% wider than that of the Hamming window.

A length-51 low-pass FIR filter is designed using this window and the responses shown in Figure 6.23. This can be compared to the one designed using Hamming window in Figure 6.21.

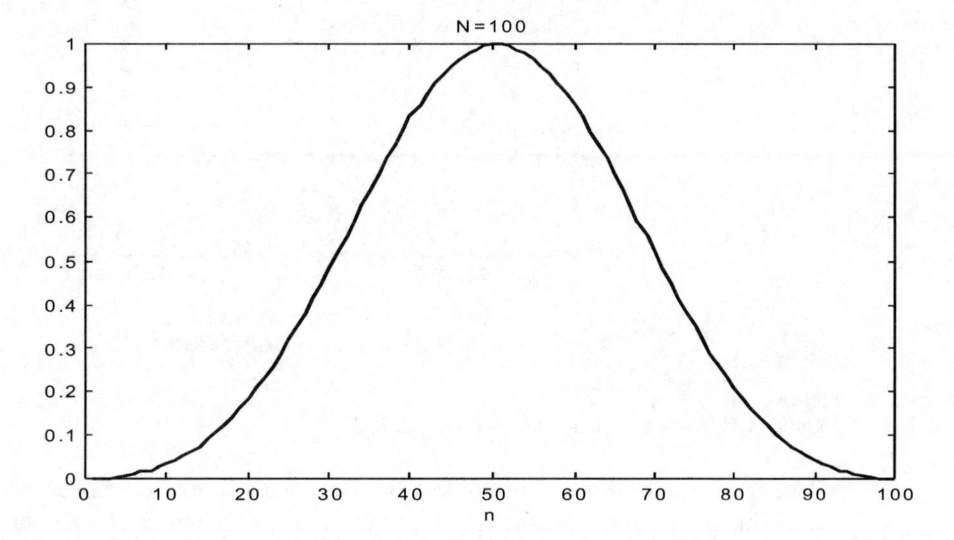

Figure 6.22
The Blackman window

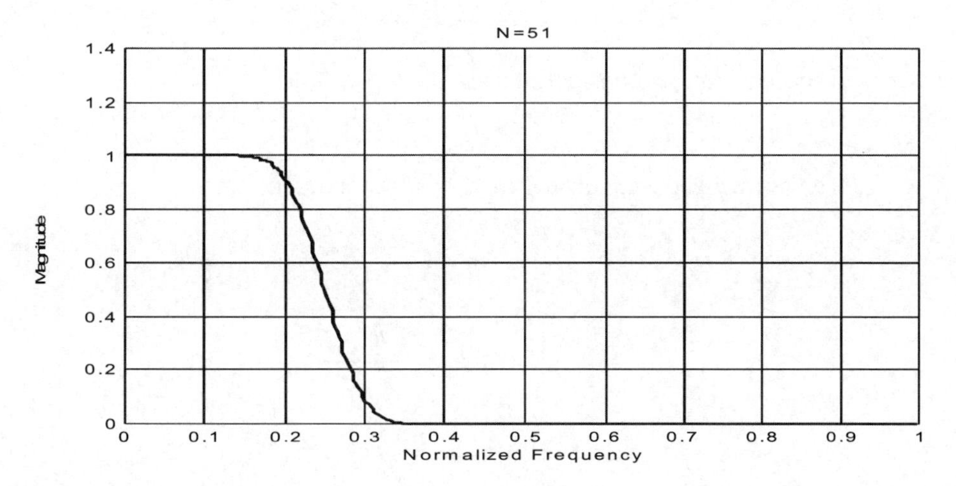

Figure 6.23
Low-pass FIR filter designed using the Blackman window

6.4.4 Kaiser window

The main advantage of the previous three window functions is that they are simple to apply and the resulting filter characteristics are reasonably good. For a large number of applications, Hamming or Blackman window designs will be sufficient to satisfy the specifications. The major drawback of these window functions is that its characteristics such as maximum stopband attenuation and the amount of overshoot are basically fixed. So if the filter specifications include the amount of overshoot and passband-to-stopband transition width, for instance and if the above window functions do not produce designs that can satisfy them, then we are stuck.

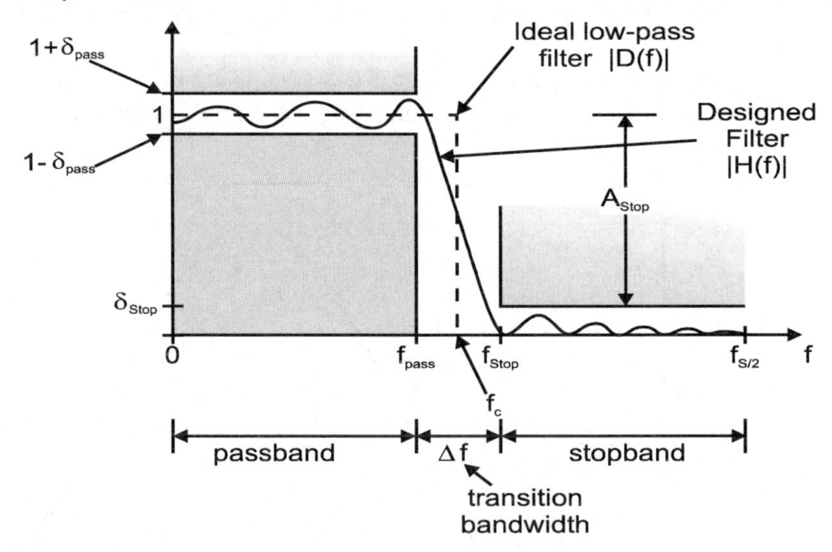

Figure 6.24
Magnitude response characteristics of a low-pass filter

Consider the magnitude response characteristics of a low-pass filter as shown in Figure 6.24. The ideal cut-off frequency is at the midpoint between the passband and stopband edge frequencies.

$$f_c = \frac{1}{2}\left(f_{\text{pass}} + f_{\text{stop}}\right)$$

The transition width is defined as

$$\Delta f = f_{\text{stop}} - f_{\text{pass}}$$

The normalized frequencies are the digital frequencies:

$$\omega_{\text{pass}} = \frac{2\pi f_{\text{pass}}}{f_s}$$

$$\omega_{\text{stop}} = \frac{2\pi f_{\text{stop}}}{f_s}$$

$$\omega_c = \frac{2\pi f_c}{f_s}$$

$$\Delta\omega = \frac{2\pi\Delta f}{f_s}$$

The maximum passband and stopband ripples are usually expressed in decibels (dB) in practice:

$$A_{\text{pass}} = 20\log_{10}\left(\frac{1+\delta_{\text{pass}}}{1-\delta_{\text{pass}}}\right)$$

$$A_{\text{stop}} = -20\log_{10}\delta_{\text{stop}}$$

These equations relate the two sets of specifications $\{f_{\text{pass}}, f_{\text{stop}}, A_{\text{pass}}, A_{\text{stop}}\}$ and $\{f_c, \Delta f, \delta_{\text{pass}}, \delta_{\text{stop}}\}$.

If δ_{pass} is small, then we can use a first order approximation to get

$$A_{\text{pass}} = 17.372\delta_{\text{pass}}$$

Note that it is a property of all window designs that δ_{pass} and δ_{stop} are equal in the filter designed. Therefore, instead of dealing with two variables, we can choose the maximum ripple to be the smaller of the two:

$$\delta = \min(\delta_{\text{pass}}, \delta_{\text{stop}})$$

Practical choices of passband and stopband attenuation will usually result in the stopband ripple being smaller than the passband one.

6.4.4.1 Kaiser window design

Kaiser has developed a flexible family of window functions. This family of window functions has adjustable shape parameters that allow the designer to achieve the specified ripple and attenuation. It is mathematically defined as

$$w(n) = \frac{I_0\left(\beta\sqrt{1-(n-M)^2/M^2}\right)}{I_0(\beta)} = \frac{I_0\left(\beta\sqrt{n(2M-n)/M}\right)}{I_0(\beta)}$$

for $n = 0,1,\ldots,N-1$ where $I_0(x)$ is the zero-th order modified Bessel function of the first kind. Here we assumed as before that N is odd. The second format is more convenient for numerical computations. β is called the shape parameter.

The Kaiser window is symmetric about its midpoint and has a maximum value of 1 at that point. Since $I_0(0)=1$, the end-points have the value $1/I_0(\beta)$. Typical values of β are in the range of

$$4 < \beta < 9$$

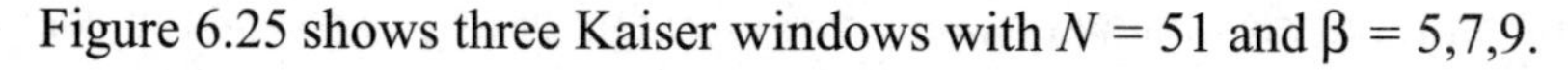

Figure 6.25 shows three Kaiser windows with $N = 51$ and $\beta = 5,7,9$.

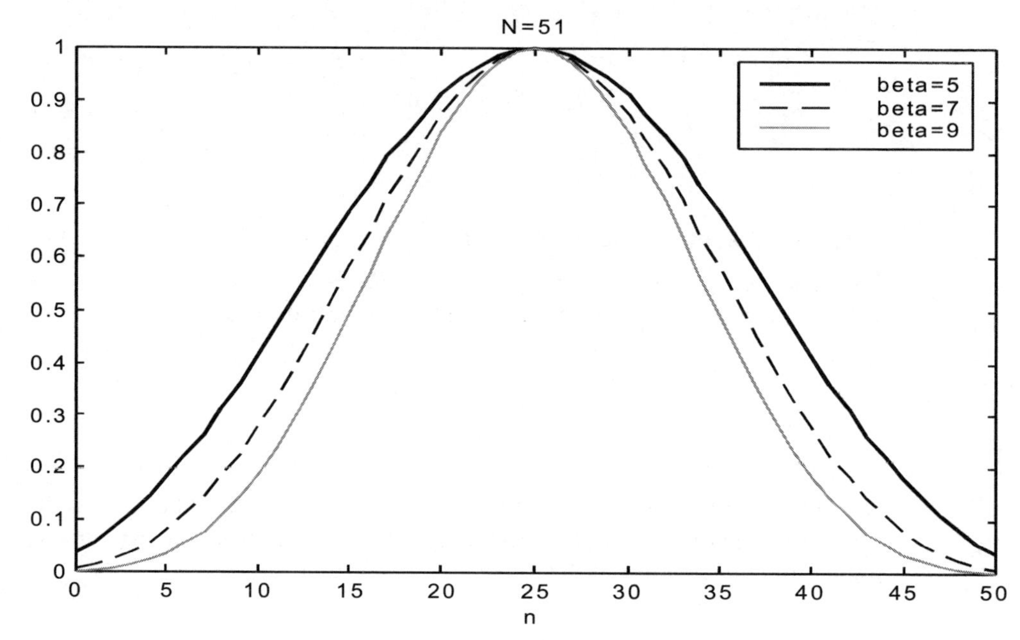

Figure 6.25
Kaiser windows

The Kaiser window is reduced to a rectangular one for $\beta = 0$. It resembles the Hamming window for $\beta = 5$, except near the end-points.

Table 6.1 compares the transition bandwidth and the maximum stopband ripple for various values of the shape parameter.

β	Transition bandwidth	Max. stopband ripple (dB)
4.0	2.6	–45
5.0	3.2	–54
6.0	3.8	–63
7.0	4.5	–72
8.0	5.1	–81
9.0	5.7	–90

Table 6.1
Comparison of transition bandwidth and maximum stopband ripple

The filter order N and the shape parameter can be calculated from the specifications. Kaiser has derived empirical design formulas as follows:

$$\beta = \begin{cases} 0.1102(A-8.7), & A \geq 50 \\ 0.5842(A-21)^{0.4} + 0.07886(A-21), & 21 < A < 50 \\ 0, & A \leq 21 \end{cases}$$

where A is the ripple in dB.

The filter length or order is inversely related to the transition bandwidth:

$$\Delta f = \frac{Df_s}{N-1} \Leftrightarrow N-1 = \frac{Df_s}{\Delta f}$$

where D is a factor computed from A:

$$D = \begin{cases} \dfrac{A-7.95}{14.36} & A > 21 \\ 0.922 & A \leq 21 \end{cases}$$

6.4.4.2 Design steps

- Compute f_c and Δf and then the normalized digital frequencies.
- Compute the passband and stopband ripples and hence δ and A.
- Compute β and D.
- Compute the filter length required and round it up to the next odd integer.
- Compute the window function $w(n)$.
- Compute the windowed impulse response $h(n)$.

Note that the parameters N and β depend only on A and Δf and not on f_c. However, $h(n)$ does depend on f_c.

Example 6.3
Design a low-pass FIR filter using Kaiser windows with the following specifications:

$$f_s = 44.1\,\text{kHz}$$
$$f_{pass} = 12\,\text{kHz}$$
$$f_{stop} = 18\,\text{kHz}$$
$$A_{pass} = 0.2\,\text{dB}$$
$$A_{stop} = 50\,\text{dB}$$

Solution:
Now we have A = 50. Therefore,

$$\begin{aligned}\beta &= 0.1102(A-8.7) \\ &= 0.1102(50-8.7) \\ &= 4.55126\end{aligned}$$

and

$$\begin{aligned}D &= \frac{A-7.95}{14.36} \\ &= 2.9283\end{aligned}$$

Since

$$\begin{aligned}\Delta f &= f_{stop} - f_{pass} \\ &= 18-12 \\ &= 6\text{ kHz}\end{aligned}$$

we have

$$N-1=\frac{Df_s}{\Delta f}$$
$$=\frac{2.9283(44.1)}{6}$$
$$=21.5$$

and so the order of the filter is

$$N=22.5$$

We can try to use $N = 22$ or $N = 23$.

MATLAB functions provide a better approximation for β and N. Figure 6.26 shows the magnitude response of the resulting filter.

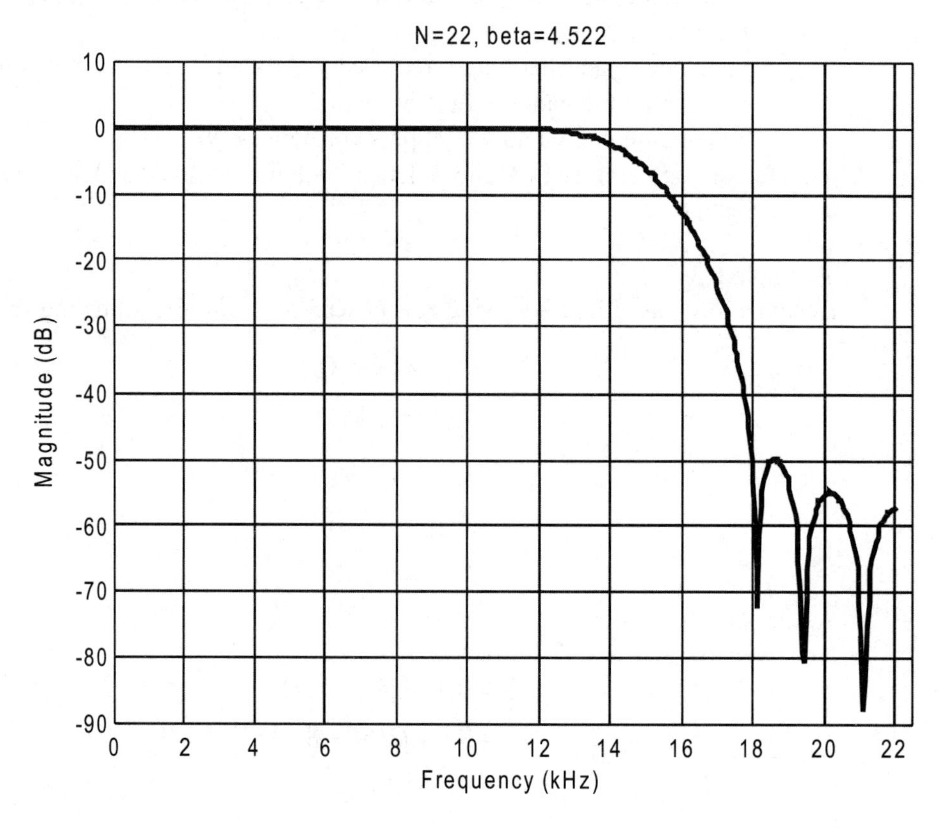

Figure 6.26
Magnitude response of Kaiser window designed FIR filter

6.4.4.3 High-pass filter design

High-pass filter design using Kaiser windows is very similar to low-pass filter design. The only change in the steps is simply define

$$\Delta f = f_{\text{pass}} - f_{\text{stop}}$$

since the role of f_{pass} and f_{stop} are interchanged.

The ideal high-pass impulse response is obtained from the inverse DTFT of the ideal high-pass frequency response. It is given by

$$d(k) = \delta(k) - \frac{\sin(\omega_c k)}{\pi k}$$

The windowed filter impulse response is therefore

$$h(n) = w(n) \cdot \left[\delta(n - M) - \frac{\sin((n - M)\omega_c)}{(n - M)\pi} \right]$$

$$= \delta(n - M) - w(n) \cdot \frac{\sin((n - M)\omega_c)}{(n - M)\pi}$$

The second term on the right-hand side of this equation is the impulse response of the low-pass filter with the same cut-off frequency.

Note that with the same value of ω_c, the low-pass and high-pass filters are complementary. That is,

$$h_{\text{LP}}(n) + h_{\text{HP}}(n) = \delta(n - M) \qquad n = 0, 1, \ldots, N - 1$$

Example 6.4

Two-way crossover filters. Conventional loudspeakers make use of an analog crossover network to split the audio signal into its low frequency and high frequency components. The low frequency signal drives the woofer and the high frequency one drives the tweeter. Digital loudspeaker systems use digital filters to perform the same function on the incoming digitized audio signal. The digital signals in the two frequency bands are then converted to analog signals, amplified and drive the corresponding parts of the loudspeaker.

Let the crossover frequency be 1 kHz. The low-pass filter has specifications:

$$f_{\text{pass}} = 800\,\text{Hz}$$

$$f_{\text{stop}} = 1200\,\text{Hz}$$

$$A_{\text{pass}} = 0.1\,\text{dB}$$

$$A_{\text{stop}} = 60\,\text{dB}$$

Note that we only need to design the low-pass filter. The high-pass filter is complementary.

The resulting low-pass and high-pass filter magnitude responses are shown in Figure 6.27.

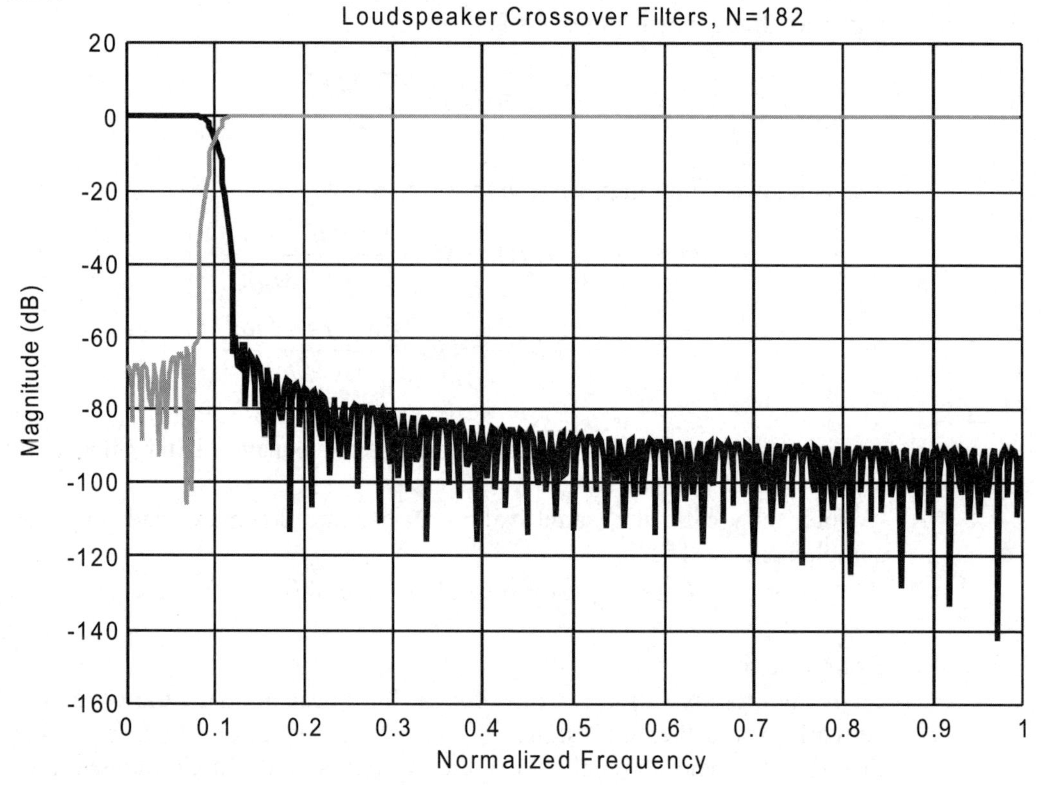

Figure 6.27
Crossover filter responses

The complementary relationship between the two filters leads to a very efficient implementation shown in Figure 6.28.

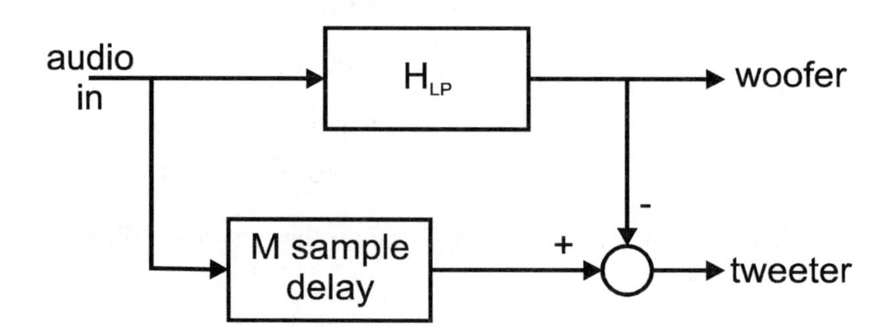

Figure 6.28
Implementation of complementary relationship between two filters

In fact, only one filter is needed; the high-pass output is obtained by combining the low-pass output and the input delayed by *M* sampling instants.

6.4.4.4 Band-pass filter design

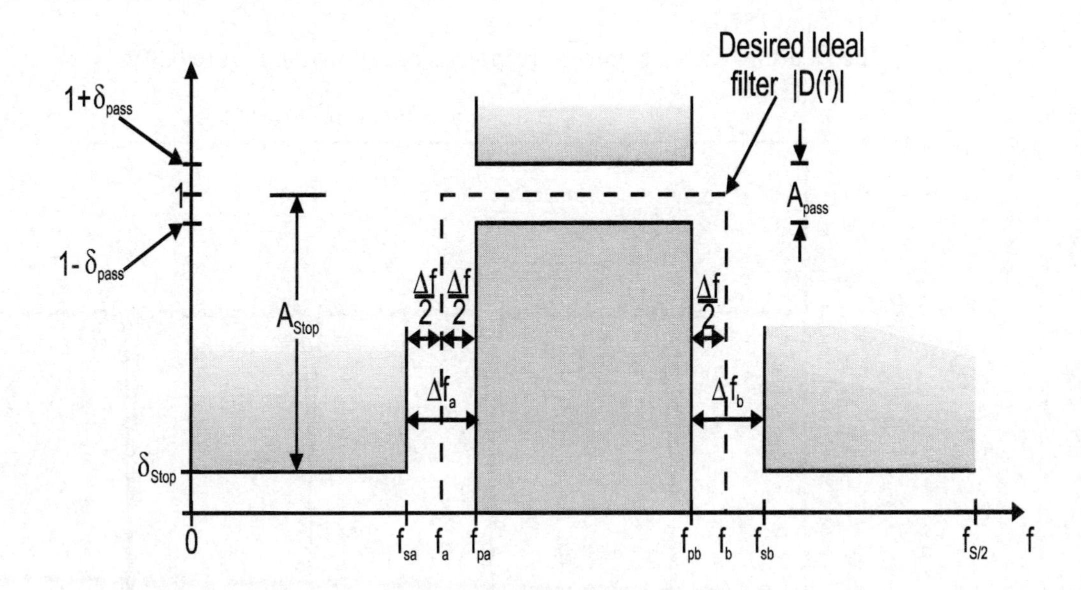

Figure 6.29
Typical specifications for a band-pass filter

Typical specifications for a band-pass filter are shown in Figure 6.29. There are two stopbands and two transition bands. The final design will have equal transition bandwidths. If the transition bandwidths of the original specifications are not equal, the smaller one will be used.

The ideal cut-off frequencies are defined in the same way as in the low-pass case:

$$f_a = f_{pa} - \frac{1}{2}\Delta f$$

$$f_b = f_{pb} + \frac{1}{2}\Delta f$$

The window parameters can then be calculated. The band-pass impulse response is given by

$$h(n) = w(n) \cdot \frac{\sin((n-M)\omega_b) - \sin((n-M)\omega_a)}{(n-M)\pi}$$

for $n = 0,1,\ldots,N-1$ where

$$h(M) = \frac{\omega_b - \omega_a}{\pi}$$

Example 6.6

Five-band graphic equalizer. The crossover frequencies of the 5 bands are 3 kHz, 7 kHz, 11 kHz, and 15 kHz.

The resulting filter magnitude responses are shown in Figure 6.30.

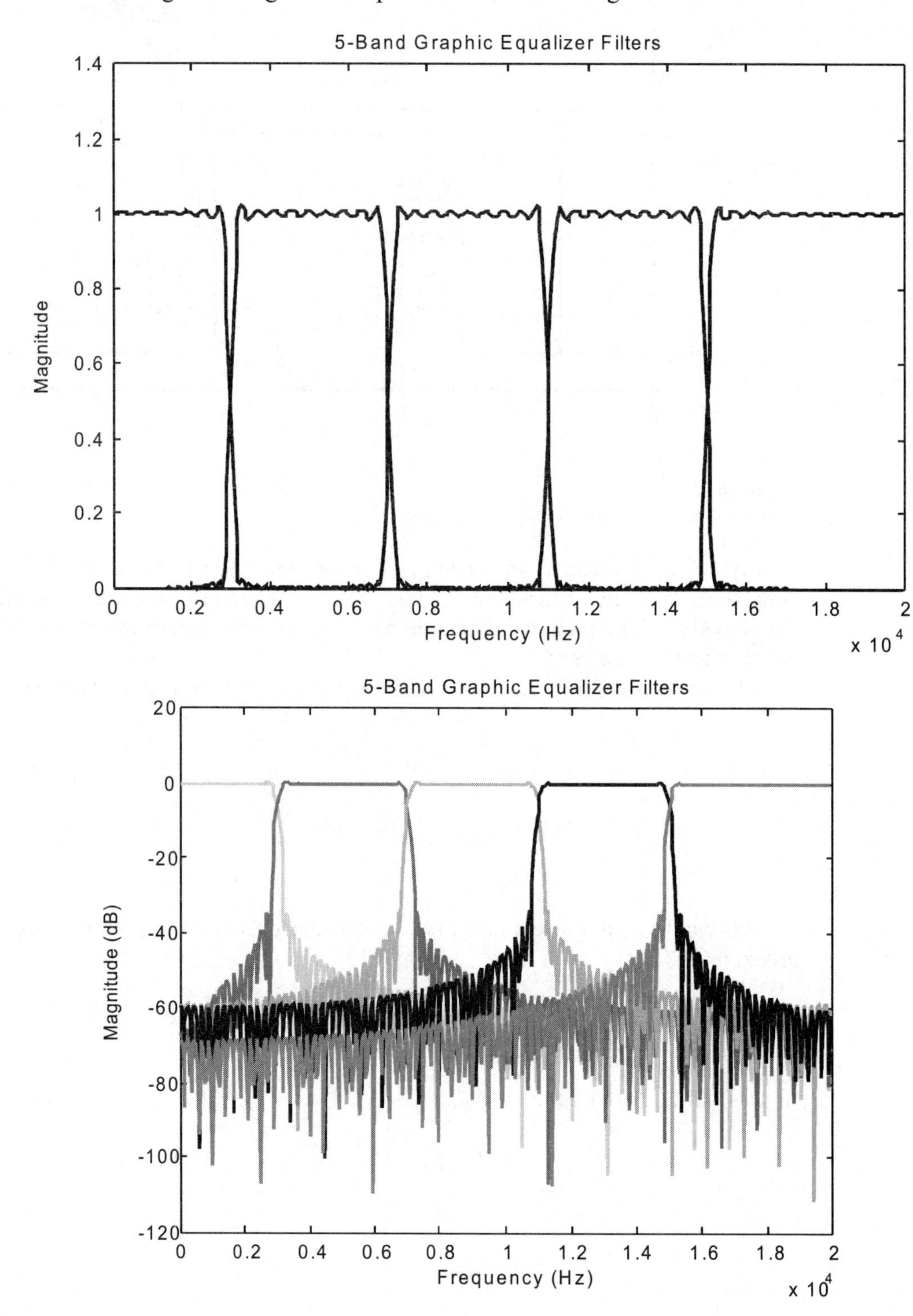

Figure 6.30
Magnitude responses of the five filters in the graphic equalizer

The final high-pass filter is complementary to the sum of the first 4 filters. An implementation using four filters is shown in Figure 6.31.

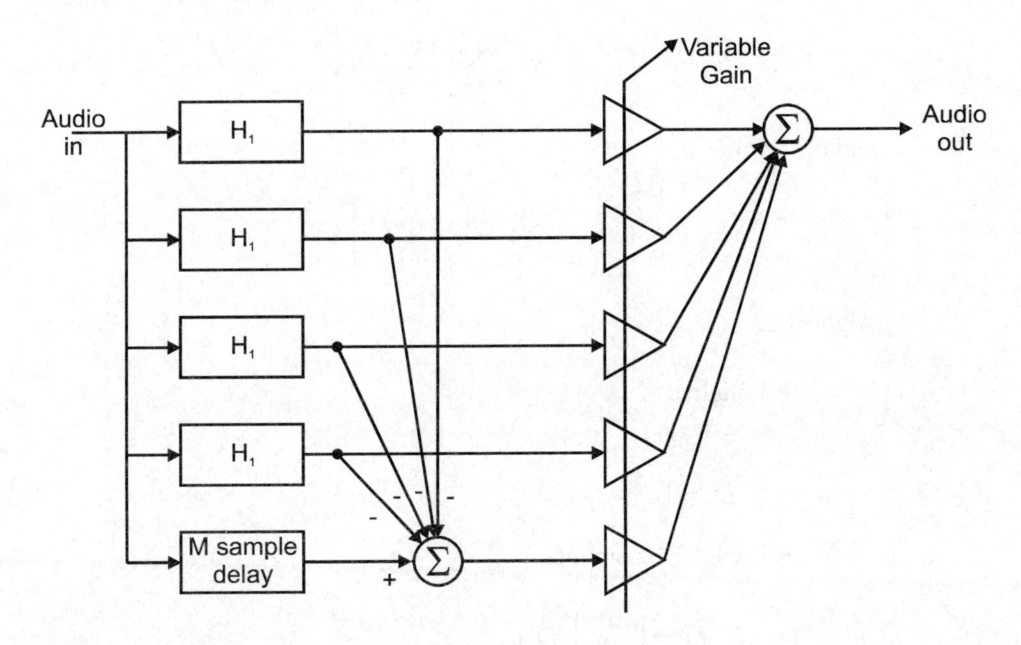

Figure 6.31
An implementation of the 5-band graphic equalizer

In practice, the digital filters employed for digital graphic equalizers are typically second order IIR filters.

6.4.4.5 Remarks

This is a particular example of a filter bank where the input is split into a number of non-overlapping frequency bands. Filter banks have been used very successfully in speech coding – a technique known as sub-band coding. The output of each filter is quantized to a different resolution. The allocation of bits in each band is usually governed by psychoacoustic perceptual criteria. To put simply, fewer bits are assigned to bands that are less audible.

There have been a lot of research activities in the design of filter banks in the last decade. This research has led to the development of wavelet transformation and wavelet filter banks, which is still a very active research area. However, it is beyond the scope of this introductory course.

6.4.4.6 Computation of Bessel functions

To complete our discussions on Kaiser window design method, we shall consider the computation of the Bessel function. The zero-th order Bessel function of the first kind can be defined by its Taylor series expansion:

$$I_0(x) = \sum_{k=0}^{\infty} \left[\frac{(x/2)^k}{k!} \right]^2$$

Evaluation of $I_0(x)$ is over the range limited by the shape parameter and

$$0 \le x \le \beta$$

The Taylor series can be recursively computed to a desired level of accuracy. Define the partial sum

$$S_n = \sum_{k=0}^{n} \left[\frac{(x/2)^k}{k!} \right]^2$$

and the term

$$D_n = \left[\frac{(x/2)^n}{n!} \right]^2$$

Algorithm:

- Initialize

  $S_0 = 1$

  $D_0 = 1$

- For $n \geq 1$, compute

  $$D_n = \left(\frac{x}{2n} \right)^2 D_{n-1}$$

  $$S_n = S_{n-1} + D_n$$

- Above step is repeated for successive values of n until the ratio

  $$\frac{D_n}{S_n} = \frac{S_n - S_{n-1}}{S_n} < \varepsilon$$

  where ε is a small number (say, 10^{-9}).

6.5 Frequency sampling method

The window method of FIR filter design requires the inverse DTFT of the desired frequency response. While the calculations may be straightforward for simple ideal low-pass, band-pass and high-pass responses, it may not be the case for an arbitrary filter response such as the ERMES pre-modulation filter specifications described in the previous chapter. Instead of considering the continuous frequency response, we can take samples of it and deal with the discrete spectrum.

Samples of the desired frequency response $D(\omega)$ are taken at N uniformly spaced frequencies ω_k within the interval $(0,2\pi)$. If N is also the order of the FIR filter to be designed, then the coefficients $h(n)$ can be found by solving the N simultaneous equations:

$$\sum_{n=0}^{N-1} h(n) e^{-j2\pi nk/N} = D\left(\frac{2\pi k}{N} \right) \qquad k = 0, 1, \ldots, N-1$$

This approach makes sure that the frequency response of the FIR filter will pass through those sampled frequency points.

The disadvantage of this direct approach is the computational complexity. Solving N simultaneous equations requires on the order of N^3 arithmetic operations. While this is acceptable for small values of N, it becomes prohibitive when N increases.

Recalling that the relationship between the uniformly spaced discrete frequency samples and the discrete-time impulse response is given by the DFT, the inverse DFT (IDFT) of the frequency samples will give the impulse response. IDFT only requires approximately N^2 arithmetic operations. If FFT is used, the number of operations is reduced to $N \log N$.

6.5.1 Design formulas

Explicit formulas can be derived for the four types of linear phase FIR filters described in section 6.3.2. These formulas are simplified from the IDFT equation by making use of the fact that the impulse responses of linear phase FIR filters are real-valued and symmetric (or anti-symmetric).

- **Type 1**

  N is odd and $M = (N-1)/2$.

  where A_k are the equally spaced DFT samples at frequencies

$$\omega_k = \frac{2\pi k}{N} \qquad k = 0, 1, \ldots, N-1$$

  If no sample at $\omega = 0$ is included, then

$$\omega_k = \frac{(2k+1)\pi}{N} \qquad k = 0, 1, \ldots, N-1$$

  and the design formula becomes

$$h(n) = \frac{1}{N}\left[\sum_{k=0}^{M-1} 2A_k \cos\left(\frac{2\pi(n-M)\left(k+\frac{1}{2}\right)}{N}\right) + A_M \cos\pi(n-M)\right]$$

- **Type 2**

  N is even.

  If a zero frequency sample is available, the design formula is:

$$h(n) = \frac{1}{N}\left[A_0 + \sum_{k=1}^{N/2-1} 2A_k \cos\left(\frac{2\pi(n-M)k}{N}\right)\right]$$

  This formula is essentially the same as that for type 1 filters except for the upper limit of the summation and

$$A_{N/2} = 0$$

  If no zero frequency sample is available, the design formula becomes

$$h(n) = \frac{1}{N}\left[\sum_{k=0}^{N/2-1} 2A_k \cos\left(\frac{2\pi(n-M)\left(k+\frac{1}{2}\right)}{N}\right)\right]$$

$$h(n)=\frac{1}{N}\left[A_0+\sum_{k=1}^{M}2A_k\cos\left(\frac{2\pi(n-M)k}{N}\right)\right]$$

- **Type 3**
  The design formulas for anti-symmetric impulse responses involve terms with the sine function instead of the cosine function. The design formulas are

$$h(n)=\frac{1}{N}\left[\sum_{k=1}^{M}2A_k\sin\left(\frac{2\pi(M-n)k}{N}\right)\right]$$

$$h(n)=\frac{1}{N}\left[\sum_{k=0}^{M-1}2A_k\sin\left(\frac{2\pi(M-n)\left(k+\frac{1}{2}\right)}{N}\right)\right]$$

  respectively for the cases where a zero frequency sample is and is not available.

- **Type 4**
  The corresponding design formulas when N is even are

$$h(n)=\frac{1}{N}\left[\sum_{k=1}^{N/2-1}2A_k\sin\left(\frac{2\pi(M-n)k}{N}\right)+A_{N/2}\sin\pi(M-n)\right]$$

$$h(n)=\frac{1}{N}\left[\sum_{k=0}^{N/2-1}2A_k\sin\left(\frac{2\pi(M-n)\left(k+\frac{1}{2}\right)}{N}\right)\right]$$

6.5.2 Transition region

Let us consider the design of a linear phase FIR filter to approximate the ideal low-pass response with a passband from 0 to 0.4π (normalized). The frequency samples are given by

$$|D(\omega_k)|=\begin{cases}1, & k=0,1,\ldots,P\\ 0, & k=P+1,\ldots,M\end{cases}$$

With $N = 40$, P = 8. The DFT samples are therefore

$$A_k=\begin{cases}(-1)^k/N, & k=0,1,\ldots,P\\ 0, & k=P+1,\ldots,M\end{cases}$$

Choosing type 2 design since N is even and a zero frequency sample is available, we arrive at a FIR filter with magnitude response shown in Figure 6.32.

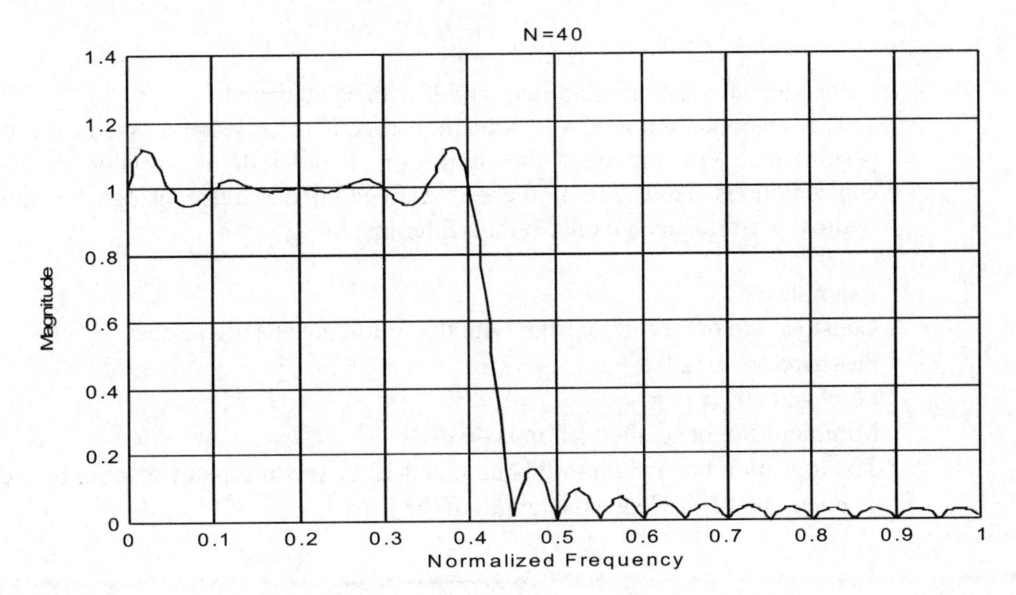

Figure 6.32
FIR filter designed using frequency sampling

The resulting filter response is very similar to the one we arrived at using the rectangular window. The amount of overshoot is relatively large near the band edge and the minimum stopband attenuation is particularly disappointing (at around –20 dB). The reason is that the transition bandwidth is too narrow. If a transition sample is added which has a magnitude that is halfway between the passband and stopband

$$A_p = 0.5(-1)^p / N$$

then the magnitude response is greatly improved as shown in Figure 6.33.

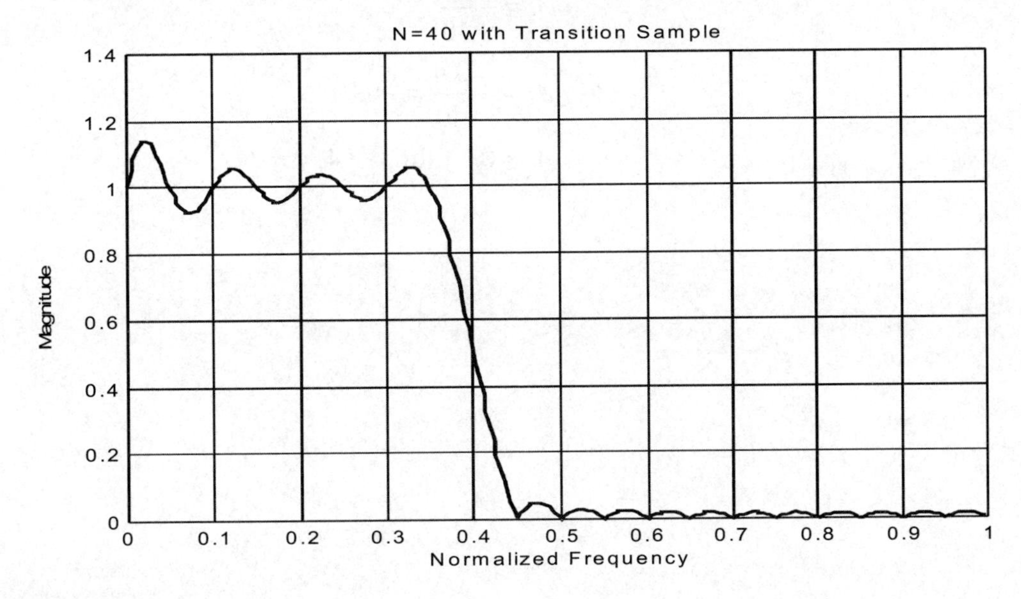

Figure 6.33
Re-design of the same filter as Figure 6.32

The stopband attenuation is now about –30 dB. Adjusting the value of A_P can make further improvements. With

$$A_p = 0.4(-1)^p / N$$

a stopband attenuation of around –40 dB can be achieved.

This example shows that a transition region is very important in the resulting filter performance. In practice, the transition bandwidth is usually dictated by other considerations. However, if there is a freedom of choice, it can be adjusted to give optimal performance given a certain filter length.

Example 6.6

Consider one more filter design with the following specifications:

Passband: 0 to 0.08π

Stopband: 0.2π to π

Minimum stopband attenuation is 40 dB

The transition bandwidth in this case is 0.12 π; the minimum spacing between frequency samples is 0.06 π. Hence the length of the filter is

$$N \geq \frac{2\pi}{0.06\pi} = 34$$

The minimum spacing is not sufficient since the stopband will have to start at the fourth sample (0.18π). In order to put a sample at 0.2π, a frequency spacing of 0.05π can be chosen which corresponds to $N = 40$. The passband will be extended to 0.1π and the transition sample will be at 0.15π.

The DFT samples are:

$$A_0 = A_2 = \frac{1}{40}$$

$$A_1 = -\frac{1}{40}$$

$$A_3 = -\frac{0.4}{40} = -0.01$$

$$A_k = 0 \quad \text{for } k \geq 4$$

Type 2 design with zero frequency samples results in a filter response shown in Figure 6.34. The filter specifications are satisfied.

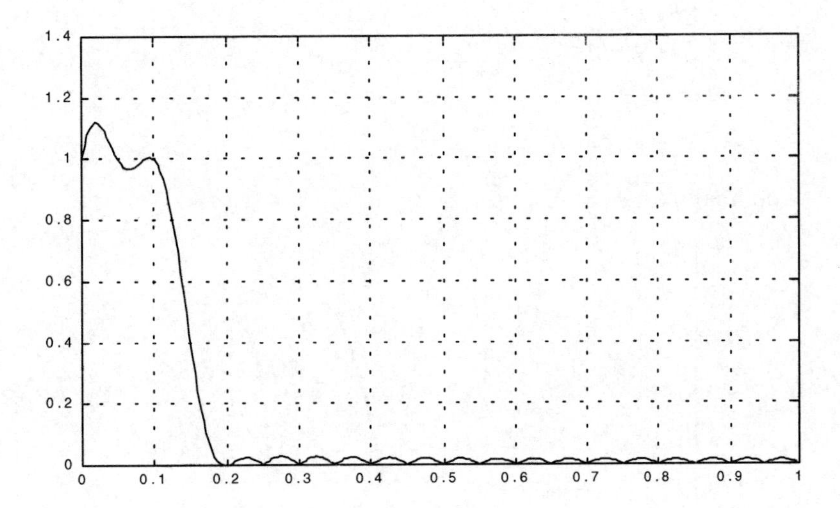

Figure 6.34
Filter response of Example 6.7

Alternatively, we can choose not to include a zero frequency sample. If the frequency sample spacing of 0.06π is used, the passband will be extended to 0.09π but the stopband edge will be at 0.21π, which is too high. Setting

$$\omega_4 = 3.5\Delta\omega = 0.2\pi$$

where $\Delta\omega$ is the sample spacing, we arrive at

$$\Delta\omega = 0.0571\pi$$

or

$$N \geq \frac{2\pi}{0.0571\pi} = 35$$

The passband edge is now at 0.0857π. The DFT samples are now

$$A_0 = \frac{1}{35}$$

$$A_1 = -\frac{1}{35}$$

$$A_2 = \frac{0.4}{35}$$

$$A_k = 0 \qquad \text{for } k \geq 3$$

The resulting filter magnitude response is shown in Figure 6.35. All specifications are satisfied.

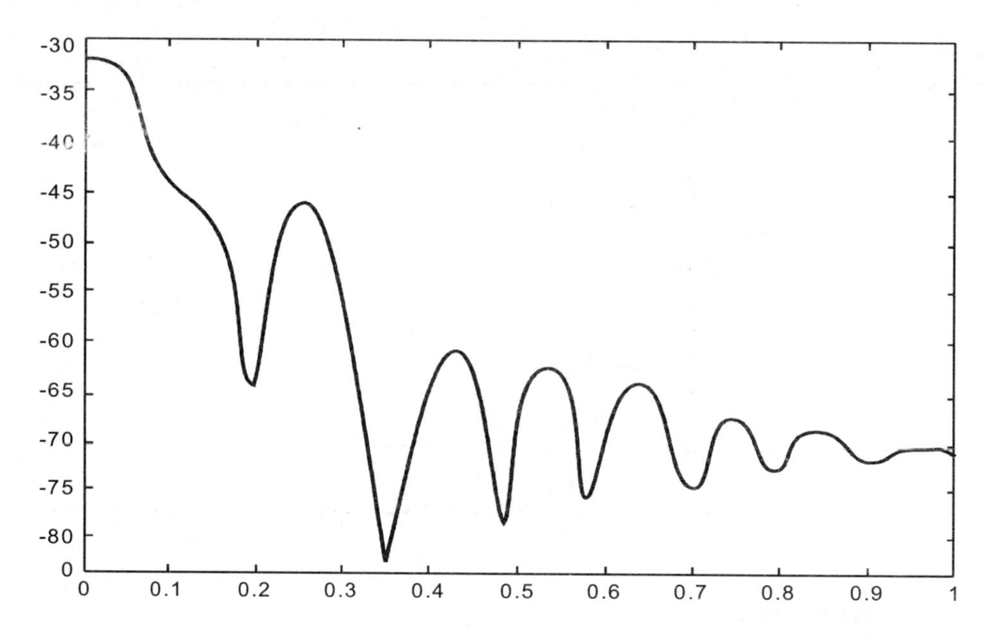

Figure 6.35
Filter response

We can see that in this case a more efficient filter (in terms of filter length) can be obtained by not including a sample at zero frequency. In general, for each design, we need to examine alternative design methods. By comparison, the most efficient windowing design is obtained through the Kaiser window with a minimum length of 38.

6.6 Parks-McClelland method

Early in the chapter we said that the first step in the filter design process is approximation. The frequency sampling design method discussed above is, strictly speaking, not an approximation approach but an interpolation approach. It produces a filter with frequency response that passes through the frequency sample points exactly but there is no constraint on the response between sample points. Consequently we cannot guarantee the behavior of the frequency response apart from that at the sample points. Peaks and overshoots can occur at various parts of the response. For low-pass filters, examples have shown that the transition bandwidth affects the resulting design to a large extent. By carefully optimizing the placement and values of the samples at the transition region, better designs are obtained. The question is how far can the maximum error be reduced?

The answer to this question lies in a technique that was used widely for analog filter approximation, known as Chebyshev approximation. This approximation, when applied to filter design, minimizes the maximum error over a set of frequencies. This type of filter exhibits equiripple behavior in the frequency responses. Thus the filter designed using this approximation are called equiripple FIR filters. They are also called optimum and minimax filters.

Closed form design formulas are not available for these filters, however. An iterative algorithm has to be used. A very efficient one is the Remez exchange algorithm. It was first developed in the early 1970s.

6.6.1 The approximation problem

A typical specification for a low-pass filter suitable for Chebyshev approximation is shown in Figure 6.36.

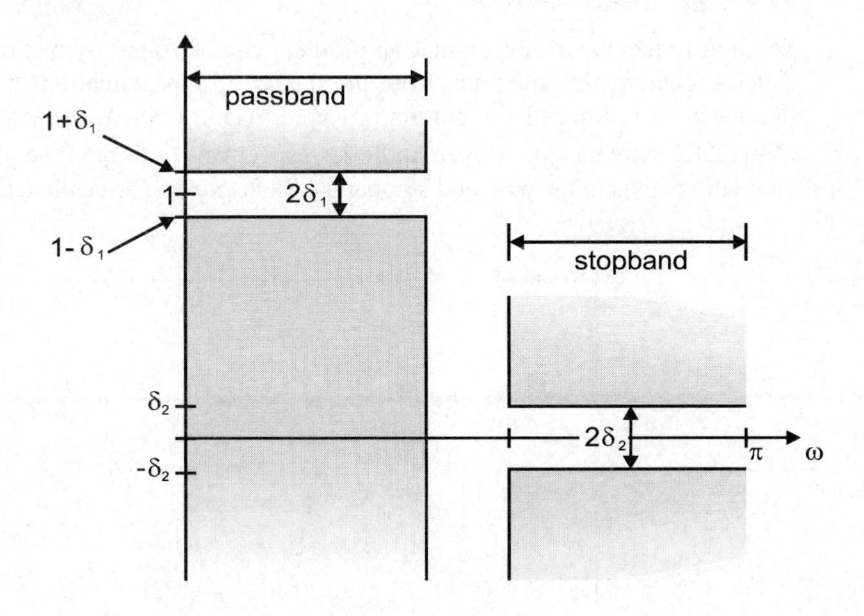

Figure 6.36
Typical low-pass filter specification for Chebyshev approximation

In the passband, the maximum deviation of the magnitude response from unity is $\pm\delta_1$. In the stopband, it is $\pm\delta_2$.

The desired frequency response $D(\omega)$ is assumed to be zero phase which means it is purely real-valued. The form of the frequency response of the final filter is

$$H(\omega) = Q(\omega)\sum_{k=0}^{N-1} h(k)\cos\left(\frac{2\pi k}{N}\right)$$

where

$$Q(\omega) = \begin{cases} 1, & \text{for Type 1 filters} \\ \cos(\omega/2), & \text{for Type 2 filters} \\ \sin\omega, & \text{for Type 3 filters} \\ \sin(\omega/2), & \text{for Type 4 filters} \end{cases}$$

The approximation problem is to minimize the maximum of the weighted error function

$$\|E(\omega)\| = \max W(\omega)|D(\omega) - H(\omega)|$$

for all $h(k)$ by choosing a suitable $H(\omega)$. Here Ω is the entire frequency range of interest, which is $[0,\pi]$. $W(\omega)$ is a user-defined weighting function so that more importance can be placed on certain frequency intervals compared with others. For instance, a zero weight can be assigned to the transition frequencies so that the shape of the response in this region will not affect the performance in the passband and stopband which are usually much more important.

6.6.2 The equiripple solution

Solution to the above approximation problem can be found by making use of a theorem, called the alternation theorem, from the theory of approximation. It basically states that the design is optimized for minimum ripple, if and only if, there are at least $N/2+2$ or $(N-1)/2+2$ extrema (for N even and odd respectively) of equal weighted amplitudes and alternating signs in the pass and stopbands. Such extrema are called alternations.

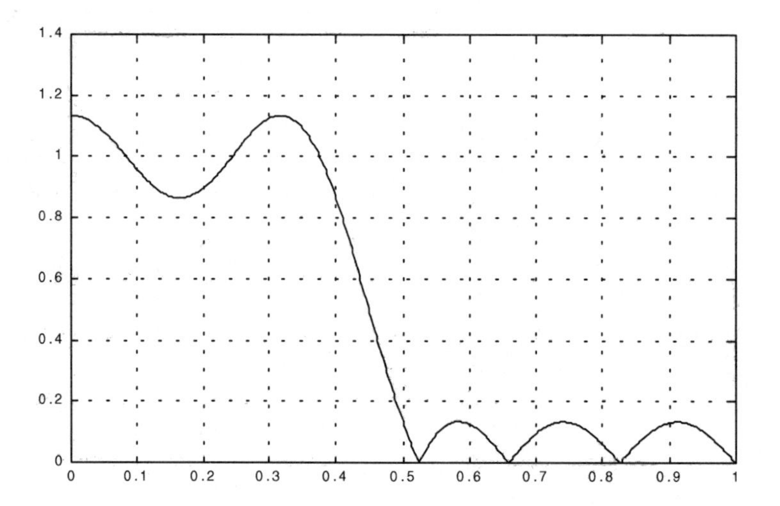

Figure 6.37
A length-13 equiripple FIR filter

Figure 6.37 shows a solution for a length-13 equiripple FIR filter. Since N=13, the number of extrema is 8. These eight extrema are also indicated in the figure.

It should be pointed out that the best equiripple design is also unique. For a given set of specifications, the unique best solution may have more than the minimum number of extrema as stated above. Let N be even and $r = (N/2+2)$. If the unique best filter has $r+2$ extrema, then there cannot be another filter with only $r+1$ or r extrema for the same set of specifications. In other words, using the Chebyshev approximation, optimality is guaranteed.

The alternation theorem is useful in that it helps us to establish the form of the optimal solution so that it can be recognized once we have found it. But it does not tell us how to arrive at the optimal solution. One approach is to identify the extremal frequencies. Once the extrema are found, the filter coefficients can be obtained by using the frequency sampling method. Thus the filter design problem becomes one of finding the extrema given a set of specifications. The Remez exchange algorithm is an efficient one for finding these extrema.

6.6.3 The Remez exchange algorithm

Given the order N of the filter, we do not know beforehand the minimum amount of ripples, δ_1 and δ_2, that can be achieved. So they become the additional variables, apart from the filter coefficients, that need to be determined. There are two main ways to handle this.

Parks and McClelland introduced a weight K to the stopband specification so that

$$K\delta_2 = \delta_1 = \delta$$

Hence instead of two variables, only δ will need to be determined together with the filter coefficients. They are evaluated iteratively for a given filter order. If the specifications are not met, then the filter order is increased and the optimization is repeated.

Another method, proposed by Hersey, Lewis and Tufts, is to let δ be equal to either δ_1 or δ_2. In this way the algorithm will still only need to deal with one additional variable but either the passband or stopband constraint will be satisfied exactly.

The Remez exchange algorithm makes use of the fact that the error function

$$E(\omega) = D(\omega) - \sum_{k=0}^{N-1} h(k)\cos k\omega$$

with $0 \le \omega \le \pi$ will always take on values of $\pm\delta$ for a given set of $(N+1)$ normalized frequency points denoted by ω_m for $m = 1,2,\ldots, N+1$. Therefore we have a set of linear equations

$$D(\omega_m) = \sum_{k=0}^{N-1} h(k)\cos k\omega_m + (-1)^m \delta \qquad m = 1,2,\ldots,N+1$$

There are $N+1$ equations with $N+1$ unknowns (N filter coefficients and the ripple amplitude) which we can solve. If the extremal frequency ω_m is known, then the equations can be easily obtained and no iteration is needed.

Obviously the algorithm cannot deal with a continuum of frequencies, even within the Nyquist interval. Parks and McClellan suggested the use of a set of frequencies, which are equally spaced, with a size of about 10 times the order of the filter. Since the number of frequencies is larger than N+1, we cannot directly solve the set of equations set out above. The Remez exchange algorithm starts with a trial set of frequencies and systematically exchange frequencies until the set of extremal frequencies is found.

Remez exchange algorithm:

- Choose an initial set of N+1 frequencies:

$$T^{(0)} = \left\{\omega_1^{(0)}, \omega_2^{(0)}, \ldots, \omega_{N+1}^{(0)}\right\}$$

- Solve the set of linear equations for $T^{(i)}$. The error function has a magnitude of $\delta^{(i)}$ for the *i*-th iteration.
- Find the frequency response at the whole set of frequencies.
- Search the entire set of frequencies to see where the magnitude of error is larger than that found in the second step. If none exists, then stop.
- Update the set of trial frequencies to be the $N+1$ frequencies where the errors are largest among the errors computed for the whole set of frequencies.

$$T^{(i+1)} = \left\{\omega_1^{(i+1)}, \omega_2^{(i+1)}, \ldots, \omega_{N+1}^{(i+1)}\right\}$$

- Repeat from the second step.

It should be pointed out that in the above discussion, we have assumed that the weight function $W(\omega)$ is unity for all frequencies. But the results apply to a general positive weight function.

6.6.4 Design formulas

For low-pass filters, Kaiser has developed some empirical formulas that helps in estimating the order of the filter required for a given set of specifications.

$$N = \frac{-10\log_{10}(\delta_1\delta_2)}{14.6\Delta f} + 1$$

where Δf is the normalized transition bandwidth given by

$$\Delta f = \frac{\omega_S - \omega_P}{2\pi}$$

and ω_p, ω_s are the passband and stopband edge frequencies respectively. This formula gives a good estimate when the bandwidth is neither extremely wide nor extremely narrow.

For filters with very narrow passbands, the stopband behavior governs the filter order and the following formula can be used for estimation:

$$N = \frac{0.22 - 20\log_{10}\delta_2 / 27}{\Delta f}$$

For very wide passband filters, such as notch filters, the following equation can be used instead:

$$N = \frac{0.22 - 20\log_{10}\delta_1 / 27}{\Delta f}$$

A more accurate estimate can be obtained by:

$$N = \frac{f(\delta_1, \delta_2) - g(\delta_1, \delta_2)(\Delta f)^2}{\Delta f}$$

where

$$f(\delta_1, \delta_2) = (0.005309x_1^2 + 0.07114x_1 - 0.4761)x_2 - (0.00266x_1^2 + 0.5941x_1 + 0.4278)$$

$$g(\delta_1, \delta_2) = 11.012 + 0.51244(x_1 - x_2)$$

and

$$x_1 = \log_{10}\delta_1$$

$$x_2 = \log_{10}\delta_2$$

Example 6.7

An FIR low-pass filter with the following specifications is designed using the Remez exchange algorithm:

Passband: $0 - 0.66\pi$

Stopband: $0.74\pi - \pi$

$\delta_1 = \delta_2 = 0.1$

Figures 6.38 and 6.39 show an odd length and even length filter response that satisfied these specifications.

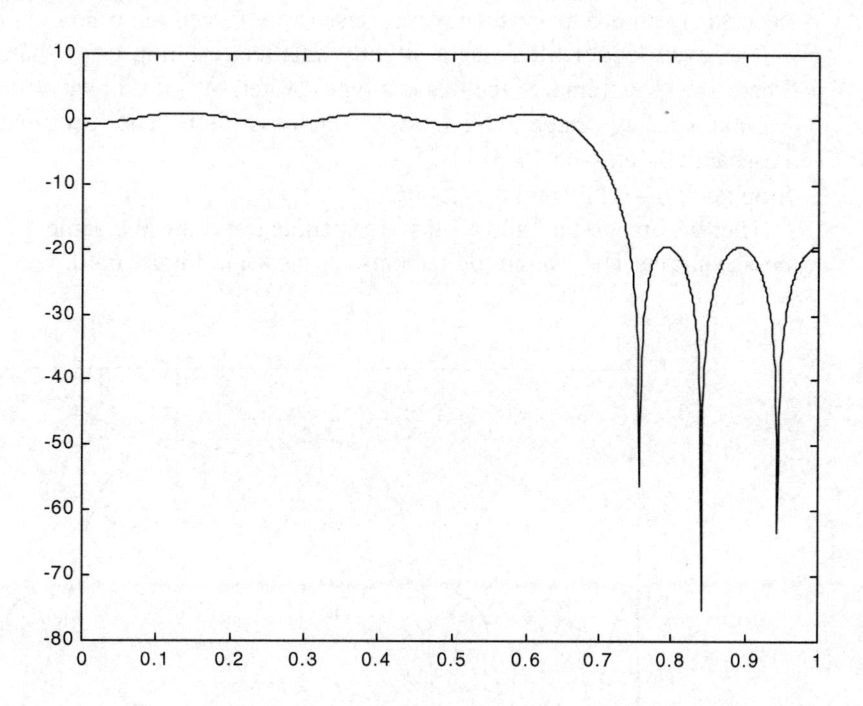

Figure 6.38
Odd length filter response

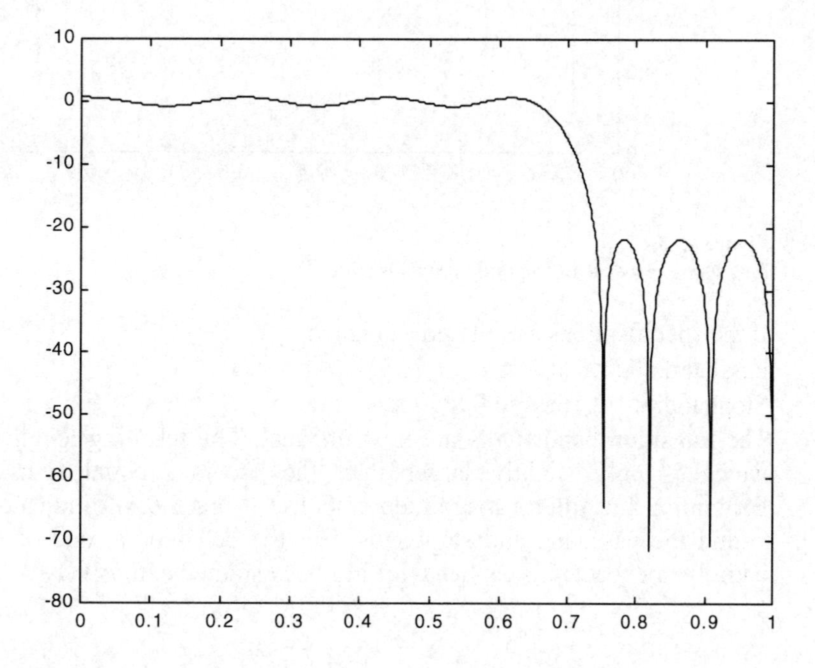

Figure 6.39
Even length filter response

For the length-21 filter, there are 12 extrema as expected, including the two band edges. Note that one of these two band edges will always be an extrema frequency, but not necessarily both. The frequency response is not forced to be zero at either $\omega = 0$ or $\omega = \pi$.

The even length filter has a slightly smaller resulting error than the length-21 filter. There are 11 extrema. Since this is a type 2 filter, $\omega = \pi$ is always zero.

Next we shall design two length-21 bandpass filters. The specifications are:

Passband: $0.36\pi - 0.66\pi$

Stopband: $0 - 0.28\pi$ and $0.74\pi - \pi$

The two transition bandwidths are identical and are the same as in the previous low-pass example. The magnitude response is shown in Figure 6.40.

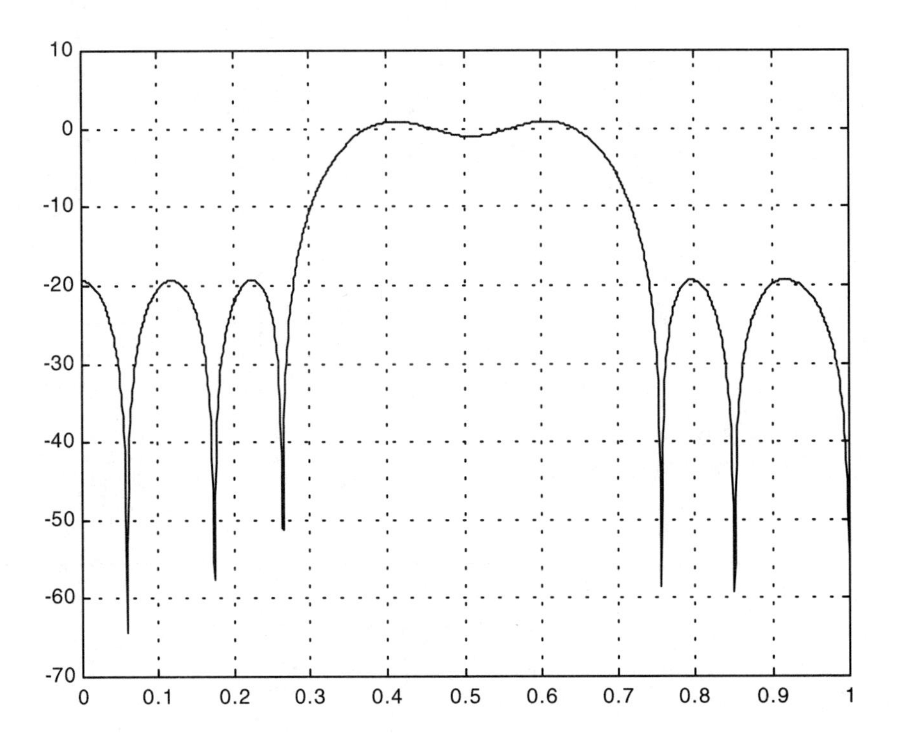

Figure 6.40
Magnitude response of length-21 bandpass filter

If the specifications are altered as below:

Passband: $0.5\pi - 0.74\pi$

Stopband: $0 - 0.16\pi$ and $0.8\pi - \pi$

The transition bandwidths are now unequal. The resulting length-21 filter has an error, which is only slightly larger than the previous equal transition bandwidth case. Examining the filter's magnitude response (Figure 6.41) indicates that it behaves well within the passband and stopbands. But the behavior in one of the transition bands is entirely unexpected. This behavior has been studied extensively.

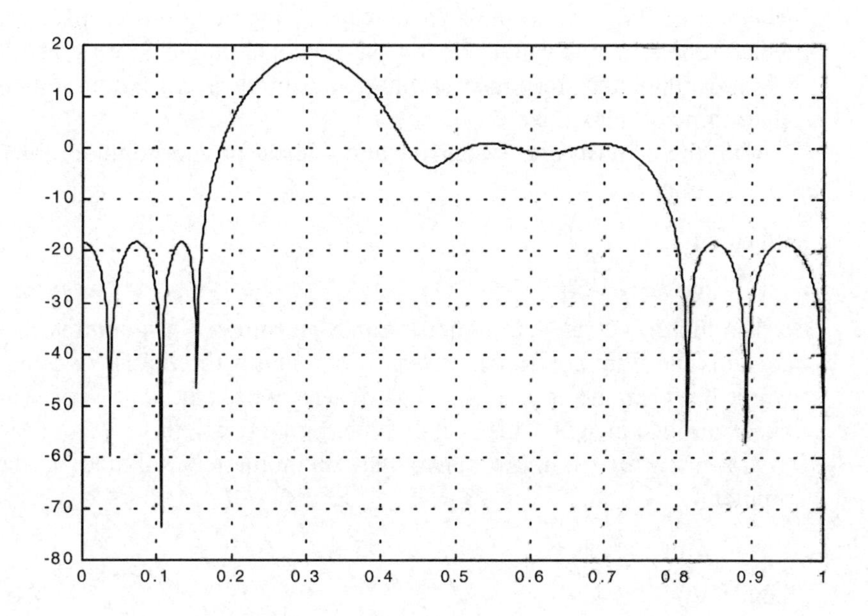

Figure 6.41
Length-21 FIR bandpass filter response

One way of reducing the possibility of transition band peaks is to calculate the following

$$N_1 = \frac{-10\log_{10}(\delta_1\delta_2)-13}{14.6\Delta f_1}+1$$

$$N_2 = \frac{-10\log_{10}(\delta_2\delta_3)-13}{14.6\Delta f_2}+1$$

where Δf_1 and Δf_2 are the two transition bandwidths. If $N_2 > N_1$, then Δf_1 can be reduced by moving the stopband edge frequency closer to the passband edge so that N_1 is approximately equal to N_2.

Alternatively, the weighing function can be used to control the maximum amount of error in the transition band. However, the appropriate amount of weighing is usually obtained by experience and trial-and-error.

A much better way to control the behavior in the transition band is to use the linear programming design method.

6.7 Linear programming method

One of the most recent approaches to linear phase FIR filter design makes use of the well-known linear programming method. In this case, the desired frequency response is composed of two parts: the upper limit function and the lower limit function. For a set of frequency points (approximately 10 times that of the filter order), the constraints are denoted as

$$H(\omega_k)+x \le U(\omega_k)$$

$$H(\omega_k)-x \ge L(\omega_k) \quad or \quad -H(\omega_k)+x \le -L(\omega_k)$$

where U and L are the upper and lower limit functions respectively. x is a parameter, which represents the distance between the upper and/or lower constraints. x can be zero if we allow the final response to 'hug' one of these two limit functions. Otherwise, the algorithm will maximize x.

With these constraints, we arrive at the linear programming problem:

$$\max \quad x$$

subject to

$$C^{\mathrm{T}}h + ax \le b$$

The matrix C is determined from the sampled trigonometric functions. Vector h contains the filter coefficients. Vector b contains the limits (or bounds) and vector a is 1 where the parameter x is used and zero where it isn't. The variables h and x are unconstrained in sign. This is called the primal problem.

A form, which is more convenient for numerical solution, is the dual of the primal problem:

$$\min \quad b^{\mathrm{T}}y$$

subject to

$$Cy = 0, \qquad a^{\mathrm{T}}y = 1, \qquad \text{and} \quad y \ge 0$$

Using the well-known simplex algorithm can easily solve the dual problem. The algorithm will terminate under one of the following conditions:

- Negative cost is obtained which implies that the original design problem is feasible.
- Optimal solution is reached with non-negative cost, which means that the design problem has a feasible solution.
- The dual is unbounded which means that the primal problem is infeasible.
- Dual is infeasible which implies that the primal problem is infeasible or unbounded.

6.8 Design examples

A computer program called METEOR is publicly available which implements this approach. A low-pass filter is designed using METEOR and the result is shown in Figure 6.42. The solid lines are the constraints.

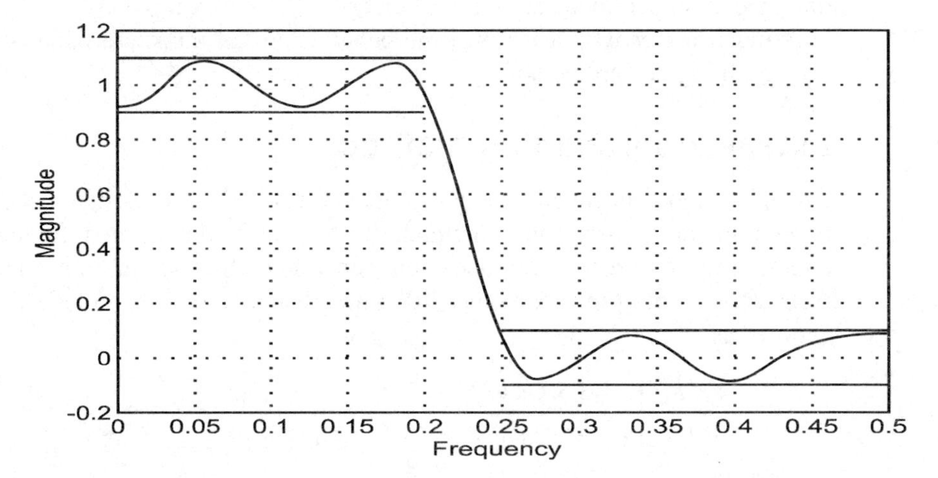

Figure 6.42
Low-pass filter designed using METEOR

Figure 6.43 shows the magnitude response of a length-25 bandpass filter with unequal transition bandwidths.

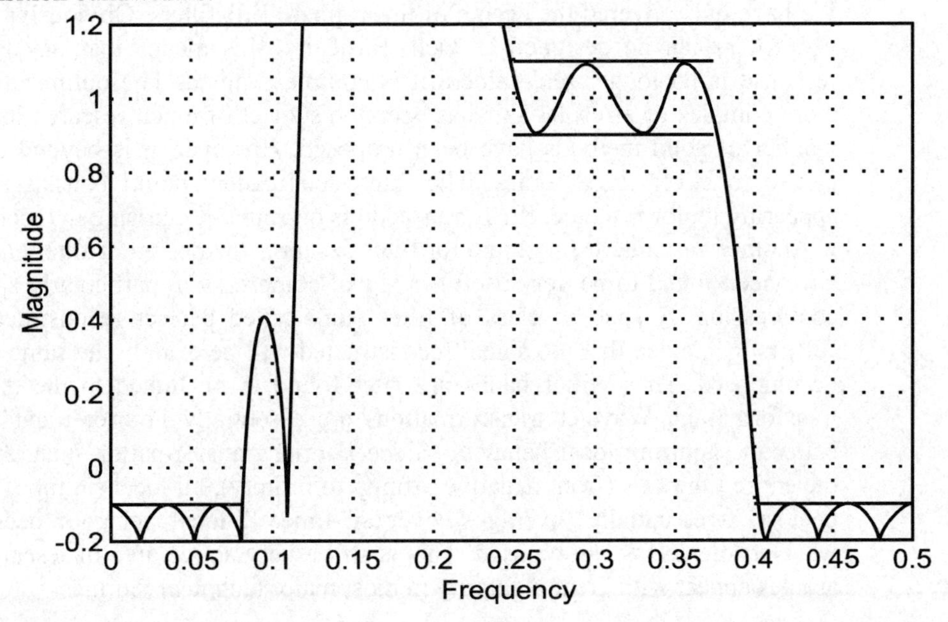

Figure 6.43
Length-25 bandpass filter designed using METEOR

This solution is essentially the same as that obtained using the Parks-McClelland method. The behavior in the first transition band is undesirable. This problem can be overcome by placing an upper limit on the first transition band. The resulting filter response is found in Figure 6.44.

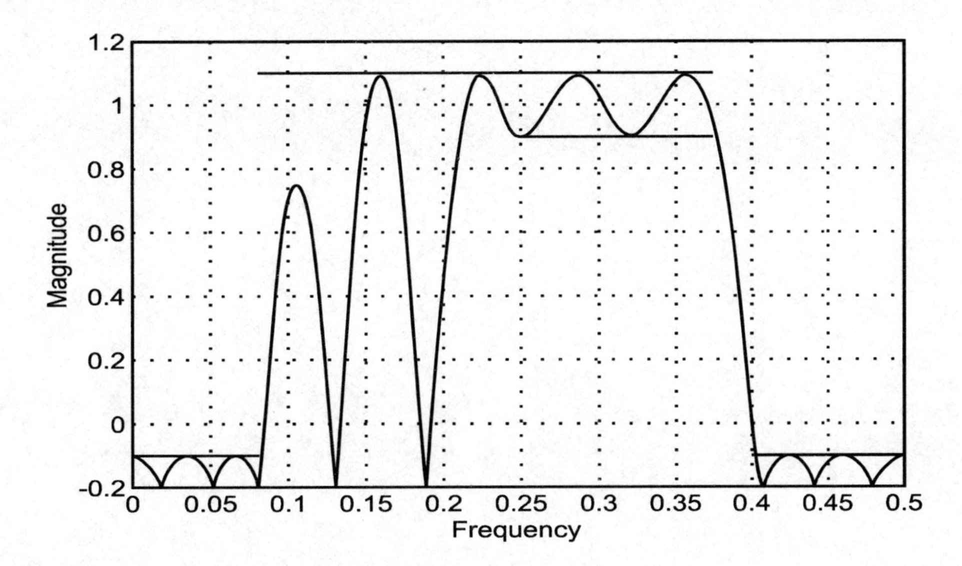

Figure 6.44
Limiting the overshoot in the first transition band

A better response can be obtained by the stopband very much similar to the solution proposed previously.

6.9 To probe further

We have only covered the design of linear phase FIR filters. Obviously, non-linear phase FIR filters can be designed as well. Basically, this means that the desired frequency response is no longer real-valued; it is complex-valued. The optimization will become more complex as a result. This has been the subject of much research in the 1980s and a number of good methods have been proposed. However, it is beyond the scope of this course to cover these topics. The interested reader should consult technical articles appearing in, for instance, IEEE transactions on signal processing in recent years.

Another very interesting topic for filter design is the design of filter banks. Filter banks have been found to be very useful in signal compression, particularly speech and image compression. A specific class of filter bank called perfect reconstruction filter banks (PRFB) guarantee that the signal reconstructed will be exactly the same as the one being decomposed. These filter banks are later found to be linked to the theory of wavelet transformation. Wavelet transformations are essentially Fourier transforms, which are better in capturing local behavior. Since Fourier transformation integrate (or sum) over the entire time axis (from negative infinity to infinity), all local (in time) information will be lost. It essentially 'averages' over all time. If local behavior becomes important, wavelet transforms can be used. This is still a very active area of research and technical articles appear with great regularity in most major technical journals.

7

Infinite impulse response (IIR) filter design

Infinite impulse response (IIR) filters have impulse responses that are infinite in duration. This is in contrast with the impulse responses of FIR filters, which are non-zero for only a finite number of samples. The relationship between the input and output of an IIR filter is given by the recursive formula:

$$y(n) = -\sum_{k=1}^{N} a(k)y(n-k) + \sum_{k=0}^{M} b(k)x(n-k)$$

The current output sample is computed from the past M inputs and the current input sample plus the previous N output samples. The second term on the right-hand side is of the same form as FIR filters. The first term is the feedback or recursive part of the equation, which causes the response to an impulse to last forever, at least theoretically. This is the reason why IIR filters are also called recursive filters.

$$y(n) = \sum_{k=0}^{\infty} h(k)x(n-k)$$

Since the IIR filter is a linear system, the output and input are related by the convolution sum.

What makes it different from FIR filters is that the upper limit of the summation is infinity because the impulse response $h(n)$ has infinite duration. Obviously computation using the convolution sum is impractical. The recursive relationship defined in the first equation is much more efficient and practical. It requires $N+M+1$ multiplications and $N+M$ additions if implemented directly.

We shall first discuss the characteristics of IIR filters with an emphasis on relating the recursive equation to its spectral properties. This will be followed by description of design methods for IIR filters. A most common approach to the IIR filter approximation problem is via classical analog filter approximations. Thus a review of some analog filter approximations are discussed. Design formulas are set out in detail. The relatively large number of equations in the later sections should not deter the reader; they are simply included for completeness. The appropriate formulas can be identified and made use of when needed.

7.1 Characteristics of IIR filters

We have seen that the input and output samples of an IIR filter are related by the recursive equation

$$y(n) = -\sum_{k=1}^{N} a(k)y(n-k) + \sum_{k=0}^{M} b(k)x(n-k)$$

This equation can be re-arranged as

$$\sum_{k=0}^{N} a(k)y(n-k) = \sum_{k=0}^{M} b(k)x(n-k)$$

with $a(0)=1$. Alternatively, we can write it as

$$\sum_{k=1}^{N} a(k)\left[z^{-k}y(n)\right] = \sum_{k=0}^{M} b(k)\left[z^{-k}x(n)\right]$$

where z^{-1} can be considered as a unit delay operator, i.e. a delay of one sample interval. So

$$z^{-k}x(n) = x(n-k)$$

In this way, we can define a transfer function in terms of z:

$$H(z) = \frac{y(n)}{x(n)} = \frac{\sum_{k=0}^{M} b(k)z^{-k}}{\sum_{k=0}^{N} a(k)z^{-k}}$$

$$= \frac{b(0) + b(1)z^{-1} + \cdots + b(M)z^{-M}}{1 + a(1)z^{-1} + a(2)z^{-2} + \cdots + a(N)z^{-N}}$$

which is a ratio of two polynomials.

The reader should note that this explanation of the transfer function of the digital system is not mathematically rigorous or strictly correct. But it is perhaps the most intuitive way of arriving at the transfer function. The proper mathematical way to arrive at this equation is through the Z transform. We shall not go into details of the Z transform in this course. The interested reader should consult any textbooks on DSP or discrete-time linear systems. All of them would have covered some aspects of the Z transform.

It suffices here to say that the z is generally a complex variable and so the polynomials are complex polynomials. The frequency response $H(\omega)$ is related to $H(z)$ by

$$z = e^{j\omega}$$

assuming that the sampling period T is normalized to 1 second. Substitution of this equation into $H(z)$ will give us the normalized frequency response.

Unlike the FIR filter, exact linear phase is impossible to achieve using an IIR filter. This is because exact linear phase implies that the filter impulse response must be symmetric. From the input/output relation of the IIR filter, it can be observed that it is impossible to have an impulse response that is zero for $k<0$ and non-zero for k from zero to infinity. However, near-linear phase IIR filters do exist.

Linear phase filters allow us to remove considerations of the phase response in our design process. As we have seen with linear phase FIR filters, the approximation problem is real-valued and thus is relatively simple to solve. Since IIR filters cannot have exact linear phase, its design is therefore more complex. The equations that need to be solved are generally non-linear. Complex optimization techniques are needed in this case. Most of the common design techniques assume that the phase response is not important and only approximate the magnitude response.

If phase response is not important, then for the same magnitude response specifications, the required order for IIR filter is normally lower than that required for FIR filters. This is a definite advantage since a lower order implies a lower computational cost and a shorter delay.

7.2 Review of classical analog filter

Analog filter approximation has been studied very extensively in the past. There is a vast amount of knowledge accumulated. One of the most successful approaches to IIR digital filter design is to make use of this knowledge base. The design approach is illustrated in Figure 7.1.

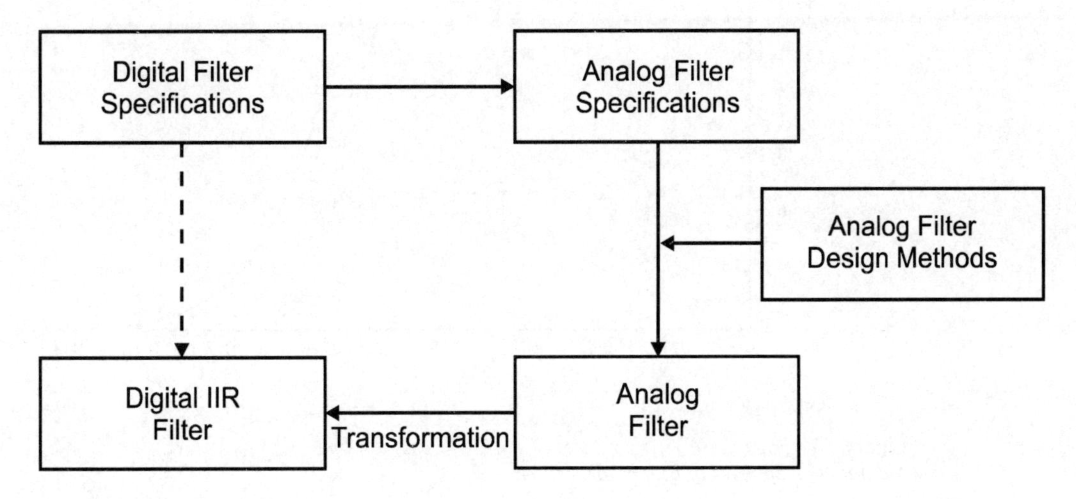

Figure 7.1
IIR filter design approach

The digital filter approximation problem is translated to an equivalent analog filter approximation problem. This analog filter design problem is then solved using well-known techniques. The resulting analog filter impulse response is then properly transformed into a digital one, thus giving us the digital filter transfer function and the filter coefficients.

Therefore in this section we shall review the most common analog filter approximation techniques. Four approximation functions are considered standard: Butterworth, Chebyshev, inverse Chebyshev and elliptic functions. They are presented in terms of a normalized low-pass filter. The low-pass filter can then be transformed to high-pass or band-pass if necessary.

7.2.1 Butterworth function

The magnitude-squared function of the Butterworth approximation is

$$|H(\omega)|^2 = \frac{1}{1+c\omega^{2n}}$$

for an n-th order filter. The constant c determines at which frequency ω the transition of the passband to stopband occurs. For normalized Butterworth filters, this point is at $\omega = 1$ and so $c = 1$. This frequency is also referred to as the cut-off frequency. Figure 7.2 shows the magnitude responses of several Butterworth filters of different orders.

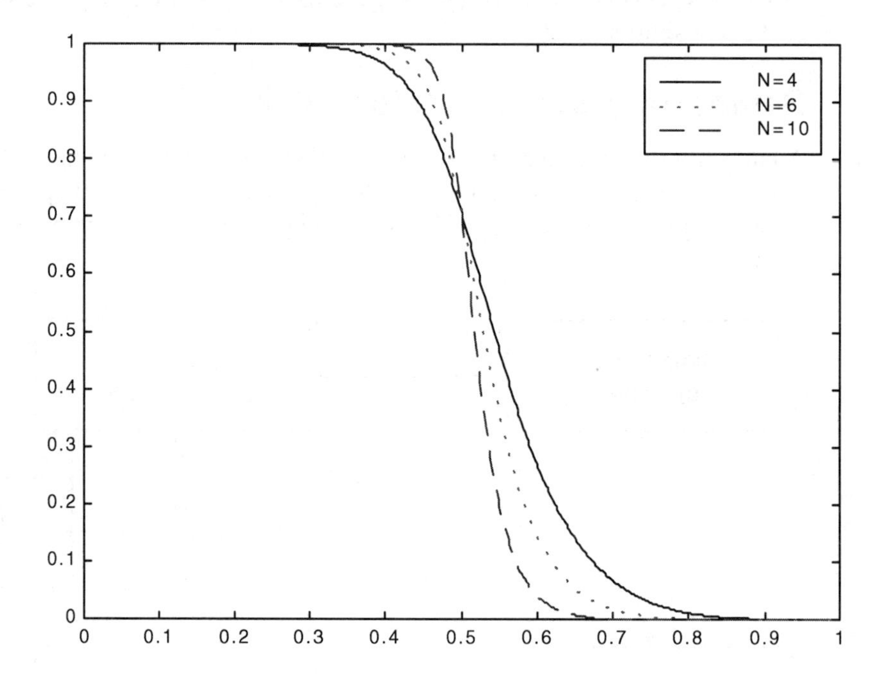

Figure 7.2
Magnitude responses of Butterworth filters

Butterworth filters are sometimes called maximally flat filters because the response at both $\omega = 0$ and $\omega = \infty$ are completely flat (or horizontal). At these frequencies, the Butterworth response is very close to the ideal. This flat passband and stopband response is achieved at the expense of the transition bandwidth, which is considerably larger than the other classical approximations. The roll-off from the passband to the stopband is relatively slow and the phase response near the cut-off frequency is non-linear. At other frequencies, the phase response is smooth.

Given the magnitude-squared function, the transfer function of a Butterworth filter can be found by standard procedures, which we shall not dwell on. The most important part of a Butterworth filter design is to determine the order of the filter required satisfying the specifications. Once the filter order is determined, the filter coefficients can be found by calculations or by simply looking up tables. Some good references on analog filter approximation are listed in the last section of this chapter.

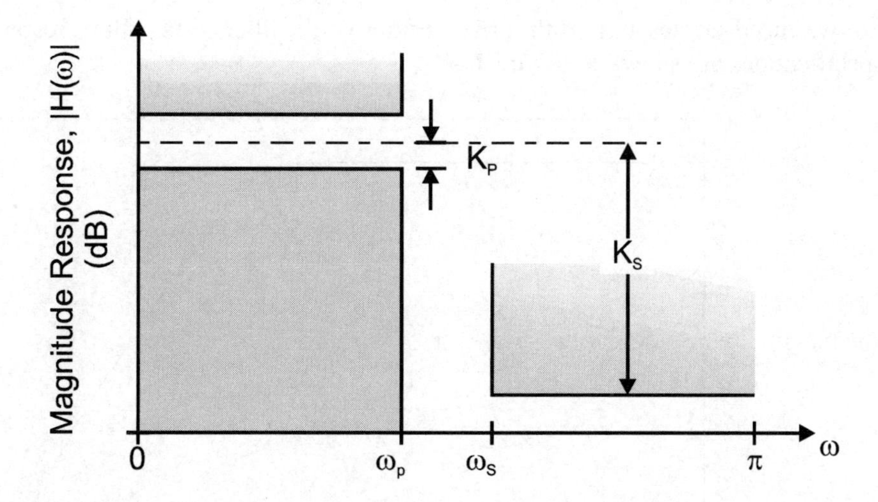

Figure 7.3
Typical specification of the magnitude response

A typical specification on the magnitude response of a filter to be designed is shown in Figure 7.3. The order of the required Butterworth filter is given by

$$n \geq \frac{\log_{10} M}{\log_{10} \Omega} = \frac{\ln M}{\ln \Omega}$$

where

$$\Omega = \frac{\omega_s}{\omega_p}$$

$$M = \sqrt{\frac{10^{0.1K_s} - 1}{10^{0.1K_p} - 1}}$$

and K_s, K_p is in decibels (dB).

Example 7.1
Determine the order of a Butterworth filter that meets the following specifications:
$\omega_p = 1$ rad/s
$\omega_s = 1.3$ rad/s

$K_s = 22$ dB

$K_p = 3.0103$ dB
Solution:
Using the above equations:

$$\Omega = \frac{1.3}{1.0} = 1.3$$

$$M = \sqrt{\frac{10^{22} - 1}{10^{0.30103} - 1}} = 12.5495$$

$$n \geq \frac{\ln(12.5495)}{\ln(1.3)} = 9.6419$$

So we need at least a 10th order Butterworth filter. The filter response and the specifications are shown in Figure 7.4.

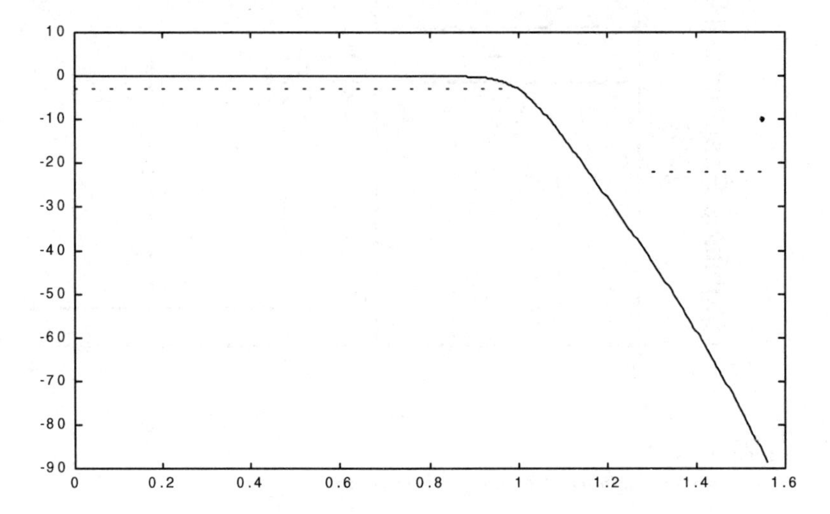

Figure 7.4
10th order Butterworth filter response

7.2.2 Chebyshev approximation

The Butterworth filter, while possessing some desirable properties, does not provide a sufficiently good approximation near the passband edge. The roll-off from passband to stopband is also relatively gradual. Hence for filters that have a narrow transition band, a very high order Butterworth filter is required.

If the application can tolerate some ripples in the passband, the Chebyshev approximation will overcome some of the problems associated with Butterworth filters stated above. The magnitude-squared response of Chebyshev filters has the form

$$|H(\omega)|^2 = \frac{1}{1+\varepsilon^2 C_N^2(\omega)}$$

where $C_N(\omega)$ is an N-th order Chebyshev polynomial and ε is a parameter that is associated with the size of the ripples. Figure 7.5 shows the magnitude response of 3 Chebyshev filters of different orders but the same ε.

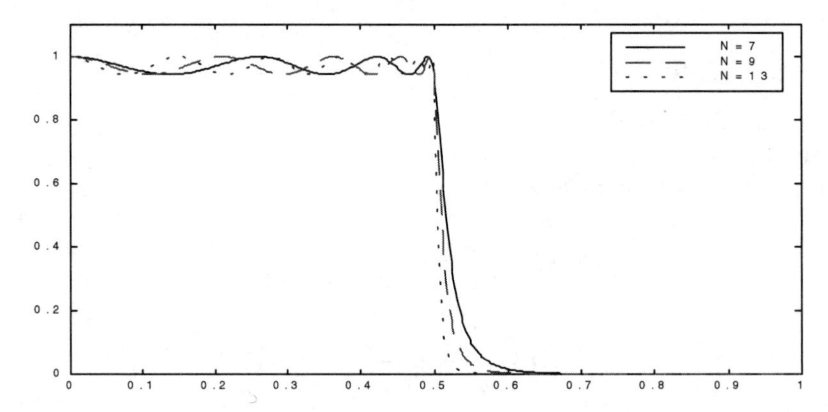

Figure 7.5
Magnitude responses of Chebyshev filters

We have already encountered Chebyshev approximation in FIR filter design in the previous chapter. The N-th order Chebyshev polynomial is given by

$$C_n(\omega) = \cos(n\cos^{-1}\omega) \qquad \text{for } 0 \le \omega \le 1$$

$$C_n(\omega) = \frac{1}{2}\left[\left(\omega + \sqrt{\omega^2 - 1}\right)^n + \left(\omega - \sqrt{\omega^2 - 1}\right)^n\right] \qquad \text{for } \omega > 1$$

Alternatively, these polynomials can be defined recursively as

$$C_1(\omega) = \omega$$

$$C_2(\omega) = 2\omega^2 - 1$$

$$C_{n+1}(\omega) = 2\omega C_n(\omega) - C_{n-1}(\omega)$$

Figure 7.6 plots some of these polynomials for $-1 \le \omega \le 1$.

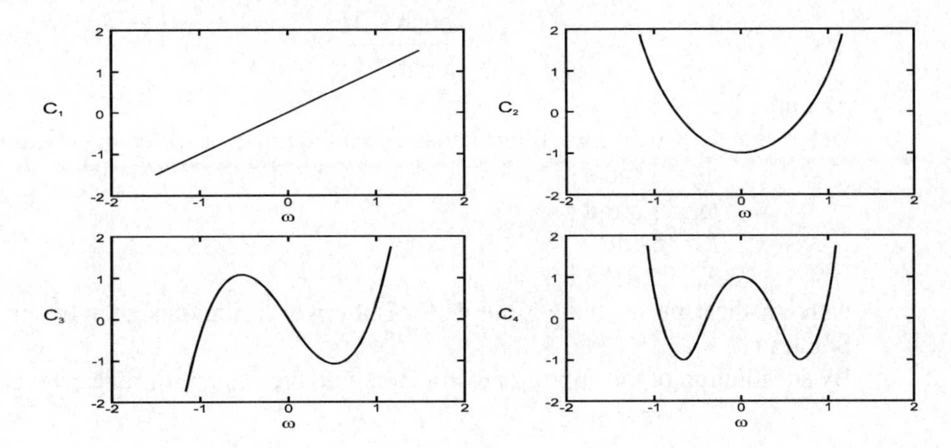

Figure 7.6
Chebyshev polynomial functions

These polynomials have the properties that

$$0 \le C_n^2(\omega) \le 1 \qquad \text{for } 0 \le \omega \le 1$$

$$C_n^2(\omega) \ge 1 \qquad \text{for } \omega \ge 1$$

That is, they oscillate between +1 and –1 for $-1 < \omega < 1$ and increase monotonically outside this interval.

As with FIR filters, the Chebyshev approximation minimizes the maximum error over the passband and is optimal in that sense. The order N of the filter determines the transition bandwidth and the number of oscillations within the passband.

Design of Chebyshev filters involves the determination of both the order N and the parameter ε. We shall refer to Figure 7.3 as the generic specifications for the low-pass filter. The procedures are as follows:

The maximum allowed passband variation is absolute value d or in decibels K_p.

$$d = 1 - 10^{-K_y/20}$$

Calculate ε from d or a.

$$K_p = 10\log(1+\varepsilon^2)$$

$$d = 1 - \frac{1}{\sqrt{1+\varepsilon^2}}$$

$$\varepsilon = \sqrt{\frac{2d - d^2}{1 - 2d + d^2}} = \sqrt{10^{K_o/10} - 1}$$

The order N of the filter is determined using the formulas:

$$\Omega = \frac{\omega_s}{\omega_p}$$

$$M = \sqrt{\frac{10^{0.1K_s} - 1}{10^{0.1K_o} - 1}}$$

$$N = \frac{\cosh^{-1} M}{\cosh^{-1} \Omega}$$

Example 7.2
Determine the Chebyshev filter that satisfies the low-pass filter specifications:

$\omega_p = 1$ rad/s
$\omega_s = 1.3$ rad/s
$K_s = 22$ dB
$K_p = 3.0103$ dB

which is the same as in the example for Butterworth filter design in the previous example.
Solution:
By substitution of the appropriate numbers into the design formulas, we have

$$\Omega = 1.3$$

$$M = 12.5495$$

$$N \geq \frac{\cosh^{-1} 12.5495}{\cosh^{-1} 1.3} = \frac{3.2212}{0.7564} = 4.2585$$

Thus the minimum order of the Chebyshev filter is 5 with a passband ripple of about 3 dB. The filter response and constraints are shown in Figure 7.7.

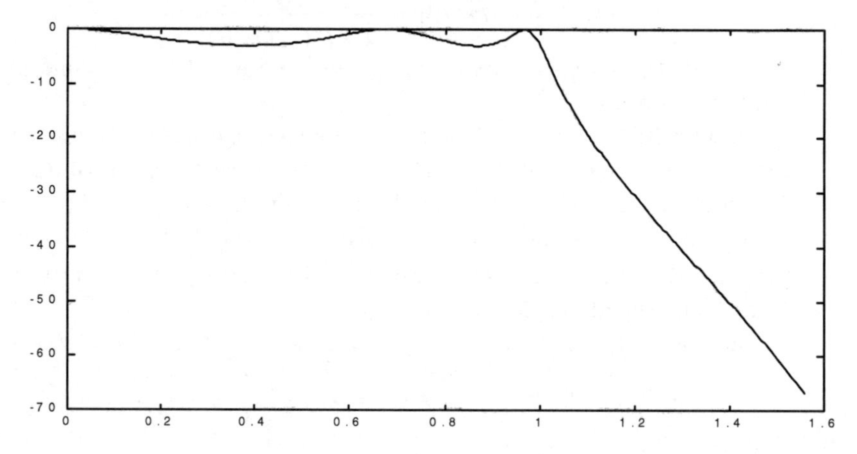

Figure 7.7
Filter response

7.2.3 Inverse Chebyshev approximation

An alternative to the Chebyshev approximation is the inverse Chebyshev approximation. As the name implies, the frequency response behavior is inverse to the Chebyshev approximation. An inverse Chebyshev filter has a flat passband and ripples in the stopband.

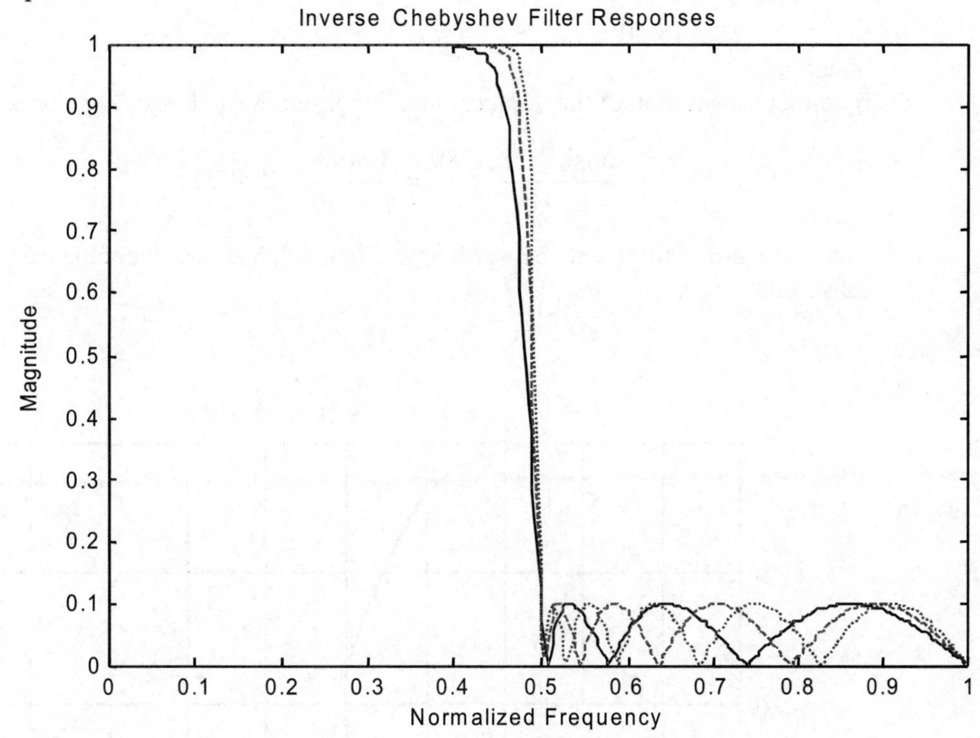

Figure 7.8
Inverse Chebyshev filter responses

The magnitude-squared responses of several filters with different orders are shown in Figure 7.8. The magnitude-squared function is given by

$$|H(\omega)|^2 = \frac{\varepsilon^2 C_N^2(1/\omega)}{1+\varepsilon^2 C_N^2(1/\omega)}$$

for an N-th order filter.

The design procedures are similar to that for Chebyshev filters except the formulas are slightly different. The design formulas are given below:

$$K_s = 10\log\left(1+\frac{1}{\varepsilon^2}\right)$$

$$\varepsilon = \frac{1}{10^{0.1K_s}-1}$$

$$N = \frac{\cosh^{-1}\sqrt{\left(10^{0.1K_s}-1\right)/\left(10^{0.1K_o}-1\right)}}{\cosh^{-1}\left(1/\omega_p\right)}$$

Example 7.3

Determine the order of an inverse Chebyshev filter that satisfies the following specifications:

$\omega_p = 0.5$ rad/s
$\omega_s = 1$ rad/s
$K_p = 0.5$ dB
$K_s = 18$ dB

Solution:

By direct substitution of the numbers into the design formula for N, we have

$$N \geq \frac{\cosh^{-1} 22.5589}{\cosh^{-1} 2} = \frac{3.8088}{1.3170} = 2.8921$$

A third order filter will be sufficient. Figure 7.9 shows the filter response and the constraints.

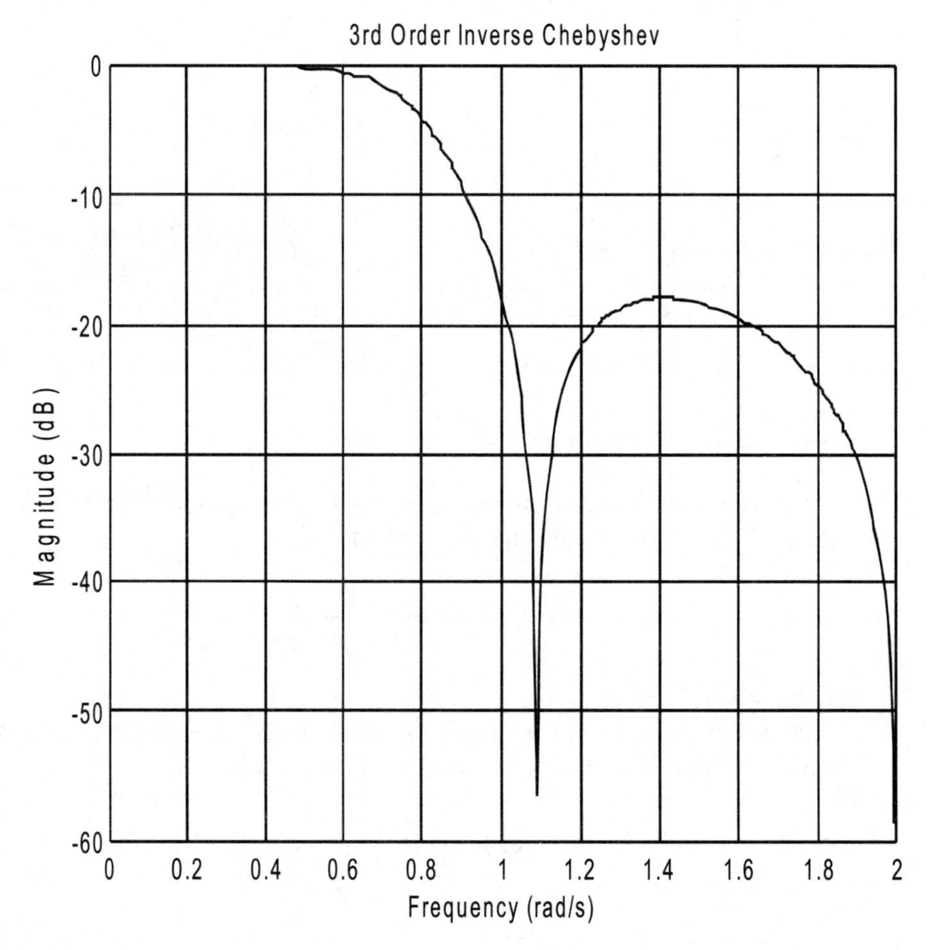

Figure 7.9
Third order inverse Chebyshev response

7.2.4 Elliptic function

For any set of low-pass filter specifications as shown in Figure 7.3, the elliptic filter is the most efficient in the sense that, compared to the previous three filter approximations, it requires the lowest order filter. It has equal ripple in the passband and in the stopband. The magnitude response of some elliptic filters is shown in Figure 7.10.

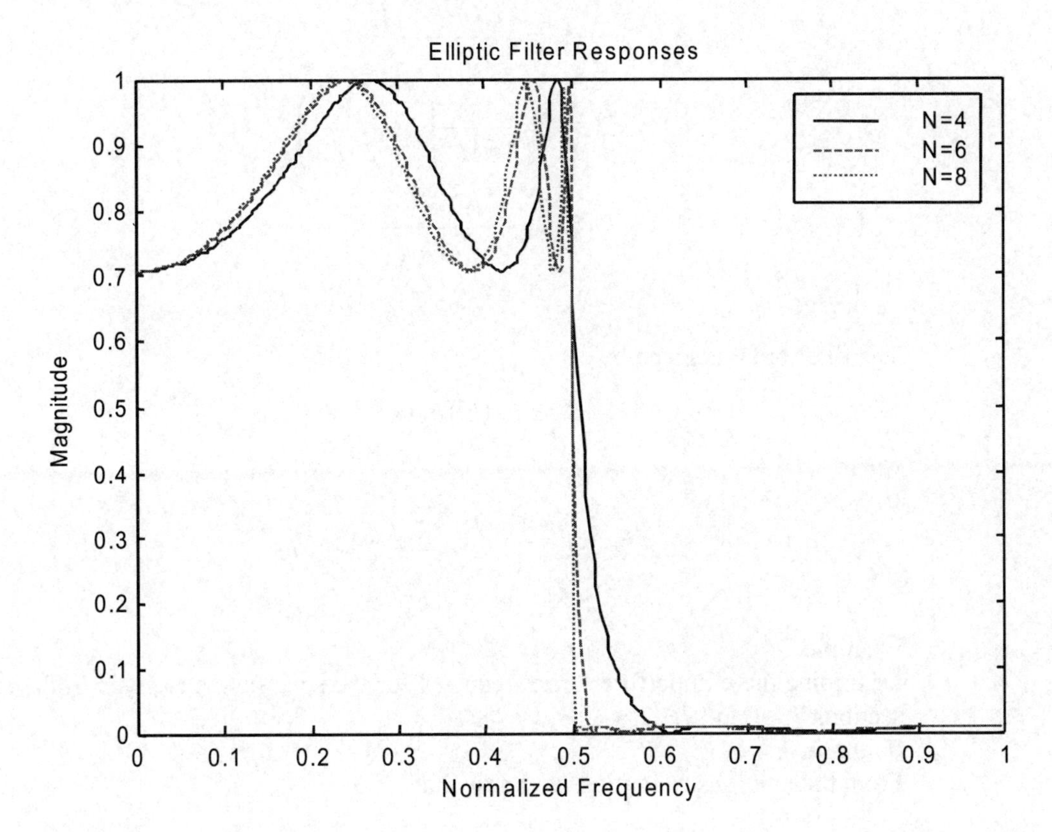

Figure 7.10
Elliptic filter responses

The design of elliptic filters is considerably more complex compared with the procedures for Butterworth and Chebyshev filters. Its magnitude-squared response is given by

$$|H(\omega)|^2 = \frac{1}{1+\varepsilon^2 R_n^2(\omega)}$$

which has the same form as a Chebyshev filter except that the function R_n is now a rational function with numerator and denominator polynomials. This rational function is called the Chebyshev rational function.

Mathematically it is based on what are called Jacobian elliptic functions and is the most complex of all the approximation functions we have discussed. Therefore we will not go too deeply into the theory and just provide the design formulas so that the order of elliptic filters can be determined.

Again, we refer to the generic low-pass filter specification in Figure 7.3. To determine the filter order N, first we calculate the quantities

$$\Omega = \frac{\omega_s}{\omega_p}$$

$$M = \sqrt{\frac{10^{0.1K_s} - 1}{10^{0.1K_o} - 1}}$$

$$C(M) = \frac{1}{16M^2}\left(1 + \frac{1}{2M^2}\right)$$

$$D(\Omega) = \frac{\sqrt{\Omega} - 1}{2\left(\sqrt{\Omega} + 1\right)}$$

The filter order is given by

$$N \geq F_E(C)F_E(D)$$

where

$$F_E(x) = \frac{1}{\pi}\ln(x + 2x^5 + 15x^9)$$

Example 7.4
Determine the elliptic filter order required for the specifications given in the examples in sections 7.2.1 and 7.2.2.
Solution:
From the previous examples, we know that

$$\Omega = 1.3$$
$$M = 12.5495$$

Now we apply these quantities to the above design formulas to get

$$C(M) = 0.0003981$$
$$D(\Omega) = 0.03275$$
$$F_E(C) = \frac{1}{\pi}\ln(0.0003981) = -2.4920$$
$$F_E(D) = \frac{1}{\pi}\ln\left[0.03275 + 2(0.03275)^5\right] = -1.0883$$
$$N \geq (-2.4920)(-1.0883) = 2.7119$$

Thus a third order elliptic filter will satisfy these specifications whereas we need a 10th order Butterworth and a 5th order Chebyshev filter for the same specifications. The responses of these three filters are plotted in Figure 7.11.

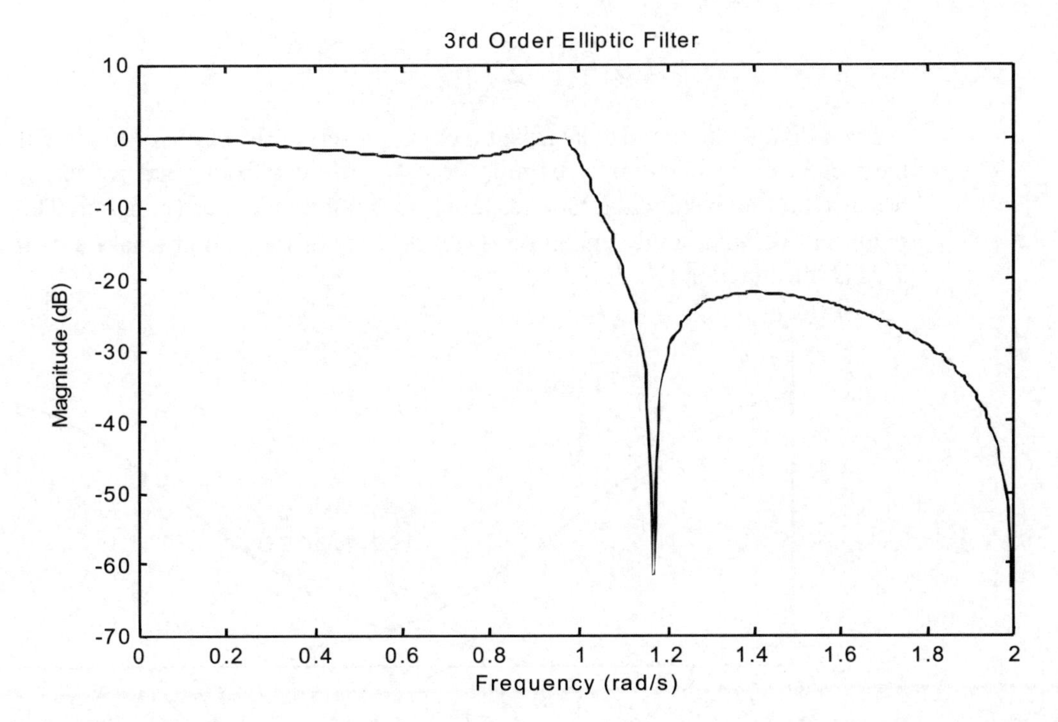

Figure 7.11
Comparison of several filters

Note that elliptic filters are also called Cauer filters and rational Chebyshev filters.

7.3 IIR filters from analog filters

One of the most efficient ways of designing IIR digital filters is via a corresponding analog filter. This approach is illustrated in Figure 7.1. The magnitude response specifications of the digital filter are translated to that for an analog filter. This analog filter is designed using the approximation methods discussed in the previous section. A suitable transformation then converts the analog filter into a digital IIR filter.

Most of the above steps are straightforward. The only step that requires more thought is the transformation from the analog filter to digital filter. A number of different methods have been proposed. We shall discuss two most common ones. They are called impulse invariance and bilinear transformation.

7.3.1 Impulse invariant method

Let $H_a(\omega)$ be the transfer function of the analog filter that has been designed. The impulse response $h_a(t)$ of this filter can be obtained through Fourier transformation. The idea behind the impulse invariance method is that the impulse response of the digital filter $h(n)$ is a sampled version of $h_a(t)$. Thus

$$h(n) = h_a(nT)$$

where T is the sampling period.

As we have discussed in Chapter 2, the magnitude spectrum of a sampled signal is a periodic continuation of the original spectrum of the analog signal. So the magnitude response of the digital filter is periodic with a period of $f_s=1/T$. It can be expressed mathematically as

$$H(\omega)=\frac{1}{T}\sum_{k=-\infty}^{\infty} H_a\left[\frac{j(\omega-2\pi k)}{T}\right]$$

Even though the impulse invariant method preserves the shape of the impulse response, the frequency response may be significantly different from what we expected. This is because the stopband magnitude response does not drop to zero at $\omega = \pi$. This means that $H(\omega)$ will be an aliased version of $H_a(\omega)$ because of the periodic nature of the response. This is illustrated in Figure 7.12.

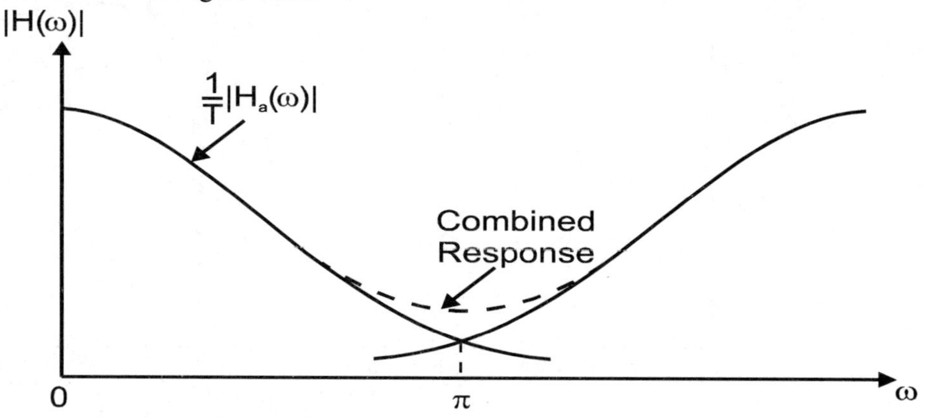

Figure 7.12
Effect of sampling an analog response

The stopband attenuation of the aliased version of $H_a(\omega)$ may not be sufficient to satisfy the original filter specifications. This will render the design useless. So we have to make sure that the aliased parts of $H(\omega)$ are small enough. The passband is also affected but the effect is usually much smaller than for the stopband. A sufficiently high sampling rate can be used to overcome this problem, the higher the sampling rate, the smaller the error introduced by aliasing.

For this reason, Butterworth and Chebyshev filters are more suitable if impulse invariant method is used. This is because both these filters do not have ripples in the stopband and the response in this region is monotonically decreasing.

The analog filter transfer function can be expressed in terms of partial fractions.

$$H_a(\omega)=\sum_{i=1}^{N}\frac{K_i}{j\omega-s_i}$$

where s_i is generally complex-valued. In fact, it is usually given in terms of a variable s:

$$H_a(s)=\sum_{i=0}^{N}\frac{K_i}{s-s_i}$$

For those readers who are familiar with laplace transformation, the variable s is usually used in the laplace transform domain. Those who are not familiar with laplace transform can simply regard s as a complex variable which is called the complex frequency variable. The usual frequency response in terms of the real frequency variable ω can be obtained from any function $H(s)$ by using the following substitution:

$$s = j\omega$$

Without going into the details of the mathematics, we shall simply state the relationship between the transfer function of the digital filter $H(z)$ and that of the analog filter $H_a(s)$:

$$H(z)=\sum_{i=1}^{N}\frac{K_i}{1-e^{s_iT}z^{-1}}$$

Thus the coefficients of the IIR filter can be obtained directly from $H_a(s)$ without having to compute the impulse response first. The relationship between $H(z)$ and the filter coefficients can be found in section 7.1.

$$H(z) = \frac{K_1}{1 - e^{s_1 T} z^{-1}} + \frac{K_2}{1 - e^{s_2 T} z^{-1}}$$

Characteristics of impulse invariance designed filters are:

- The IIR filter is stable if the analog filter is stable.
- The frequency response of the IIR filter is an aliased version of the analog filter. So the optimal properties of the analog filter are not preserved.
- The cascade of two impulse invariance designed filters is not impulse invariant with the cascade of the two analog filter prototypes.
- The step response (the response of the filter to a unit step input) is not a sampled version of the step response of the analog filter. Therefore, if the step response of the final filter is important, then the impulse invariant method may not be a suitable choice.

Example 7.5
A second order Butterworth filter has the following normalized transfer function:

$$H(s) = \frac{1}{s^2 + \sqrt{2}s + 1}$$

Design an IIR filter based on this analog filter using the impulse invariant method.
Solution:
First expand $H(s)$ in terms of partial fractions:

$$H(s) = \frac{K_1}{s - s_1} + \frac{K_2}{s - s_2}$$

$$s_1 = -\frac{1}{\sqrt{2}}(1 - j)$$

$$s_2 = -\frac{1}{\sqrt{2}}(1 + j)$$

$$K_1 = -\frac{j}{\sqrt{2}}$$

$$K_2 = \frac{j}{\sqrt{2}}$$

Then

$$H(z) = \frac{K_1}{1 - e^{s_1 T} z^{-1}} + \frac{K_2}{1 - e^{s_2 T} z^{-1}}$$

Using a normalized sampling period of $T=1$, we have

$$H(z) = \frac{0.2265z^{-1}}{1 - 0.7497z^{-1} + 0.2431z^{-2}}$$

The frequency response is obtained by substituting

$$z = e^{j\omega}$$

and is plotted in Figure 7.13.

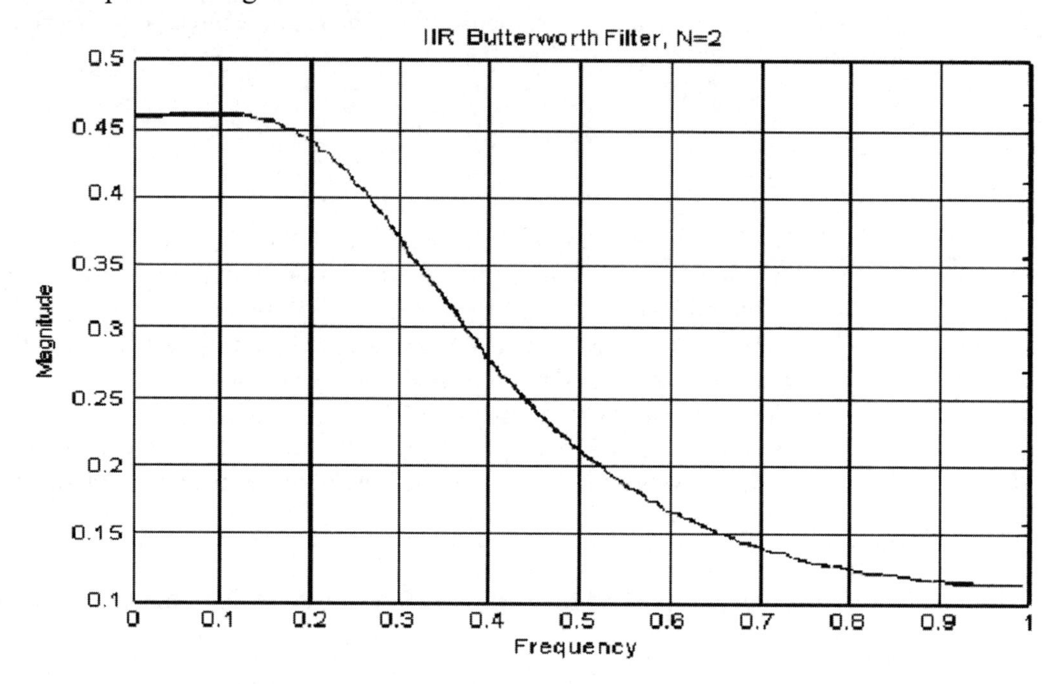

Figure 7.13
Frequency response of filter design using impulse invariant

7.3.2 Bilinear transformation method

Bilinear transformation is a frequency domain method of converting the analog filter transfer function $H(s)$ into a digital one $H(z)$. The transformation is performed by a change of variables

$$s = \frac{2}{T}\frac{z-1}{z+1}$$

It is called bilinear because both the numerator and denominator of this transformation equation are linear. This transformation is reversible in that $H(s)$ can be obtained from $H(z)$ by the substitution

$$z = \frac{2/T+s}{2/T-s}$$

In order to understand the effects of this transformation let ω′ and ω denote the analog and digital frequencies respectively. The frequency response of the digital filter can be obtained from $H(z)$ by the substitution

$$z = e^{j\omega T}$$

and that of the analog filter by the substitution

$$s = j\omega'$$

into $H(s)$. The analog and digital frequencies are related by

$$\omega' = \frac{2}{T}\tan\left(\frac{\omega T}{2}\right)$$

which is plotted in Figure 7.14.

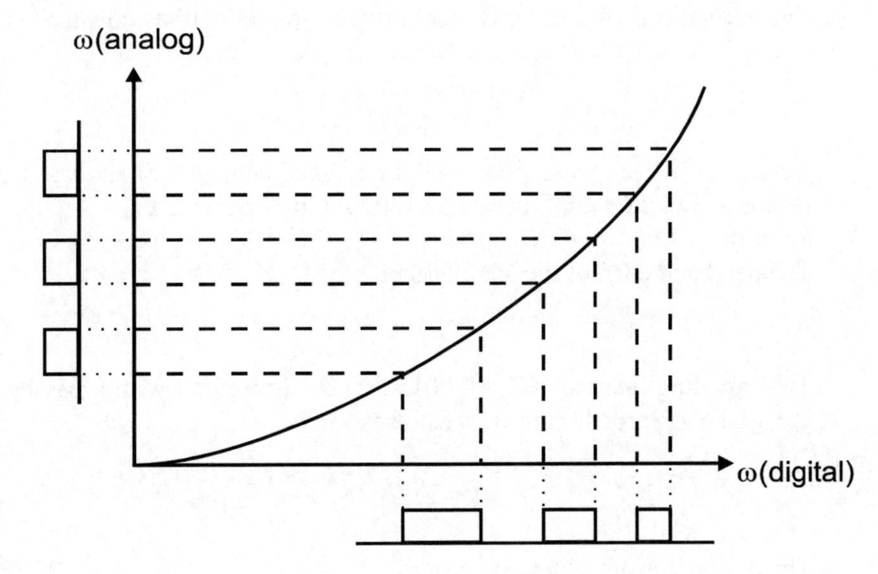

Figure 7.14
Frequency warping

This figure shows that the relationship between ω′ and ω are approximately linear for small values of ω but becomes increasingly non-linear as ω increases. This non-linearity leads to a distortion (or warping) of the digital frequency response.

The figure also shows three passbands in analog frequency that is of constant width and is regularly spaced. After applying the bilinear transformation, the passbands in digital frequency are no longer of equal width and are not equally spaced. This effect can be overcome by pre-warping the analog filter before applying the bilinear transformation.

Pre-warping or scaling of the analog frequency scale is done by replacing s with K_s where K is some constant. Since bilinear transformation is itself a change of variables, the two steps, pre-warping and bilinear transform, can be performed in a single step. The pre-warping required is

$$u_0 = \frac{2}{T}\tan\left(\frac{\omega_0 T}{2}\right)$$

where u_0 is the critical frequency of the analog filter and ω_0 is the desired critical frequency of the digital filter.

Combine the pre-warping with bilinear transformation, we have

$$u_0 = \frac{2K}{T}\tan\left(\frac{\omega_0 T}{2}\right)$$

and so

$$K = \frac{u_0 T}{2\tan\left(\omega_0 T/2\right)}$$

The bilinear transformation with pre-warping is therefore given by

$$s = \frac{u_0}{2\tan\left(\omega_0 T/2\right)}\frac{z-1}{z+1}$$

Example 7.6
The normalized third order Butterworth low-pass filter has transfer function given by

$$H(s) = \frac{1}{(s+1)(s^2+s+1)}$$

Design an IIR low-pass filter with a passband edge at 200 Hz based on this analog filter response. Let the sampling rate be 1000 samples per second.
Solution:
The passband edge of the analog filter is

$$u_0 = 1$$

The sampling interval is T = 0.001 seconds. Therefore, with a passband edge at 200 Hz, the total pre-warped bilinear transformation is

$$s = 1.376382\frac{z-1}{z+1}$$

The digital transfer function is

$$H(z) = \frac{0.09853116(z+1)^3}{(z-0.158384)(z^2-0.418856z+0.355447)}$$

$$= \frac{0.09853116(1+z^{-1})^3}{(1-0.158384z^{-1})(1-0.418856z^{-1}+0.355447z^{-2})}$$

The filter coefficients can be readily obtained by expanding *H*(*z*) and comparing it with the expression given in section 7.1.

Characteristics of the bilinear transformation are summarized below:

- Provided that the analog filter is stable, the resulting digital filter is guaranteed to be stable.
- The order of the digital filter is the same as the prototype analog filter.
- Optimal approximations to the analog filter transform into optimal digital filters.
- The cascade of sections designed by the bilinear transformation is the same as that obtained by transforming the total system.

The last characteristic is what makes bilinear transformation more useful than the impulse invariant method. The cascade of sections that are of lower orders (typically first and second order sections) are important for the realization of higher-order IIR filters to maintain numerical stability. This point will be discussed in more detail in the next chapter.

7.3.3 Frequency transformation

So far we have only considered low-pass filter designs. Designing high-pass and band-pass filters requires a transformation to be performed on the low-pass filter. Two approaches can be used for IIR digital filters. They are illustrated in Figure 7.15.

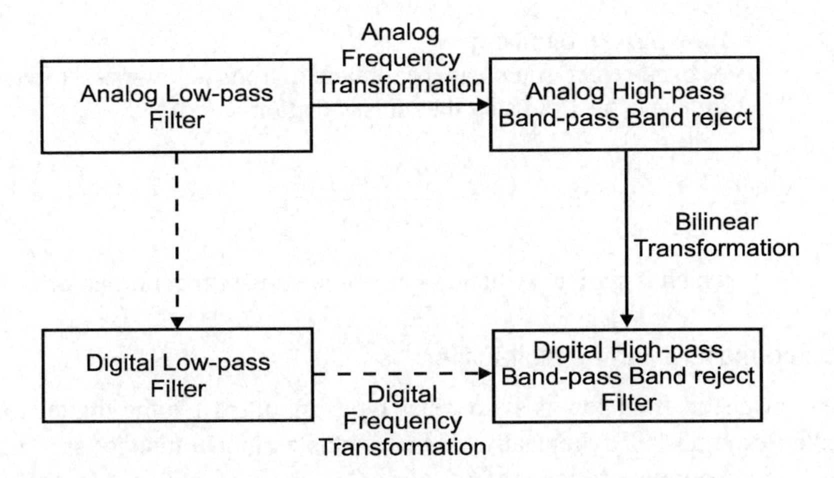

Figure 7.15
Frequency transformation approaches for IIR filters

In the first approach, the analog prototype filter is transformed to be appropriate high-pass or band-pass analog filter. Then this analog filter is turned into an IIR filter by bilinear transformation. Alternatively, the analog low-pass filter is first converted to a digital low-pass filter. The digital high-pass or band-pass filter is then obtained by a spectral transformation of the digital low-pass filter.

7.3.3.1 Transformations for the analog filter

The following substitutions should be made for the analog filter transfer function $H(s)$.

- **Low-pass to low-pass**
  If we have a low-pass filter with cut-off frequency at ω_p and we wish to convert it to another low-pass filter with a different cut-off frequency ω_p', then the transformation

$$s \rightarrow \frac{\omega_p}{\omega_p'} s$$

  is needed.

- **Low-pass to high-pass**
  To convert a low-pass filter to a high-pass filter with cut-off frequency ω_p', we need the transformation

$$s \rightarrow \frac{\omega_p \omega_p'}{s}$$

- **Low-pass to band-pass**
  If the low-pass filter has a cut-off frequency of 1 rad/s, then it can be converted to a band-pass filter by the transformation

$$s \rightarrow \frac{s^2 + \omega_u \omega_l}{s(\omega_u - \omega_l)}$$

  where ω_l and ω_u are the lower and upper cut-off frequencies of the passband respectively.

- **Low-pass to band-reject**

  A band-reject filter can be obtained from a low-pass filter with a cut-off frequency at 1 rad/s by the transformation

$$s \rightarrow \frac{s(\omega_u - \omega_u)}{s^2 + \omega_u \omega_l}$$

which is similar to the low-pass to band-pass transformation.

7.3.3.2 Transformations for the digital filter

The conversion from low-pass to other types of filters for the digital transfer function $H(z)$ is accomplished by replacing z^{-1} by a suitable rational function $g(z^{-1})$.

- **Low-pass to low-pass**

$$z^{-1} \rightarrow \frac{z^{-1} - a}{1 - az^{-1}}$$

where

$$a = \frac{\sin\left[(\omega_p - \omega'_p)/2\right]}{\sin\left[(\omega_p + \omega'_p)/2\right]}$$

and ω_p, ω_p' are the original and new cut-off frequencies respectively.

- **Low-pass to high-pass**

$$z^{-1} \rightarrow -\frac{z^{-1} + a}{1 + az^{-1}}$$

where

$$a = -\frac{\cos\left[(\omega_p - \omega'_p)/2\right]}{\cos\left[(\omega_p + \omega'_p)/2\right]}$$

- **Low-pass to band-pass**

$$z^{-1} \rightarrow -\frac{z^{-2} - a_1 z^{-1} a_2}{a_2 z^{-2} - a_1 z^{-1} + 1}$$

where

$$a_1 = -\frac{2aK}{K+1}$$

$$a_2 = -\frac{K-1}{K+1}$$

$$\alpha = \frac{\cos[(\omega_u + \omega_l)/2]}{\cos[(\omega_u - \omega_l)/2]}$$

$$K = \cot\left(\frac{\omega_u - \omega_l}{2}\right)\tan\frac{1}{2}$$

The cut-off frequency of the low-pass filter is at ω_c.

- **Low-pass to band-reject**

$$z^{-1} \rightarrow \frac{z^{-2} - a_1 z^{-1} a_2}{a_2 z^{-2} - a_1 z^{-1} + 1}$$

where

$$a_1 = -\frac{2\alpha}{K+1}$$

$$a_2 = -\frac{1-K}{1+K}$$

$$\alpha = \frac{\cos[(\omega_u + \omega_l)/2]}{\cos[(\omega_u - \omega_l)/2]}$$

$$K = \tan\left(\frac{\omega_u - \omega_l}{2}\right)\tan\frac{\omega_c}{2}$$

The cut-off frequency of the low-pass filter is at ω_c.

7.4 Direct design methods

The design methods based on transformation from analog filters are appropriate only if the classical approximations (Butterworth, Chebyshev, etc) provide adequate solutions. The desired filter characteristics are necessarily simple. If filters with arbitrary responses are needed, then these transform methods cannot be applied.

There are a number of existing methods that design the digital filter directly. However, some of these methods are mathematically much more involved. We shall attempt to describe two of them in this section.

7.4.1 Frequency sampling method

The concept of the frequency sampling method for design IIR filters is basically the same as that for FIR filters. The frequency response of the IIR filter will pass through the given samples of the desired response. The difference here is that since IIR filters cannot have linear phase, the samples have to be obtained from both the magnitude and phase responses.

7.4.1.1 Calculation of IIR filter response

First, let us consider how the frequency response of an IIR filter can be calculated. We know that L equally spaced samples of the frequency response $H(\omega)$ can be approximately calculated from a length-L DFT of the impulse response $h(n)$. But this direct calculation requires that the infinitely long $h(n)$ be truncated to a length of L samples.

If we have the IIR filter transfer function, then a better approach is to use DFT to calculate the numerator and denominator separately. Recall that the transfer function is given by a ratio of two polynomials

$$H(z) = \frac{\sum_{n=0}^{M} b(n) z^{-n}}{\sum_{n=0}^{N} a(n) z^{-n}}$$

DFT of the numerator of $H(z)$ is performed by first appending L-M zeros to $b(n)$ followed by a length-L DFT. For the denominator, L-N zeros are appended to $a(n)$ before performing the transform. The sampled frequency response is given by

$$H_k = H\left(\frac{2\pi k}{L}\right) = \frac{DFT\{b(n)\}}{DFT\{a(n)\}} = \frac{B_k}{A_k}$$

where the division is performed term by term for each of the L values of the DFTs as a function of k. Since the order of IIR filter is usually low, direct DFT computations are not computationally expensive.

7.4.1.2 Frequency sampling design

For the purpose of design, we can choose the number of equally spaced samples to be the same as the number of unknown coefficients, i.e.

$$L = M + N + 1$$

for $(M+1)$ numerator coefficients and N denominator coefficients ($a_0=1$).

The equation

$$H_k = \frac{B_k}{A_k}$$

can be expressed as

$$B_k = H_k A_k$$

In other words, the sequence $\{b(n)\}$ can be obtained by the cyclic convolution of the sequences $\{h(n)\}$ and $\{a(n)\}$ where $\{h(n)\}$ is the length-L inverse DFT of $\{H_k\}$.

From Chapter 3, we understand that cyclic convolution can be expressed in matrix form

The lower L-M-1 equations are

$$\begin{bmatrix} 0 \\ \vdots \\ 0 \end{bmatrix} = \begin{bmatrix} h_{M+1} \\ \vdots \\ h_{L-1} \end{bmatrix} + \begin{bmatrix} h_M & \cdots & h_{M+N} \\ \vdots & \vdots & \vdots \\ h_{L-2} & \cdots & h_{L+N-2} \end{bmatrix} \begin{bmatrix} a_1 \\ \vdots \\ a_N \end{bmatrix}$$

This matrix equation can be denoted in more compact form by

$$0 = \mathrm{h}_1 + \mathrm{H}_2\mathrm{a}$$

or equivalently by

$$\mathrm{h}_1 = -\mathrm{H}_2\mathrm{a}$$

where H_2 is a $(L–M–1)$ by N matrix, h_1 is a vector of length $(L–M–1)$ and a is of length N. If $L=N+M+1$, then the matrix H_2 is square. Provided H_2 is non-singular, a can be solved exactly using this set of equations.

The upper M+1 equations are

$$\begin{bmatrix} b_0 \\ \vdots \\ b_M \end{bmatrix} = \begin{bmatrix} h_0 & \cdots & h_{L-N+1} \\ \vdots & \vdots & \vdots \\ h_M & \cdots & h_{2N-1} \end{bmatrix} \begin{bmatrix} 1 \\ \vdots \\ a_N \end{bmatrix}$$

or more compactly as

$$\mathrm{b} = \mathrm{H}_1\mathrm{a}$$

Since a has already been evaluated from the previous step, b can be easily calculated.

This method is relatively simple and can be applied to arbitrary frequency responses. The main disadvantage is that we have no control over the stability of the designed filter. It also suffers the same problem as the corresponding method for FIR filter design, which sometimes gives poor approximation between the frequency sample points. A solution is to increase the number of samples, which makes the solution of the system of simultaneous equations more difficult to arrive at. The final design is also very sensitive to the choice of frequency sample points.

7.4.2 Least squared equation error design

The above frequency sampling design method can be modified to allow us to minimize the equation error. This converts the interpolation problem into an approximation problem.

If number of frequency samples used, L, is made to be larger than the number of coefficients to be determined. Therefore the matrix H_2 will not be square and we cannot solve the equation

$$\mathrm{h}_1 = -\mathrm{H}_2\mathrm{a}$$

for the vector a. However, by introducing a length-L error vector e, we have

$$\begin{bmatrix} \mathrm{b} \\ 0 \end{bmatrix} = [\mathrm{H}_0][\mathrm{a}] + [\mathrm{e}]$$

and the above equation then becomes

$$h_1 - e = -H_2 a$$

Without going into the details of the mathematics, we shall simply state that the solution that minimizes *e* is given by

$$a = -\left[H_2^T H_2\right]^{-1} H_2^T h_1$$

provided the equations are not singular. In practice, this equation is not directly used for solving for *a* since the equations are often numerically ill conditioned. Special algorithms are commonly used.

With the values of *a* solved, *b* can be obtained in the same way as for the frequency sampling method:

$$b = H_1 a$$

This makes the upper $M+1$ terms in *e* zero and the total squared error a minimum.

Note that minimizing the equation error is different from minimizing the squared error of the frequency responses. The former is based on time-domain equations while the latter is a frequency domain method. In general, with IIR filters, minimizing the squared error of frequency responses leads to non-linear equations. However, minimizing the equation error only requires the solution to linear simultaneous equations. If the desired frequency response is close to what can be achieved by an IIR filter, then both error minimization methods will lead to similar solutions. But if the specifications are not consistent with the IIR filter order and characteristics, then large errors will result. Sometimes the resulting filter may even be unstable.

Since it is not easy to set appropriate frequency responses, especially phase responses, for IIR filters, trial-and-error is sometimes the best option. In this case, a good design environment will allow the designer to experiment different specifications and designs efficiently.

Example 7.7

A sixth order IIR low-pass filter is designed using the least square equation error method. 41 frequency samples are used. The desired response has a magnitude of 1 for 0 to 0.4π and the magnitude is zero for frequencies larger than 0.4π. The phase response was adjusted by trial-and-error to give a good final magnitude response. The filter magnitude response is shown in Figure 7.16.

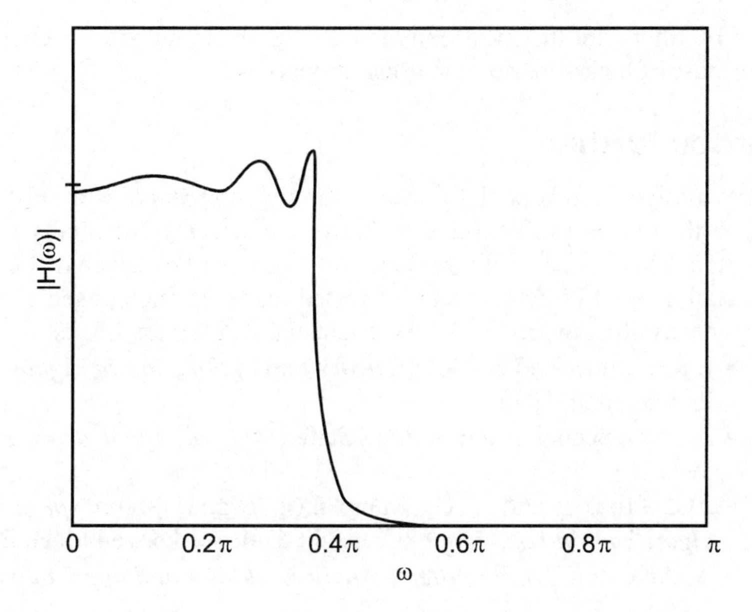

Figure 7.16
Sixth order IIR response designed using LS method

7.5 FIR vs IIR

At this point it would be useful to summarize the relative merits of FIR and IIR filters. Some of the advantages have been discussed in this and the previous chapter; others will be covered in the following chapter on realization and implementation.

- FIR filters can achieve exact linear phase. This means that the filter will not introduce any distortion in the phase of the signal. Linear phase response is important for applications in data transmission and image processing, for instance. IIR filters generally have non-linear phase response, especially near the band edges.
- FIR filters are guaranteed to be stable while no such guarantee exists for IIR filters.
- IIR filters are more suitable if sharp cut-off (small transition bandwidth) is required. The order of FIR filter needed for sharp cut-off can be very high. A high order filter also implies long delays. The implication for implementation is that higher order filters have more coefficients and therefore require more storage and are computationally more expensive. It should be emphasized that DSP chips are optimized to perform operations required by FFT and convolutions and thus can implement FIR filtering very efficiently.
- It is relatively easy to design FIR filters with arbitrary frequency responses. However, analog filters can be readily transformed into equivalent IIR digital filters with similar specifications. This is an advantage for designers who want to convert existing analog applications to digital. FIR filters have no analog counterpart.
- IIR filters are more susceptible to round-off errors and quantization errors than FIR filters.

Therefore, if we need a sharp cut-off filter with a high throughput (low delay), the IIR filter is more suitable. It should be designed using analog elliptic approximation to give fewer coefficients. On the other hand, if exact linear phase is very important, use FIR filters. FIR filters are the most common choice if the number of coefficients is not too large because of their superior numerical properties.

7.6 To probe further

IIR filter design is substantially more difficult compared with FIR filter design. To formulate the design problem and to solve it requires substantially more mathematical background. This is one of the reasons why we have not covered IIR filter design in as much detail as for FIR filters. The interested reader is encouraged to pursue the subject further by consulting the following excellent classic DSP textbooks:

- L.R. Rabiner and B. Gold, *Theory and application of digital signal processing*, Prentice-hall, 1975.
- A.V. Oppenheim and R.W. Schafer, *Digital signal processing*, Prentice-hall, 1975.
- J.G. Proakis and D.G. Manolakis, *Digital signal processing: principles, algorithms and applications*, second edition, Maxwell Macmillan, 1988.
- A. Antoniou, *Digital filters: Analysis, design and applications*, second edition, McGraw-Hill, 1993.
- More advanced design techniques continue to appear in the literature. The reader should consult IEEE transactions on signal processing for the latest in digital filter design.

8

Digital filter realizations

After the coefficients of a digital filter have been determined, the approximation problem is solved. The next stage of the filter design process is called realization or implementation of the filter. When we talk about realization here, we are talking about a structure that relates the input and output of the filter, illustrated by block diagrams. They are called filter structures. The blocks within these block diagrams can be implemented either as a piece of digital hardware or as a program to be executed by a DSP chip.

We shall describe several filter structures. Considerations for the choice of a filter structure include ease of programming on a particular DSP chip, and the regularity of the VLSI (very large scale integrated) design. Some structures are more sensitive to (quantization) errors in the coefficients. In some cases, such as IIR filters, the stability of the filter may depend on an appropriate realization.

8.1 Direct form

8.1.1 IIR filters

Consider a simple second order IIR filter with transfer function

$$H(z) = \frac{B(z)}{A(z)} = \frac{b^0 + b_1 z^{-1} + b_2 z^{-2}}{1 + a_1 z^{-1} + a_2 z^{-2}}$$

The input and output samples are related by

$$y(n) = -a_1 y(n-1) - a_2 y(n-2) + b_0 x(n) + b_1 x(n-1) + b_2 x(n-2)$$

Direct form realization is simply a realization based on the direct implementation of this difference equation. This is illustrated in Figure 8.1.

In the figure, the z^{-1} symbol represents the delay of one sample. In actual implementations, it would represent shift registers or a memory location in RAM. The three basic elements of the structure are illustrated in Figure 8.2.

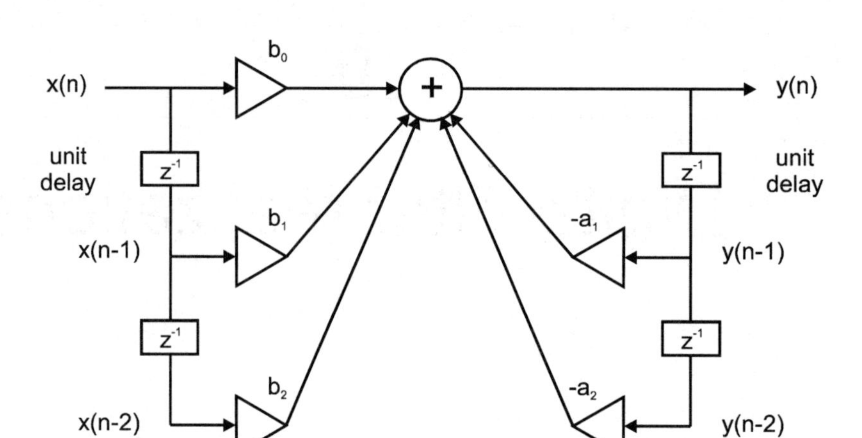

Figure 8.1
Direct form realization of the IIR filter

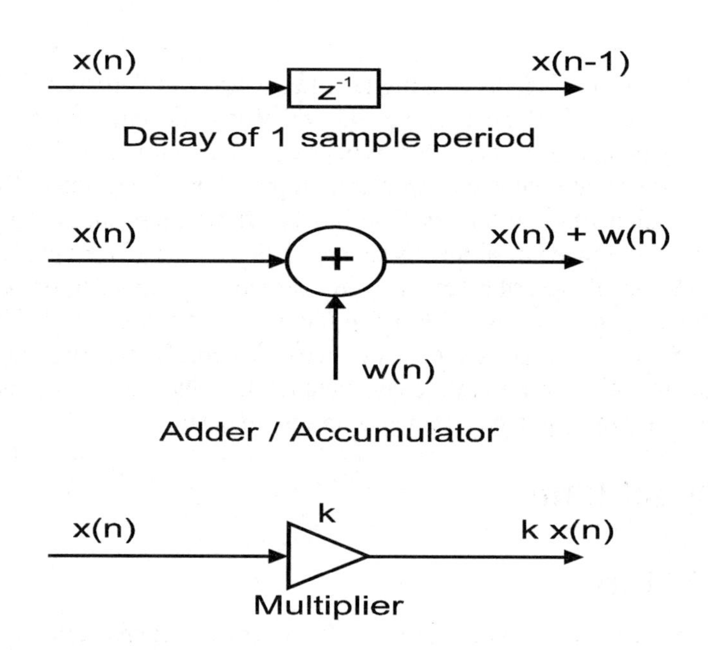

Figure 8.2
Three basic elements in the realization

This structure involves 4 unit delays, 5 multiplications and 4 additions. Notice that the right-hand side of the equation consists of two main operations: multiplications and additions. Each sample of the output y or input x is multiplied by a coefficient. The result is then stored or accumulated for addition. We call these two basic operations of the direct form structure multiply-and-accumulate (MAC).

Figure 8.1 also illustrates the two parts of the filter structure. All the numerator terms shown on the left-hand side of the adder block are the feed-forward elements. The denominator terms that depend on the previous output samples are feeding back. This direct form is called direct form I or simply direct form.

8.1.1.1 Canonical form

There is also a direct form II structure. It is also known as the canonical form. To see how we can arrive at a different direct form structure, consider the difference equation of the second order filter. The terms on the right-hand side of the equation is being regrouped as

$$y(n) = [b_0 x(n) + b_1 x(n-1) + b_2 x(n-2)] + [-a_1 y(n-1) - a_2 y(n-2)]$$

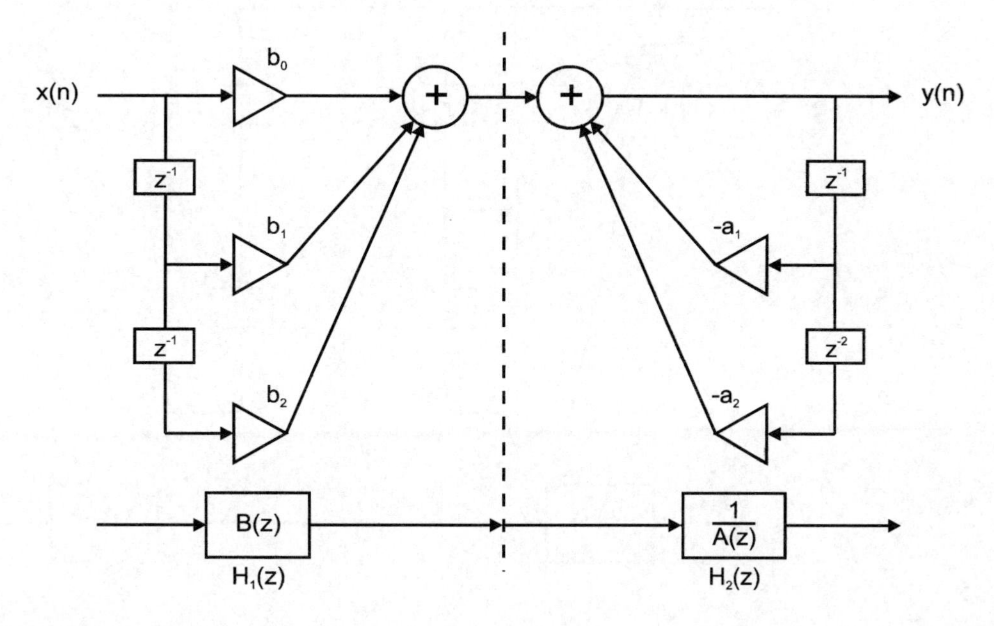

Figure 8.3
Alternative direct form structure

This regrouping is depicted in Figure 8.3. There are now two adders in the diagram instead of one as in Figure 8.1. We can view this structure as a cascade of two filters: one with only feed-forward terms and one with only feedback terms. The digital transfer functions of these two filters are then

$$H_1(z) = B(z)$$

$$H_2(z) = \frac{1}{A(z)}$$

so that their cascade is

$$H_1(z) \cdot H_2(z) = B(z) \cdot \frac{1}{A(z)} = H(z)$$

which is the original transfer function.

The order of the two filters can be interchanged without affecting the overall transfer function since they are both linear systems. Figure 8.4 depicts the cascade of the filters with the order interchanged.

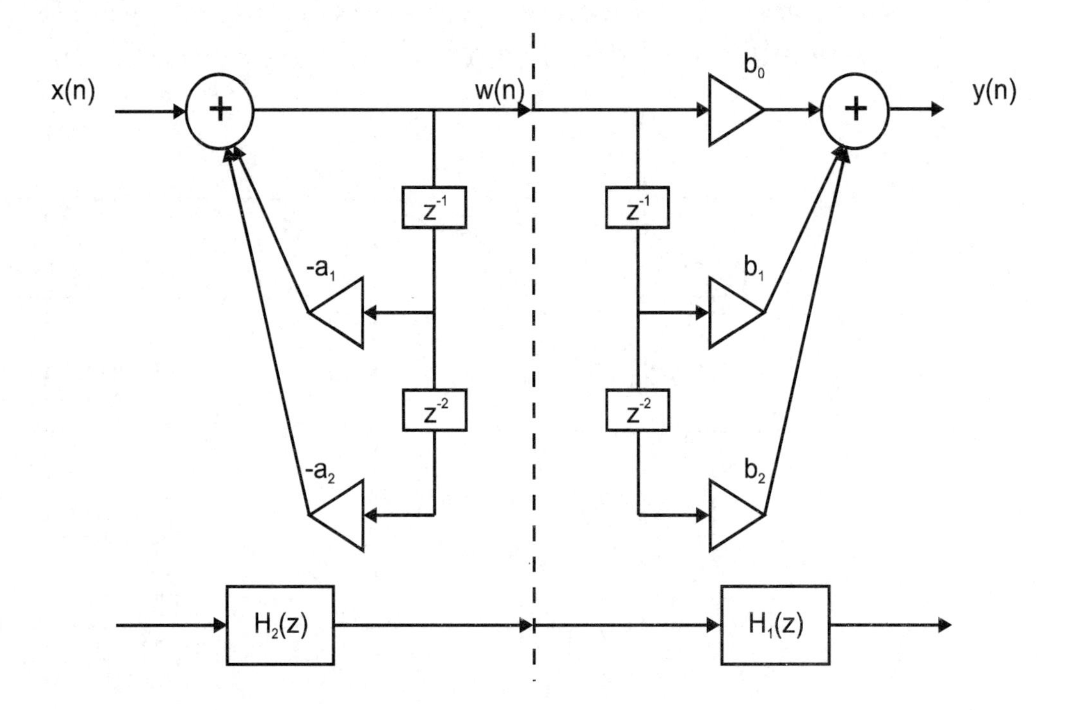

Figure 8.4
Interchanging the cascade

The output of the filter $H_2(z)$ is now the input to the filter $H_1(z)$. The output of $H_2(z)$, denoted by $w(n)$, is being delayed in the same way by the two filters. Therefore we do not need two separated sets of delays; they can be merged into one as shown in Figure 8.5. This is the canonical form.

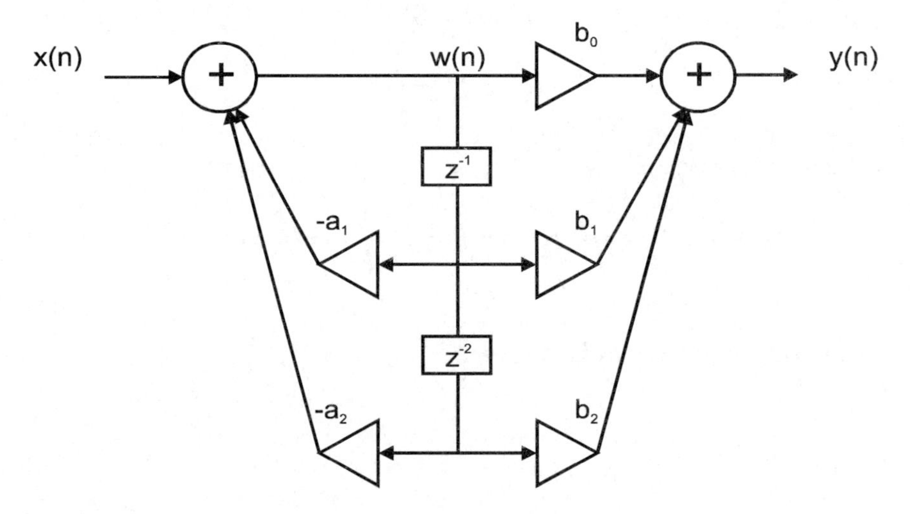

Figure 8.5
The canonical form

It is not difficult to see that the canonical form implements the original IIR difference equation. We have, from the first adder on the left,

$$
\begin{aligned}
w(n) &= -\sum_{k=1}^{2} a_k w(n-k) + x(n) \\
&= -a_1 w(n-1) - a_2 w(n-2) + x(n)
\end{aligned}
$$

The output of the second adder is

$$
\begin{aligned}
y(n) &= \sum_{k=0}^{2} b_k w(n-k) \\
&= b_0 w(n) + b_1 w(n-1) + b_2 w(n-2)
\end{aligned}
$$

Substituting the expressions for $w(n)$, $w(n-1)$ and $w(n-2)$ into the above equation, we have

$$
\begin{aligned}
y(n) = b_0&\left[-a_1 w(n-1) - a_2 w(n-2)\right] + b_0 x(n) \\
&+ b_1\left[-a_1 w(n-2) - a_2 w(n-3)\right] + b_1 x(n-1) \\
&+ b_2\left[-a_1 w(n-3) - a_2 w(n-4)\right] + b_2 x(n-2) \\
=& \sum_{k=0}^{2} b_k x(n-k) \\
&- a_1\left[b_0 w(n-1) + b_1 w(n-2) + b_2 w(n-3)\right] \\
&- a_2\left[b_0 w(n-2) + b_1 w(n-3) + b_2 w(n-4)\right] \\
=& \sum_{k=0}^{2} b_k x(n-k) - a_1 y(n-1) - a_2 y(n-2)
\end{aligned}
$$

which is the original difference equation.

Although we have been using the second order IIR filter as example, the structures can easily be generalized to filter transfer functions of higher orders.

Example 8.1
Given the digital filter transfer function

$$
H(z) = \frac{2 - 3z^{-1} + 4z^{-3}}{1 + 0.2z^{-1} - 0.3z^{-2} + 0.5z^{-4}}
$$

Draw the direct form I and II realizations of this filter.
Solution:
See Figures 8.6 and 8.7.

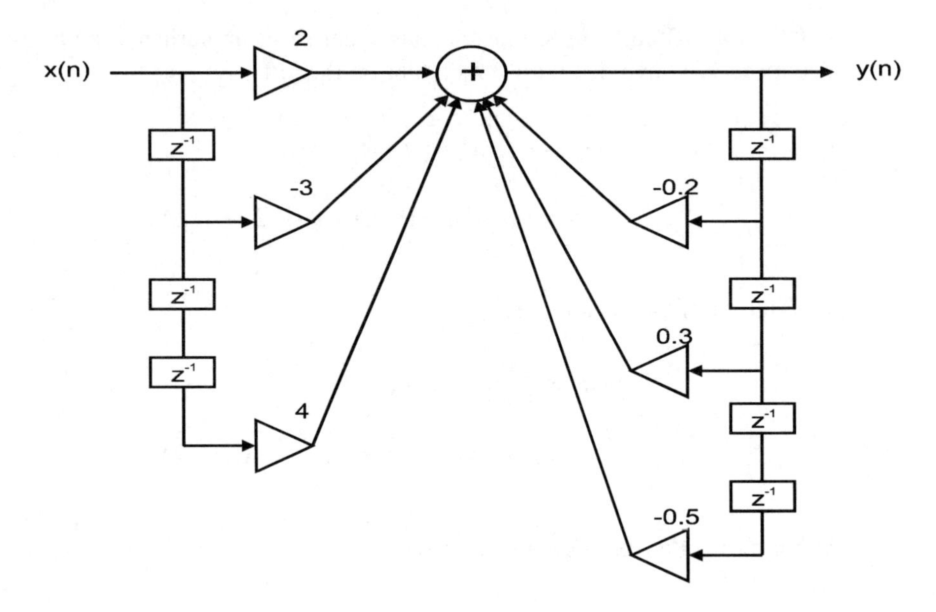

Figure 8.6
Direct form I realization

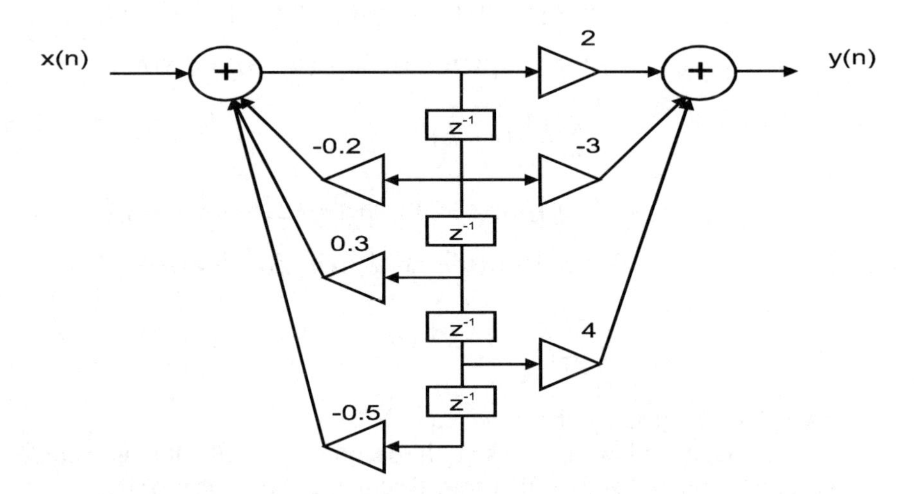

Figure 8.7
Direct form II realization

8.1.1.2 Relative merits

The canonical form is commonly used because of the following properties:

- Requires a minimum of storage space
- Good round off noise property.

The disadvantage is that it is susceptible to internal numeric overflow. The input to the filter can be scaled to avoid overflows. On the other hand, the direct form I structure does not require scaling because there is only one adder. Thus if scaling is not desirable, direct form I realization is preferred.

8.1.1.3 Transposed structures

An alternative structure based on the canonical form can be obtained by a transposition. The result is a transposed structure as shown in Figure 8.8.

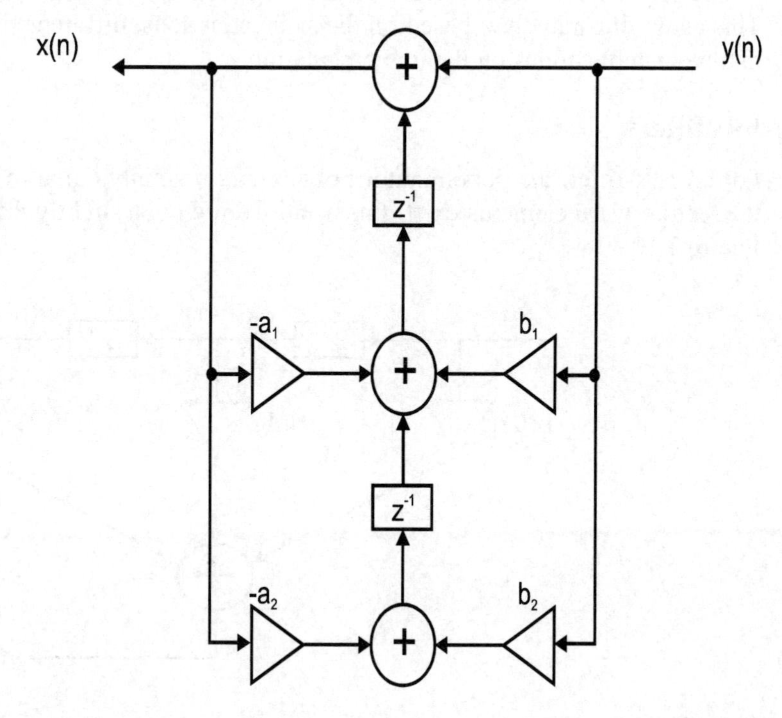

Figure 8.8
Transposed canonical structure

It is obtained by:

- Reversing all the signal flow directions;
- Change nodes (connection points) into adders and adders into nodes
- Exchanging the input and output.

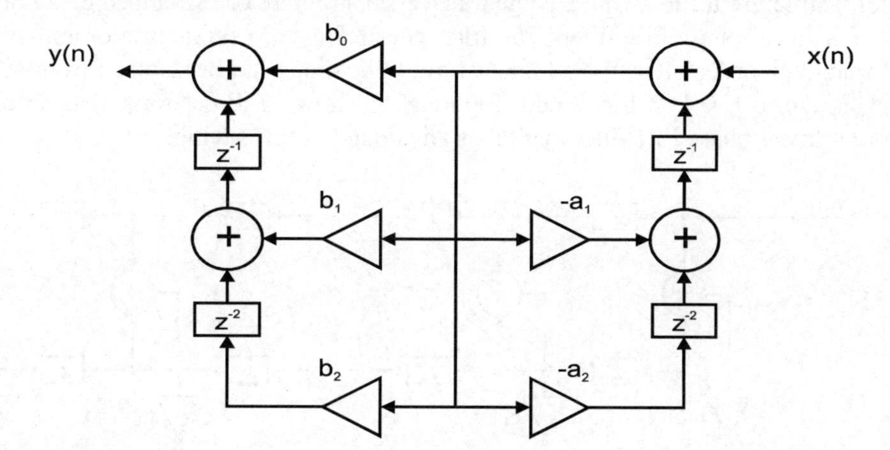

Figure 8.9
Transposed direct form I

A transposed direct form I structure can also be obtained in the same way. The result is shown in Figure 8.9. In both cases, the transfer function remains the same after transposition.

The finite wordlength effects of the original and transposed structures are different. This again illustrate that, based on the same equations, different structures can be derived that have implications on the implementation.

8.1.2 FIR filters

For an FIR filter, the denominator polynomial is simply equal to 1, i.e. $A(z)=1$. So only the feed forward elements exist. It is usually drawn in a slightly different way as shown in Figure 8.10.

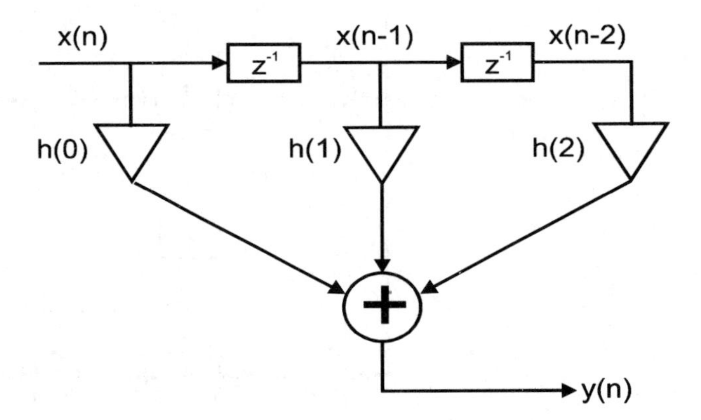

Figure 8.10
FIR filter structure

This corresponds to the FIR equation we have been using for a second order filter

$$y(n) = \sum_{m=0}^{2} h(m)x(n-m)$$

where $h(n)$ is the impulse response or the filter coefficient. Another name for the direct form structure for FIR filters is the transversal structure or the tapped delay line structure.

For linear phase FIR filters, the filter coefficients are symmetric or anti-symmetric. So for an N-th order filter, the number of multiplications can be reduced from N to $N/2$ for N even and to $(N+1)/2$ for N odd. Figure 8.11 shows a direct form realization of an odd order linear phase FIR filter that takes advantage of this saving.

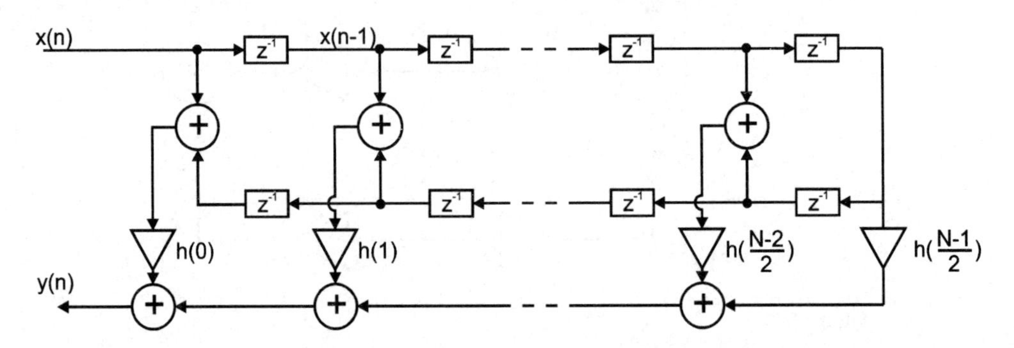

Figure 8.11
Direct form realization of odd order linear phase FIR filter

It is interesting to note that a transposed FIR direct structure can also be obtained using the method discussed for transposing IIR structures. The resulting structure for a second order filter is shown in Figure 8.12.

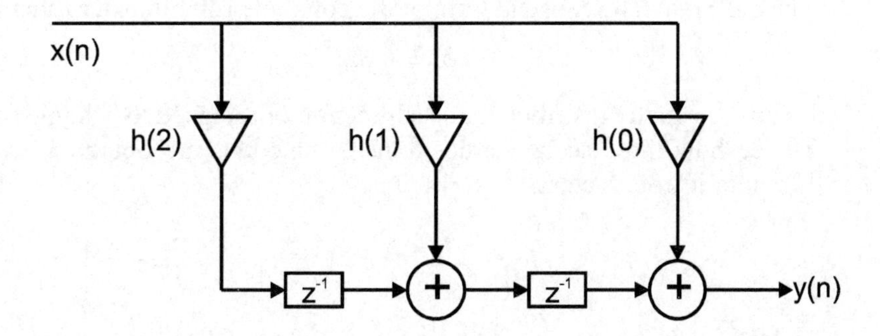

Figure 8.12
Transposed FIR direct structure

8.2 Cascade form

A general transfer function $H(z)$ of order $N>2$ can be factorized into K second order functions so that

$$H(z) = \prod_{i=1}^{K} H_i(z)$$

$$= \prod_{i=1}^{K} \frac{b_{i0} + b_{i1}z^{-1} + b_{i2}z^{-2}}{1 + a_{i1}z^{-1} + a_{i2}z^{-2}}$$

where K is $N/2$ for even N or is $(N+1)/2$ for odd N. We may view each second order section with transfer function $H_i(z)$ as a subsystem of the whole system $H(z)$ and so the full system is made up of a cascade of these subsystems as depicted in Figure 8.13.

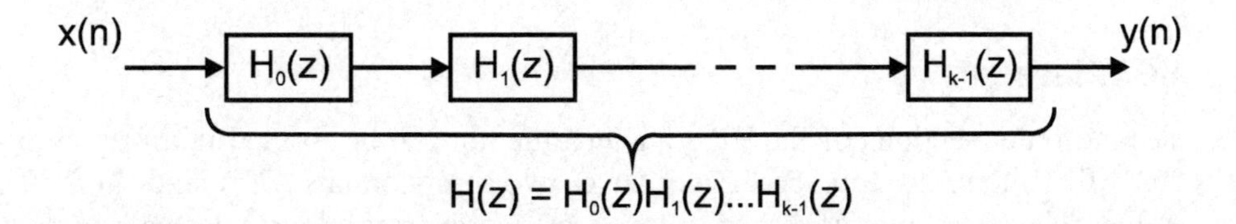

Figure 8.13
Cascade of transfer functions

Note that when we say ‘second order section’ we really mean ‘up to second order section’. Some of the b_{i2} and a_{i2} can be zero. So the actual numerator and denominator polynomial orders can be less than $2K$. Also, the coefficients of each section are real-valued.

Each second order section can be implemented using direct, canonical or transposed forms we discussed in the previous section.

8.2.1 FIR filters

For FIR filters, the transfer functions of the second order sections have the form

$$H_i(z) = b_{i0} + b_{i1}z^{-1} + b_{i2}z^{-2}$$

since $A(z)=1$. The constant term of the complete filter transfer function is given by

$$b_0 = b_{10}b_{20}...b_{K0}$$

It may be equally distributed to each section or assigned to a single section.
To see how these second order transfer functions are obtained, we shall first factorize $H(z)$ into its root factors.

$$H(z) = b_0\left(1 + \frac{b_1}{b_0}z^{-1} + \frac{b_2}{b_0}z^{-2} + \cdots + \frac{b_N}{b_0}z^{-N}\right)$$
$$= b_0(1 - r_1z^{-1})(1 - r_2z^{-2})\cdots(1 - r_Nz^{-N})$$

Some of the roots r_i are complex-valued while others are real-valued. The real-valued roots can be combined in pairs or left as they are. The complex-valued roots will always occur in complex conjugate pairs. For example, if r_1 is a complex root, then

$$r_2 = r_1^+$$

the complex conjugate of r_1 must also be a root. When we form the second order sections, it is desirable to group pairs of these complex conjugate roots so that the coefficients b_{i1} and b_{i2} are real-valued.

$$(1 - r_1z^{-1})(1 - r_1^+z^{-1}) = 1 - (r_1 + r_1^+)z^{-1} + r_1r_1^+z^{-2}$$
$$= 1 - 2\,\mathrm{Re}(r_1)z^{-1} + |r_1|^2\, z^{-2}$$

Here the real part of r_1 is necessarily real and the magnitude squared of r_1 is also real-valued.

For exact linear phase FIR filters, some computational savings can be achieved by using fourth order sections with symmetrical coefficients in each section.

$$H_i(z) = c_{i0} + c_{i1}z^{-1} + c_{i2}z^{-2} + c_{i1}z^{-3} + z^{-4}$$

By doing this, the number of multiplications for each section is reduced by half.

8.2.2 IIR filters

Second order sections of the IIR transfer function can be formed in a way similar to that for FIR filters. In this case, we have two polynomials $B(z)$ and $A(z)$. These two polynomials can be factorized into second order (quadratic) terms separately. Each quadratic term from the numerator can be paired with a quadratic term from the denominator to form a second order section.

Note that the second order transfer functions $H_i(z)$ formed are not unique. But the overall transfer function $H(z)$ remains the same. In practice, the pairing and ordering of the second order sections may affect the numeric accuracy of the resulting filter. The internal multiplication in each section may generate a certain amount of round-off error. This error is then propagated to the next section. The round-off error of the overall output is different for each combination of second order sections. Naturally we want to achieve a minimum amount of round-off error. This is a difficult problem to solve. In practice, some trial-and-error would be needed. Fortunately, most IIR filters do not have a high order so the number of possible combinations is not too large.

A rule of thumb is to pair the quadratic pairs with roots that are closest to one another. Another one is to put the section with the denominator root having magnitudes that are closest to one as the last section.

8.3 Parallel form

An alternative to cascade form for IIR filters is the parallel form. Here the transfer function $H(z)$ is expanded using partial fractions as

$$H(z) = C + \sum_{k=1}^{N} H_k(z)$$

where

$$C = \frac{b_{\mathrm{N}}}{a_{\mathrm{N}}}$$

$$H_k(z) = \frac{A_k}{1 - p_k z^{-1}}$$

for an N-th order function. In this case, the individual subsystem transfer functions are summed to form the overall transfer function. Thus the subsystems are connected in parallel in contrast with the cascade form. This is shown in Figure 8.14. The whole IIR filter now consists of a parallel bank of first order filters.

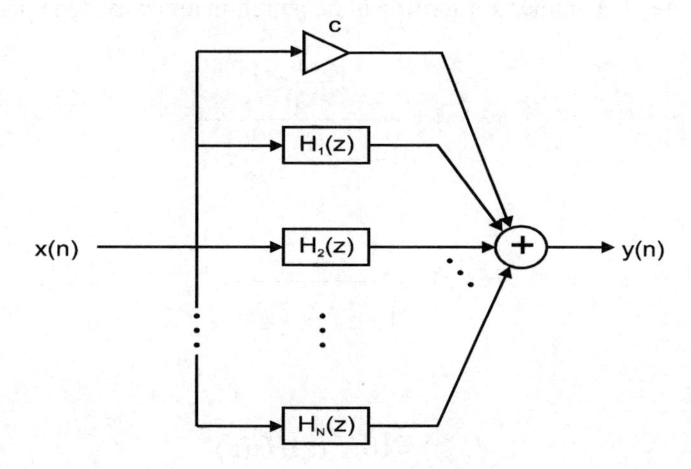

Figure 8.14
Parallel form realization

Both A_k and p_k can be complex valued. If p_k is complex, then its complex conjugate will also appear. We can combine the pair of complex conjugate terms to avoid having to deal with complex numbers. So the transfer function of the subsystems are second order sections and can be implemented in the same way as discussed in the previous sections.

$$H(z) = C + \sum_{k=1}^{k} H_k(z)$$

$$H_k(z) = \frac{b_{k0} + b_{k1}z^{-2}}{1 + a_{k1}z^{-1} + a_{k2}z^{-2}}$$

where K is $N/2$ for even N and $(N+1)/2$ for odd N.
Note that in this case the numerator coefficient for z^{-2} in each second order section is zero.

The advantage of the parallel form compared with the cascade form for IIR filters is that the ordering of the subsystems is unimportant since they are in parallel. Scaling is easier as it can be carried out for each block independently. Furthermore, round-off errors in each block are not propagated to the next block.

The cascade realization, however, is quite often still the preferred implementation method. If the IIR filter is derived from classic analog filters using the bilinear transform, then between 25% and 50% of the filter coefficients are actually simple integers (0, ±1, etc). This makes the computation a lot easier, especially if the processor is not very powerful.

Example 8.2
Determine the cascade and parallel realizations of the IIR filter with transfer function

$$H(z) = \frac{10\left(1 - \frac{1}{2}z^{-1}\right)\left(1 - \frac{2}{3}z^{-1}\right)\left(1 + 2z^{-1}\right)}{\left(1 - \frac{3}{4}z^{-1}\right)\left(1 - \frac{1}{8}z^{-1}\right)\left[1 - \left(\frac{1}{2} + j\frac{1}{2}\right)z^{-1}\right]\left[1 - \left(\frac{1}{2} - j\frac{1}{2}\right)z^{-1}\right]}$$

Solution:
The cascade realization can be obtained by pairing the complex conjugate terms in the denominator. The numerator terms can be paired in any way. One possible solution is:

$$H_1(z) = \frac{1 - \frac{2}{3}z^{-1}}{1 - \frac{7}{8}z^{-1} + \frac{3}{32}z^{-2}}$$

$$H_2(z) = \frac{1 + \frac{3}{2}z^{-1} - z^{-2}}{1 - z^{-1} + \frac{1}{2}z^{-2}}$$

with

$$H(z) = 10H_1(z)H_2(z)$$

For parallel realization, we need to expand the transfer function in partial fractions. It has the form

$$H(z) = \frac{A_1}{1 - \frac{3}{4}z^{-1}} + \frac{A_2}{1 - \frac{1}{8}z^{-1}} + \frac{A_3}{1 - \left(\frac{1}{2} + j\frac{1}{2}\right)z^{-1}} + \frac{A_3^{\neq}}{1 - \left(\frac{1}{2} - j\frac{1}{2}\right)z^{-1}}$$

The numerator constants are found to be

$$A_1 = 2.93$$
$$A_2 = -17.68$$
$$A_3 = 12.25 - j14.57$$

They can be combined into second order sections as

$$H(z) = \frac{-14.75 - 12.90z^{-1}}{1 - \frac{7}{8}z^{-1} + \frac{3}{32}z^{-2}} + \frac{24.5 + 26.82z^{-1}}{1 - z^{-1} + \frac{1}{2}z^{-2}}$$

Obviously there are more coefficients in the cascade realization that are integers.

8.4 Other structures

There are some more structures and forms by which FIR and IIR filters can be implemented. We shall look briefly at some of them in this section.

8.4.1 Lattice structure

The lattice structure is most widely used in digital speech processing and in adaptive filtering. To develop the lattice structure, let us consider a first order FIR filter. The output of this filter is given by

$$y(n) = h(0)x(n) + h(1)x(n-1)$$
$$= x(n) + \alpha_1(1)x(n-1)$$

Here we have assumed that $h(0) = 1$ without loss of generality. The output can be obtained by a single stage of a lattice structure as shown in Figure 8.15.

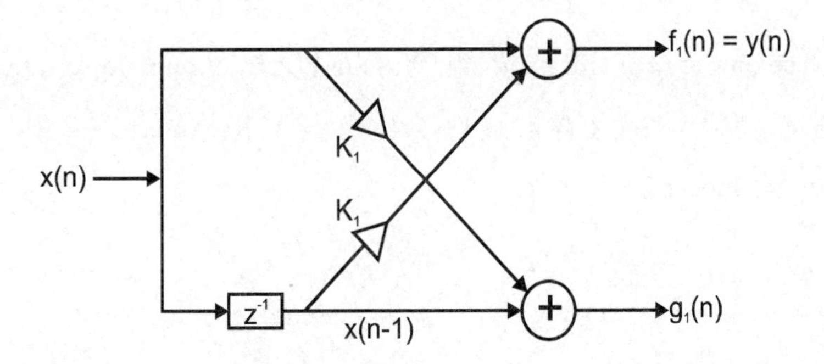

Figure 8.15
Lattice form realization

Notice that the lattice structure provides two outputs

$$f_1(n) = x(n) + K_1 x(n-1)$$
$$g_1(n) = K_1 x(n) + x(n-1)$$

The first output $f_1(n)$ is the same as the output $y(n)$ of the first order FIR filter if we choose

$$K_1 = \alpha_1(1)$$

This parameter is called the reflection coefficient. It has the property that for a stable filter.

$$|K_1| \leq 1$$

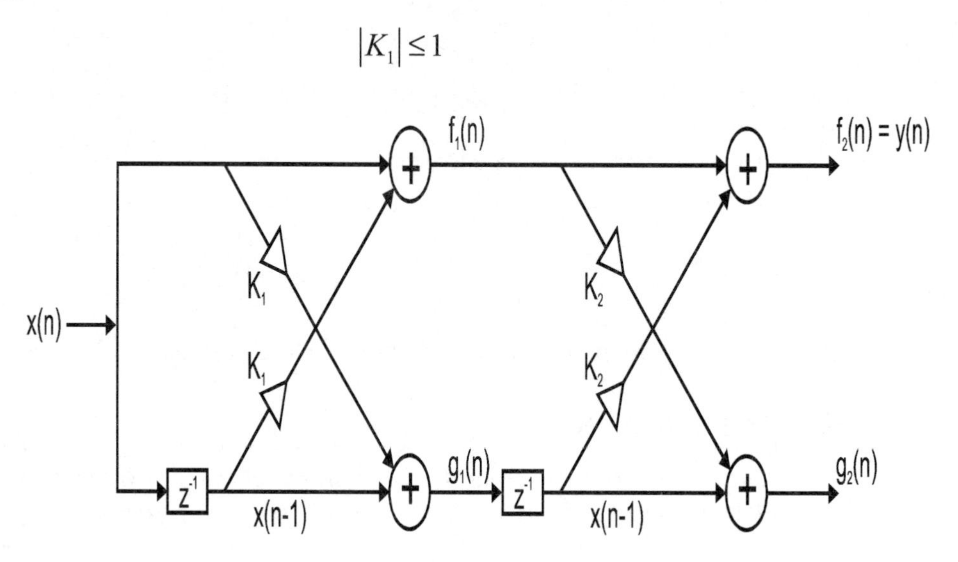

Figure 8.16
Cascade of two lattice stages to form a second order filter

If we need a second order FIR filter, we can use two lattice stages as shown in Figure 8.16. The outputs at the second stage are

$$\begin{aligned} g_2(n) &= K_2 f_1(n) + g_1(n-1) \\ f_2(n) &= f_1(n) + K_2 g_1(n-1) \\ &= x(n) + K_1 x(n-1) + K_2 \left[K_1 x(n-1) + x(n-2)\right] \\ &= x(n) + K_1(1+K_2)x(n-1) + K_2 x(n-2) \end{aligned}$$

The output $f_2(n)$ will be identical to the FIR filter output given by

$$y(n) = x(n) + \alpha_2(1)x(n-1) + \alpha_2(2)x(n-2)$$

if we choose

$$K_2 = \alpha_2(2)$$

$$K_1 = \frac{\alpha_2(1)}{1+\alpha_2(2)}$$

The subscript to the coefficient α indicates the order of the filter.

We can similarly extend this to an *N*-th order FIR filter by additional lattice stages and choosing the correct values for the reflection coefficients. These reflection coefficients have to be calculated recursively. If we denote the *N*-th order FIR filter equation as

$$y(n) = 1 + A_N(z)$$

then we know immediately that

$$K_N = \alpha_N(N)$$

To obtain K_{N-1}, we need the polynomial $A_{N-1}(z)$. So we have the following algorithm:

For $m = N - 1$ down to 1

begin

$K_m = \alpha_m(m)$

$\alpha_{m-1}(0) = 1$

for $k = 1$ to $m - 1$

begin

$$\alpha_{m-1}(k) = \frac{\alpha_m(k) - \alpha_m(m)\alpha_m(m-k)}{1 - \alpha_m^2(m)}$$

end

end

This algorithm is generally known as the Durbin-Levinson recursive algorithm.

A similar lattice structure can be derived for the IIR digital filter. If the numerator coefficient of the IIR filter transfer function is equal to 1, then we have what is called an all-pole filter. Figure 8.17 shows the lattice structure for an N-th order all-pole IIR filter.

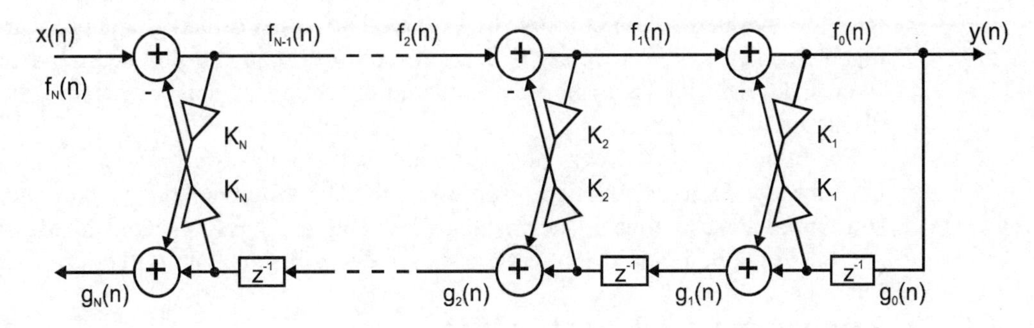

Figure 8.17
Lattice structure for IIR filters

Notice that we now have an upper path, which is a forward path, and a lower path, which is a reverse path.

The more general IIR transfer function can be realized using a lattice-ladder structure as shown in Figure 8.18.

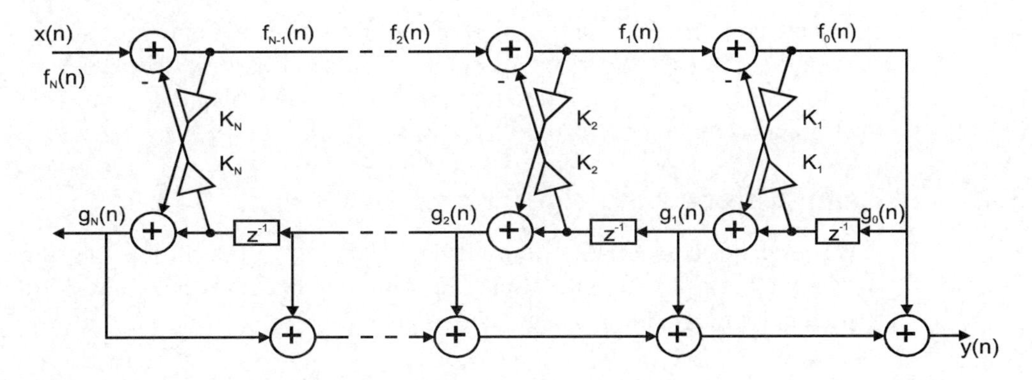

Figure 8.18
A more general IIR lattice filter

The upper half is a lattice structure for the denominator polynomial and the lower half is a ladder structure that realizes the numerator polynomial.

Lattice structures are generally less sensitive to coefficient quantization errors than the direct forms. Inspecting the reflection coefficients can easily test the stability of the filter.

8.4.2 Wave digital filter

The wave digital filter structure is derived from analog lossless LC filters. There are a variety of different types of wave digital filters with varying computational and storage requirements. It is characterized by low sensitivity to coefficient quantization errors.

Generally speaking, the more complex the structure of the digital filter, the less sensitive it is to coefficient errors. Wave digital filters are capable of operating with very few bits for coefficient representation. If the number of bits available were at a premium, then a complex structure would be needed to achieve the same level of accuracy.

8.4.3 Fast convolution

The fast convolution method applies to FIR filters. The FIR difference equation is in fact a convolution between the input sample sequence and the filter impulse response sequence. Instead of performing this convolution we can transform the two sequences into the frequency domain by FFT. The two transformed sequences are multiplied together and then the inverse FFT performed to obtain the output time sequence of the filter. Both the impulse response and the input sequences must be suitably zero padded as discussed in Chapter 4.

For high order FIR filters, the computational savings are quite substantial. This point has already been discussed in some detail. The disadvantage is that there will be a substantial delay between the instant when the input is presented to the filter and the output is obtained. Some applications can tolerate this delay but others cannot.

8.4.4 Frequency sampling structure

The frequency sampling structure is another alternative for FIR filters. Instead of using the impulse response $h(n)$ of the digital filter, samples of the desired frequency response $D(\omega)$ are used. For FIR filters with a narrow passband, most of the samples of the desired frequency response will be zero. So fewer computations will be needed.

8.5 Software implementation

We shall now briefly discuss some of the details of implementing the structures in the previous section as software algorithms. We shall assume that we have a continuous stream of input samples and describe the sample-processing algorithm – the algorithm for processing a single input sample.

8.5.1 Sample processing algorithms

The sample-processing algorithm for the direct form I structure is the simplest. Consider a general IIR filter transfer function with a numerator polynomial of order M and denominator polynomial of order N.

for each input sample x do:

$$
\begin{aligned}
v_0 &= x \\
w_0 &= -a_1 w_1 - \cdots - a_M w_M + b_0 v_0 + b_1 v_1 + \cdots + b_N v_N \\
y &= w_0 \\
v_i &= v_{i-1}, \qquad i = N, N-1, \ldots, 1 \\
w_i &= w_{i-1}, \qquad i = M, M-1, \ldots, 1
\end{aligned}
$$

The input sample is denoted as x and output sample as y. b_i and a_i are the numerator and denominator coefficients respectively. All the internal variables v_i and w_i are initialized to zero. Note that the updating of the internal variables must be done in reverse order to avoid overwriting of the previous values.

The sample-processing algorithm for the direct form II structure is also very straightforward. Using the same notations as before, we have:

for each input sample x do:

$$
\begin{aligned}
w_0 &= x - a_1 w_1 - a_2 w_2 - \cdots - a_M w_M \\
y &= b_0 w_0 + b_1 w_1 + \cdots + b_N w_N \\
w_i &= w_{i-1}, \quad i = K, K-1, \ldots, 1
\end{aligned}
$$

Here K is equal to M or N, whichever is greater. That is,

$$K = \max(M, N)$$

Again, the internal variables w_i are all initialized to zero and the updating must be performed in reverse order. Notice that only one set of internal variables is needed because of the simpler structure.

The way these sample-processing algorithms can be optimized on a DSP chip will be deferred to the next chapter.

8.6 Representation of numbers

A brief review of some simple fixed-point representation of numbers has been presented in Appendix B. In this section, we shall expand on that and discuss the fixed point and floating point representation of numbers. The fixed number of bits allocated to represent each number leads to finite numerical precision in computations. This implies that round-off and truncation errors are unavoidable. The effects are particularly severe in fixed-point implementations.

8.6.1 Fixed-point representation

The general fixed-point format is basically the same as the usual familiar decimal representation of numbers. It consists of a string of digits with a decimal point. The digits to the left of the decimal point are the integer part and those to the right are the fractional part of the number.

$$
\begin{aligned}
X &= \left(b_B b_{B-1} \cdots b_1 b_0 \cdot b_{-1} b_{-2} \cdots b_{-A}\right)_r \\
&= \sum_{i=-A}^{B} b_i r^i \qquad 0 \le b_i \le r-1
\end{aligned}
$$

where b_i are the digits and r is the *radix* or *base*.

Example 8.3

$$(12.34)_{10} = 1\times10^{1} + 2\times10^{0} + 3\times10^{-1} + 4\times10^{-2}$$

$$(110.01)_{2} = 1\times2^{2} + 1\times2^{1} + 0\times2^{0} + 0\times2^{-1} + 1\times2^{-2}$$

We shall focus on binary representations, as this is the format we need to deal with in DSP. In this case, the digits are bits. b_{-A} is the least significant bit (LSB) and b_B is the most significant bit (MSB). Naturally the binary point (as opposed to decimal point) between b_0 and b_{-1} does not physically exist and is up to the user to interpret.

Non-negative integers can easily be represented by an n-bit pattern ($B = n-1,\ A = 0$). Since we need to deal with fractional numbers, the fraction format ($B = 1, A = n-1$) is normally used. This allows us to represent numbers in the range 0 to $1-2^{n}$. This is because multiplication of two numbers that are less than 1 will give us a result that is less than 1.

Positive fractions are given by

$$X = 0.b_{-1}b_{-2}\cdots b_{-n} = \sum_{i=-1}^{-n} b_i 2^{i}$$

$$= 0.b_1 b_2 \cdots b_n = \sum_{i=1}^{n} b_i 2^{-i} \qquad X \geq 0$$

denoting the bits by a positive index for convenience.

Negative fractions can be represented by one of the following:

Sign-magnitude

$$X = 1.b_1 b_2 \cdots b_n \qquad X \leq 0$$

One's-complement

$$X = 1.\overline{b}_1 \overline{b}_2 \cdots \overline{b}_n \qquad X \leq 0$$

If X is a positive number, then the corresponding negative number is determined by complementing all the bits.

Two's-complement

$$X = 1.\overline{b}_1 \overline{b}_2 \cdots \overline{b}_{n-1} \left(\overline{b}_n \oplus 1\right)$$

This is the same as one's-complement except the least significant bit is exclusive-OR'ed (XOR) with 1. Alternatively, the LSB is added 1 modulo-2.

Example 8.4

$$\frac{7}{8} = 0.111$$

$$-\frac{7}{8} = 1.111 \qquad \text{sign-magnitude}$$

$$= 1.000 \qquad \text{one's-complement}$$

$$= 1.001 \qquad \text{two's-complement}$$

8.6.1.1 Arithmetic operations

Addition (and hence subtraction) in one's and two's-complements are straightforward. With 2's complement, if a carry bit affects the MSB, then it is dropped. With 1's complement addition, the carry in the MSB, if it is present, is carried around to the LSB. An important property of 2's complement addition is that if the final sum of a sequence of numbers X_1, X_2, ..., X_N is within the range of representation, it will be computed correctly even though some of the individual partial sums result in overflows. We can tell there is an overflow if the sum of two 2's complement numbers with the same sign has a result with an opposite sign. For instance,

$$\begin{aligned}\frac{1}{2}+\frac{3}{8}&=0100+0011\\&=0111\\&=\frac{7}{8}\end{aligned}$$

$$\begin{aligned}\frac{1}{2}-\frac{3}{8}&=0100+0011\\&=0001\\&=\frac{1}{8}\end{aligned}$$

Addition using the sign-magnitude format requires sign checking, complementing and the generation of a carry. This is significantly more complex than that for the previous two formats. This is the reason why 2's complement is commonly used.

Multiplication of two fixed-point numbers each n bits in length will generally give a product, which is $2n$ bits in length. The product therefore has to be truncated or rounded off to n bits, producing truncation or round-off errors.

8.6.2 Floating-point representation

Floating-point representations cover a much wider range of numbers. They normally consist of a mantissa M, which is the fractional part of the number, and an exponent E, which can be either positive or negative. Hence a number X is given by

$$X = M \cdot 2^{E}$$

with

$$\frac{1}{2} \le M < 1$$

Both the mantissa and the exponent require their individual sign bit.

Given a total number of bits available for representing a number, a number of different floating-point formats can result. In the past, individual computer manufacturers used their own format for their own products. A common standard floating point format has been adopted by the Institute of Electrical and Electronic Engineers (IEEE), which is usually referred to as the IEEE 754 standard. It defines the way zero is represented, the choice of M and E, the handling of overflows and other issues. For a 32-bit representation, the single precision floating point number is defined as

$$X = (-1)^{S} \cdot 2^{E-127}(M)$$

where S is the sign bit, E occupies 8 bits, and M is 23 bits long in a format as shown in Figure 8.19.

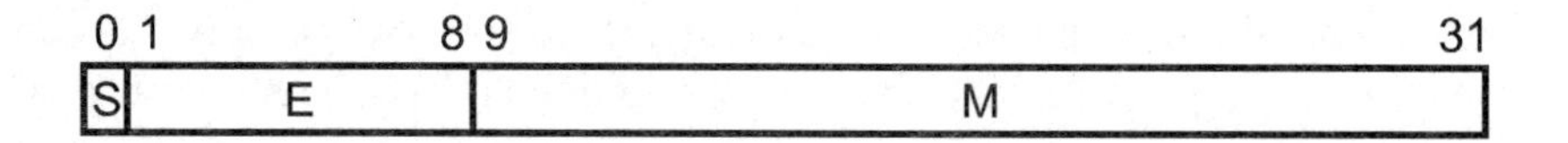

Figure 8.19
IEEE -754 floating point format

The following rules apply:

- If $E = 255$ and $M \neq 0$, then X is not a number (denoted *NaN*).
- If $E = 255$ and $M = 0$, then X is infinity (denoted *InF*).
- if $0<E<255$, then $X = (-1)^S \cdot 2^{-126}(1.M)$.
- If $E = 0$ and $M \neq 0$, then $X = (-1)^S \cdot 2^{-126}(0.M)$.
- If $E = 0$ and $M = 0$, then X is zero.

Here $0.M$ is a fraction and $1.M$ is a number with one integer bit and 23 fractional bits.

Example 8.5

0	1 0 0 0 0 0 1 0	1 0 1 0 • • • 0 0
S	E	M

The representation in above has the value

$$X = -1^0 \times 2^{130-127} \times 1.1010...0$$
$$= 2^3 \times \frac{13}{8}$$
$$= 13$$

Floating point representations can naturally represent a much larger range of numbers than a fixed point one with the same number of bits. However, it should be noted that the resolution does not remain the same throughout this range. This means that the distance between two floating point numbers increases as the number is increased. On the other hand, the resolution of fixed-point numbers is constant throughout the range.

8.6.2.1 Arithmetic operations

When two floating-point numbers are multiplied, the mantissas are multiplied and the exponents are added. But if we want to add two floating-point numbers, the exponents of the two numbers must be equal. The one with the small exponent is adjusted by increasing the exponent and reducing the mantissa. This adjustment could result in a loss in precision in the mantissa.

Overflow occurs in multiplication when the sum of the two exponents exceeds the dynamic range of the representation for the exponent.

8.7 Finite word-length effects

The number of bits that is used to represent numbers, called word-length, is often dictated by the architecture of DSP processor. If specialized VLSI hardware is designed, then we have more control over the word-length. In both cases we need to tradeoff the accuracy with the computational complexity. No matter how many bits are available for number representation, it will never be enough for all situations and some rounding and truncation will still be required.

For FIR digital filters, the finite word-length affects the results in the following ways:

- **Coefficient quantization errors.**
  The coefficient we arrived at in the approximation stage of filter design assumes that we have infinite precision. In practice, however, we have the same word-length limitations on the coefficients as that on the signal samples. The accuracy of filter coefficients will affect the frequency response of the implemented filter.
- **Round-off or truncation errors resulting from arithmetic operations.**
  Arithmetic operations such as addition and multiplication often give results that require more bits than the word-length. Thus truncation or rounding of the result is needed. Some filter structures are more sensitive to these errors than others.
- **Arithmetic overflow.**
  This happens when some intermediate results exceed the range of numbers that can be represented by the given word-length. It can be avoided by careful design of the algorithm scale.

For IIR filters, our analysis of finite word-length effects will need to include one more factor: product round-off errors. The round-off or truncation errors of the output sample at one time instant will affect the error of the next output sample because of the recursive nature of the IIR filters. Sometimes limit cycles can occur.

We shall look at these word-length effects in more detail.

8.7.1 Coefficient quantization errors

Let us consider the low-pass FIR filter that has been designed by using Kaiser windows in section 6.4.4. The coefficients obtained are listed in Table 8.1. The table also shows the values of the coefficients if they are being quantized into 8 bits. The magnitude responses of the filter, both before and after coefficient quantization, are shown in Figure 8.20.

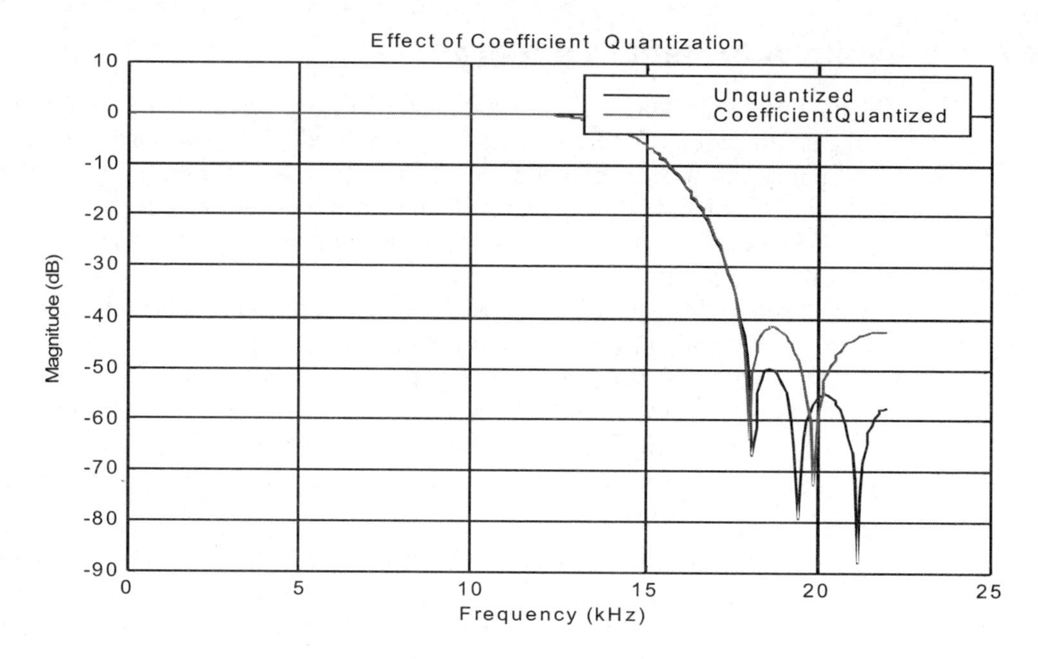

Figure 8.20
Magnitude response of a low-pass FIR filter before and after coefficient quantization

N	*Coefficient*
0	–0.0016
1, 21	0.0022
2, 10	0.0026
3, 19	–0.0117
4, 18	0.1270
5, 17	0.0071
6, 16	–0.0394
7, 15	–0.0466
8, 14	0.0117
9, 13	–0.1349
10, 12	0.2642
11	0.6803

Table 8.1

The quantized filter has violated the specification of the stopband. Clearly in this case, more than 8 bits are required for the filter coefficients.

The minimum number of bits needed for the filter coefficients can be found by computing the frequency response of the coefficient-quantized filter. A trial-and-error approach can be used. However, it will be useful to have some guideline for estimating the word-length requirements of a specific filter.

The quantized coefficients and the unquantized ones are related by

$$h_q(n) = h(n) + e(n), \qquad n = 0, 1, \ldots, N-1$$

and shown in Figure 8.21.

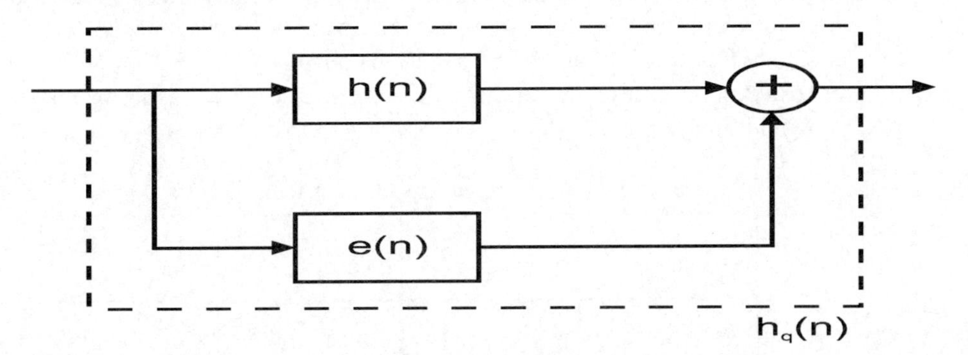

Figure 8.21
Model of coefficient quantization

This relationship can also be established in the frequency domain as in Figure 8.22.

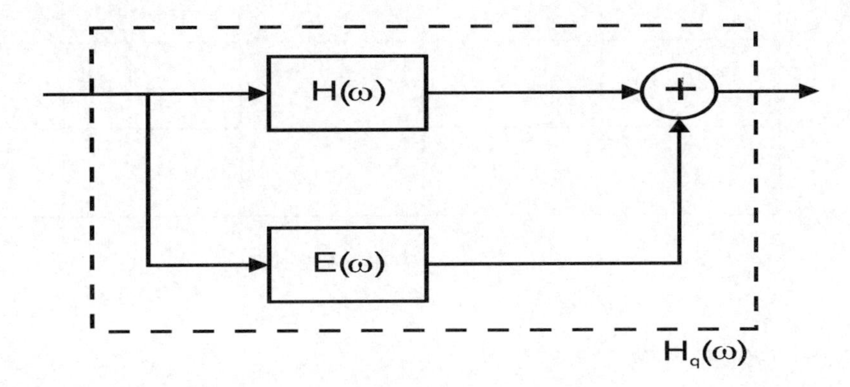

Figure 8.22
Frequency domain model of coefficient quantization

Here $H(\omega)$ is the frequency response of the original filter and $E(\omega)$ is the error in the frequency response due to coefficient quantization.

If the direct form structure is used, assuming rounding, the following bounds for the magnitude of the error spectrum are most commonly used:

$$|E(\omega)| = N2^{-B}$$

$$|E(\omega)| = 2^{-B}(N/3)^{1/2}$$

$$|E(\omega)| = 2^{-B}\left[\frac{1}{3}(N \ln N)\right]^{1/2}$$

where B is the number of bits used for representing the filter coefficients. The first bound is a worst case absolute bound and is usually over pessimistic. The other two are based on the assumption that the error $e(n)$ is uniformly distributed with zero mean. They generally provide better estimates for the word-length required.

Example 8.6
For the low-pass filter designed in section 6.4.4, we have shown that 8-bit word-length is not sufficient for the coefficients. The order of the filter is $N = 22$. The stopband attenuation is specified to be at least 50 dB down.

So we need at least 10 bits for the coefficients. The resulting magnitude response is shown in Figure 8.23.

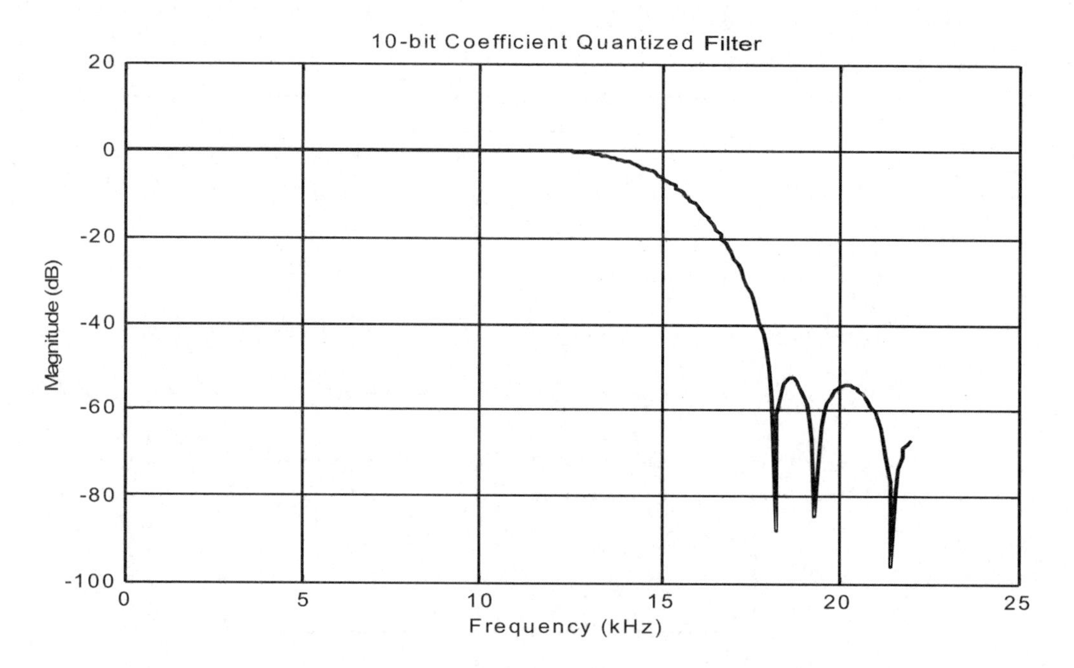

Figure 8.23
Coefficient quantized filter response

For IIR filters, the coefficient quantization error may have one more effect: instability. The stability of a filter depends on the location of the roots of the denominator polynomial in the transfer function. Consider a second order section of an IIR filter (since it is the basic building block of higher order filters) with transfer function

$$H(z) = \frac{b_0 + b_1 z^{-1} + b_2 z^{-2}}{1 + a_1 z^{-1} + a_2 z^{-2}}$$

The roots of the denominator polynomial, or poles of the transfer function, are located at

$$p_1 = \frac{1}{2}\left[-a_1 + \sqrt{a_1^2 - 4a_2}\right]$$

$$p_2 = \frac{1}{2}\left[-a_1 - \sqrt{a_1^2 - 4a_2}\right]$$

They may either be complex conjugate pairs or are both real. If they are complex conjugate pairs, they can be represented as having a magnitude and an angle:

$$p_1 = r\angle\theta$$
$$p_2 = r\angle-\theta$$

where

$$r = \sqrt{a_2}$$
$$\theta = \cos^{-1}\left(-\frac{a_1}{2r}\right)$$

For stability, the magnitude of these poles must be less than 1. This applies to both real and complex poles. So the test for stability for the second coefficient is

$$0 \le |a_2| < 1$$

From the above equation for the angle, the arguments to the arc-cosine function must have a magnitude that is less than or equal to 1. So the test for stability for the first coefficient is

$$|a_1| \le 1 + a_2$$

Both these tests must be satisfied at the same time for the IIR filter to remain stable.

8.7.2 Rounding and truncation

It is inevitable that some numbers are being represented by less number of bits than what is required to represent them exactly. For instance, when we multiply two n-bit numbers we have a result that is at most $2n$-bits long. This result will still have to be represented using n-bits. Thus rounding or truncation will be needed. The characteristics of the errors introduced depend on how the numbers are represented.

Consider the n-bit fixed point representation of a number x, which requires n_u bits to represent exactly, with $n < n_u$. For positive numbers, both sign-magnitude and 2's complement representations are identical. The error introduced by truncation is

$$-(2^{-n} - 2^{-n_u}) \le E_t \le 0$$

with the largest error discarding all $(n_u - n)$ bits, (all being 1s).

For negative numbers represented using sign-magnitude format, truncation reduces the magnitude of the numbers and hence the truncation error is positive.

$$0 \le E_t \le (2^{-n} - 2^{-n_u})$$

So the truncation error for sign-magnitude format is in the range

$$-(2^{-n} - 2^{-n_u}) \le E_t \le (2^{-n} - 2^{-n_u})$$

With 2's complement representation, truncation of a negative number will increase the magnitude of the number and so the truncation error is negative.

$$-(2^{-n} - 2^{-n_u}) \le E_t \le 0$$

So the truncation error for 2's complement format is still in the range

$$-(2^{-n} - 2^{-n_u}) \le E_t \le 0$$

Now consider round-off errors of the above fixed-point representations. In this case, the error is independent of the type of representation and it may either be positive or negative and is symmetrical about zero.

$$-\frac{1}{2}(2^{-n} - 2^{-n_u}) \leq E_t \leq \frac{1}{2}(2^{-n} - 2^{-n_u})$$

In floating point representations, the resolution is not uniform as discussed before. So the truncation and round-off errors are proportional to the number. It is more useful to consider the relative error defined as

$$e = \frac{Q(x) - x}{x}$$

where $Q(x)$ is the number after truncation or rounding.

If the mantissa is represented by 2's complement using n-bits, then the relative truncation errors has the bounds:

$$-2^{-n+1} < e_t \leq 0, \qquad x > 0$$

$$0 \leq e_t < 2^{-n+1}, \qquad x < 0$$

The relative round-off error is in the range:

$$-2^{-n} < e_r \leq 2^{-n}$$

8.7.3 Overflow errors

Overflow occurs when two large numbers of the same sign are added and the result exceeds the word-length. If we use 2's complement representation, as long as the final result is within the word-length, overflow of partial results is unimportant. If the final result does cause overflow, it may lead to serious errors in the system. Overflow can be avoided by detecting and correcting the error when it occurs. However, this is a rather expensive approach. A better way is to try and avoid it by scaling the data or the filter coefficients.

Consider an N-th order FIR filter. The output sample at time instant n is given by

$$y(n) = \sum_{m=0}^{N-1} h(m)x(n-m)$$

Assume that the magnitudes of the input and the filter coefficients are less than 1. Then the magnitude of the output is

$$|y(n)| \leq \sum_{m=0}^{N-1} |h(m)||x(n-m)|$$

In the worst case,

$$y(n) = \sum_{m=0}^{N-1} |h(m)|$$

A scaling factor G can be chosen as

$$G_1 = \|h\|_1 = \sum_{n} |h(n)|$$

The filter coefficients are all scaled by this factor

$$\tilde{h}(n) = \frac{h(n)}{G}$$

In this way, the maximum output magnitude of the filter will always be less than or equal to 1 and overflow is completely avoided.

However, this scaling method is very conservative. The worst case signal will hardly ever occur in practice. Two other less conservative scaling factors are usually chosen instead. The first one is defined by

$$G_2 = \|h\|_2 = \left[\sum_n h^2(n)\right]^{1/2} \le \|h\|_1$$

This scaling improves the signal to quantization noise ratio. The trade-off is that there is a possibility of overflow. Another scaling factor, given by

$$G_3 = \|H\|_C = \max_{\omega} |H(\omega)| \le \|h\|_1$$

guarantees that the steady-state response of the system to a sine wave will not overflow. This frequency domain-scaling factor is often the preferred method.

Scaling is even more important for IIR filters because an overflow in the current output sample affects many output samples following that one. In some cases, overflow can cause oscillation and seriously impair the usefulness of the filter. Only by resetting the filter can we recover from these oscillations.

The same set of scaling factors described above can be used for IIR filters. However, the frequency domain measure is more useful because the duration of the impulse response in this case is infinite. The procedure is illustrated by an example.

Example 8.7

An IIR filter is designed based on a 4th order elliptic low-pass filter. The filter transfer function is given by

$$H(z) = \frac{1+1.621784z^{-1}+z^{-2}}{1-0.04030703z^{-1}+0.2332662z^{-2}} \cdot \frac{1+0.7158956z^{-1}+z^{-2}}{1+0.0514214z^{-1}+0.7972861z^{-2}}$$
$$= H_1(z)H_2(z)$$

which is decomposed into a cascade of two second order sections.

The transpose structure of section 1 is shown in Figure 8.24.

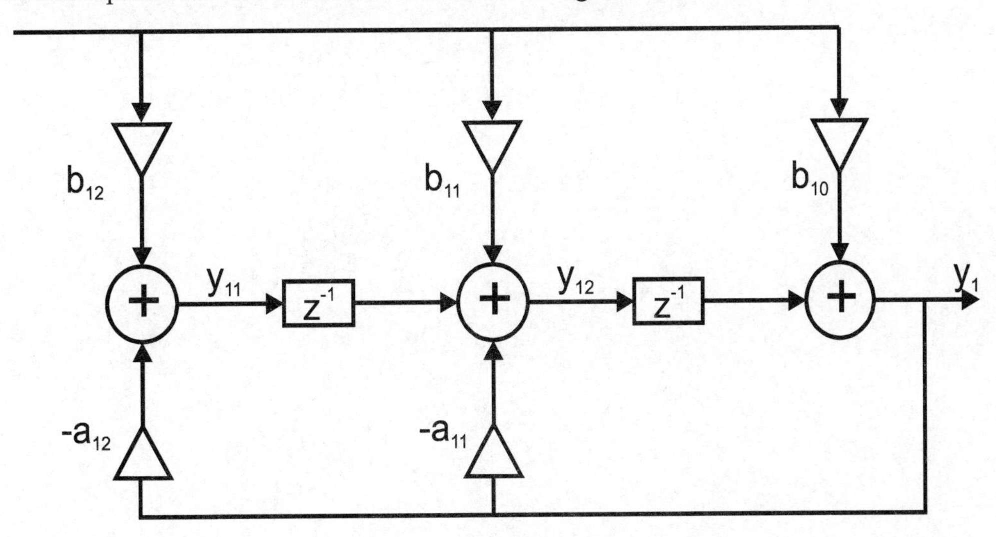

Figure 8.24
Transpose structure of section 1

The impulse response and magnitude response of this section are as shown in Figure 8.25.

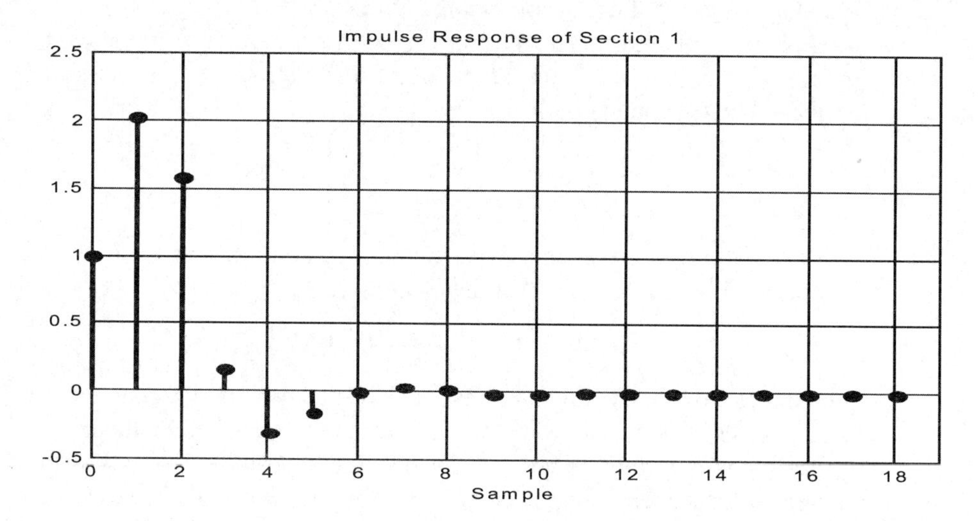

Figure 8.25
The impulse and magnitude responses of section 1

Table 8.2 shows the three different scaling factors that can be used.

Apart from the output of the section, the output of each internal adder within the section has to be examined as well. The impulse response at y_{11} and y_{12} and their frequency responses

$$H_{11}(z) = \frac{Y_{11}(z)}{X(z)}$$

$$= \frac{0.766733805 - 0.781377681z^{-1}}{1 - 0.4030702997z^{-1} + 0.766733805z^{-2}}$$

$$H_{12}(z) = \frac{Y_{12}(z)}{X(z)}$$

$$= \frac{2.024854296 + 0.766733805z^{-1}}{1 - 0.4030702997z^{-1} + 0.2332661953z^{-2}}$$

are shown in Figures 8.26 and 8.27 respectively.

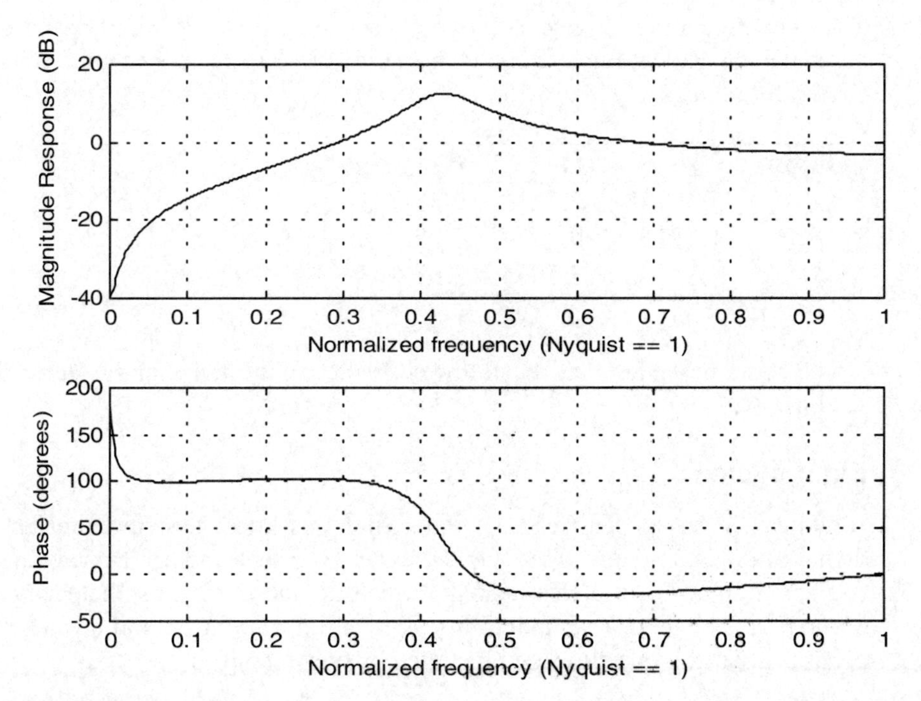

Figure 8.26
Frequency response $H_{11}(z)$

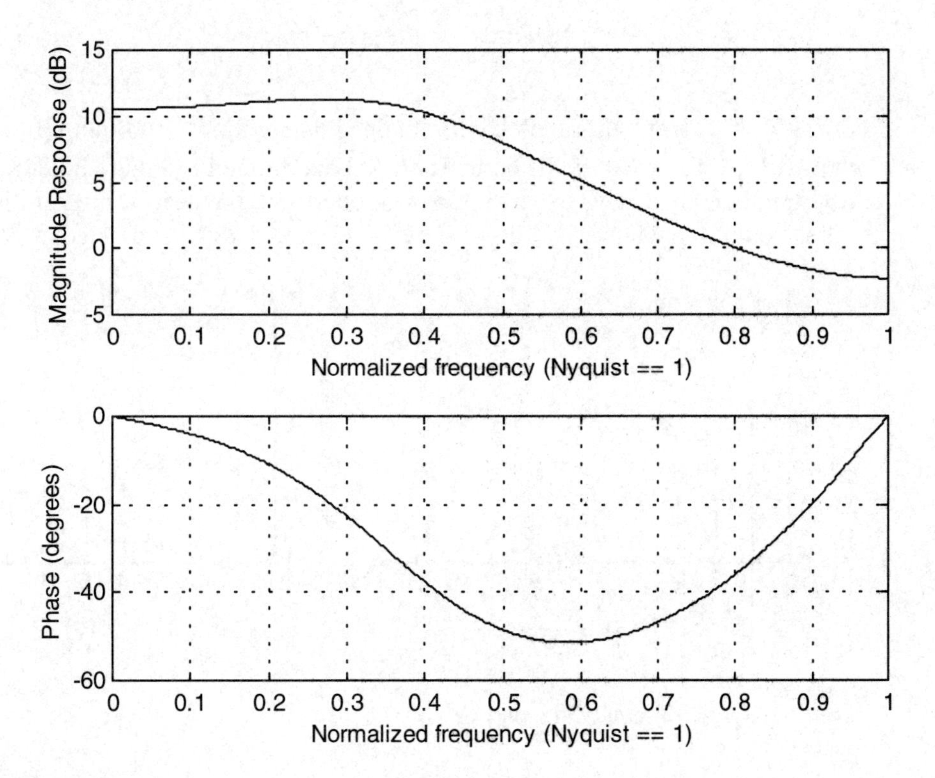

Figure 8.27
Frequency response $H_{12}(z)$

Their corresponding scaling factors are also given in Table 8.2. Having decided on the particular type of scaling factor to be used, the largest one in that column should be chosen for scaling.

Output	G_1	G_2	G_3
y_1	5.30748	2.7843	4.38
y_{11}	1.7715	0.9774	1.28
y_{12}	4.30748	2.5985	3.67

Table 8.2

The same procedure can be followed to determine the scaling factor for the second section.

8.7.4 Limit cycles

Although we have been treating digital filters as linear systems, the fact is that a real digital filter is non-linear. This is due to quantization, round off, truncation, and overflow. A filter designed as a linear system, which is stable, may oscillate when an overflow occurs. This type of oscillation is called a limit cycle. Limit cycles due to round-off/truncation and overflow are illustrated by two examples.

Example 8.8
A first order IIR filter has the following different equation:

$$y(n) = x(n) + \alpha y(n-1)$$

For $\alpha = 0.75$, the output samples $y(n)$ obtained using initial condition $y(0) = 6$ and a zero input $x(n) = 0$ for $n \geq 0$ are listed in Table 8.3 and plotted in Figure 8.28(a). It shows that the output quickly decays to zero. If $y(n)$ is rounded to the nearest integer, then after some time the output remains at 2. This is shown in Figure 8.28(b) and listed in Table 8.3.

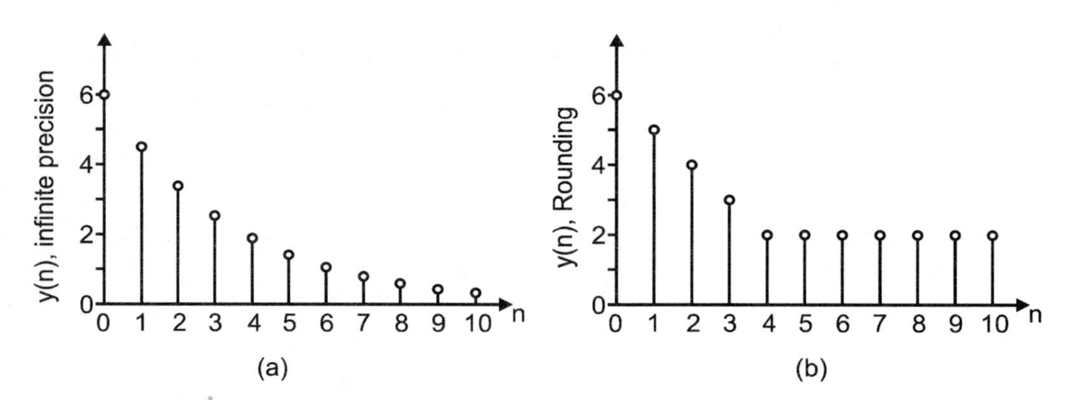

Figure 8.28
Output values before and after rounding

n	y(n), infinite precision	y(n), rounding
0	6	6
1	4.5	5
2	3.38	4
3	2.53	3
4	1.90	2
5	1.42	2
6	1.07	2
7	0.80	2
8	0.60	2
9	0.45	2
10	0.3375	2

Table 8.3

n	y(n), infinite precision	y(n), rounding
0	6	6
1	–4.5	–5
2	3.38	4
3	–2.53	–3
4	1.90	2
5	–1.42	–2
6	1.07	2
7	–0.80	–2
8	0.60	2
9	–0.45	–2
10	0.3375	2

Table 8.4

For $\alpha = -0.75$, the output oscillates briefly and decays to zero for the infinite precision version. If the result is rounded to the nearest integer, then the output oscillates between –2 and +2. These signals are plotted in Figure 8.29 and listed in Table 8.4.

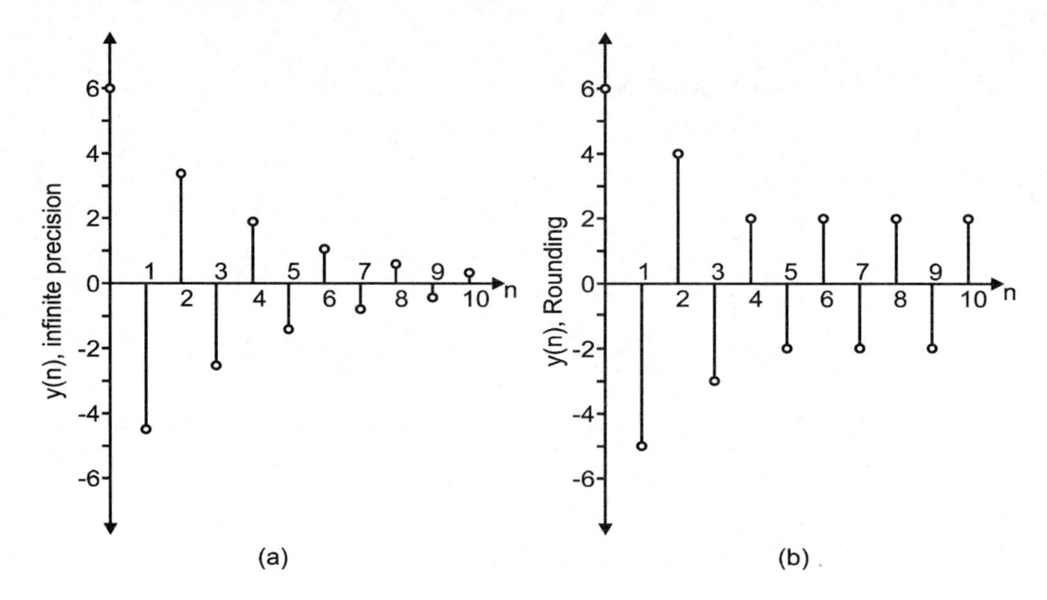

Figure 8.29
Output with and without rounding

Example 8.9
A filter with transfer function

$$H(z) = \frac{1}{z^2 - z + 0.5}$$

is stable filter. A 2's complement overflow non-linearity is added to the filter structure as shown in Figure 8.30.

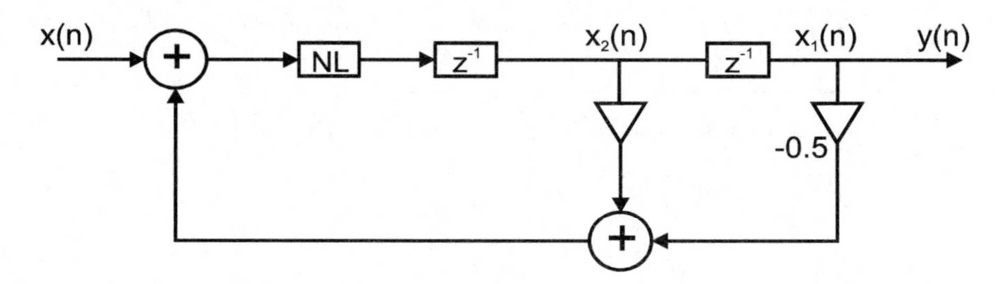

Figure 8.30
FIR filter structure with overflow non-linearity added

The transfer function of the non-linearity itself is shown in Figure 8.31.

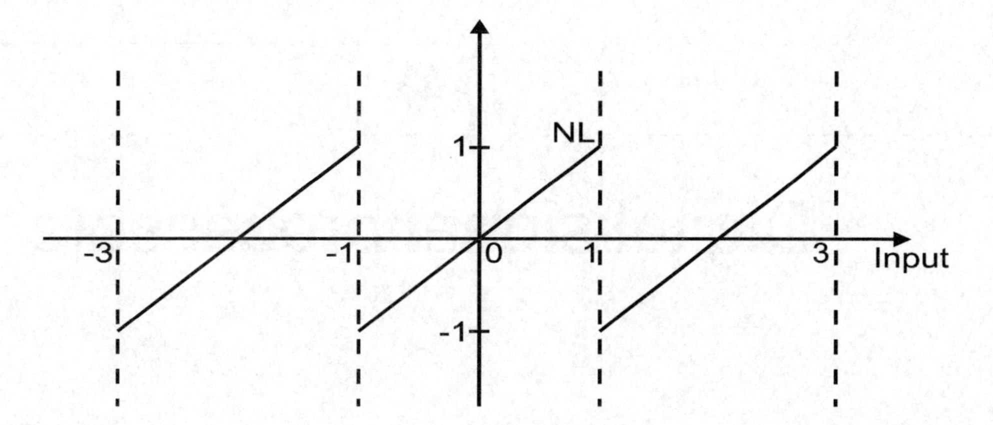

Figure 8.31
Transfer function of overflow non-linearity

According to the notations of Figure 8.30, if the input is zero,

$$x_1(n+1) = x_2(n)$$
$$x_2(n+1) = NL\left[-0.5x_1(n) + x_2(n)\right]$$

With an initial condition of

$$x_1(0) = 0.8$$
$$x_2(0) = -0.8$$

At the next time instant, we have

$$x_1(n) = (-1)^{n}0.8$$
$$x_2(n) = (-1)^{n+1}0.8$$

In fact, for $n \geq 1$,

$$\begin{aligned} x_1(1) &= -0.8 \\ x_2(1) &= NL\left[-0.5x_1(0) + x_2(0)\right] \\ &= NL\left[-1.2\right] \\ &= +0.8 \end{aligned}$$

Thus the output oscillates between 0.8 and –0.8.

Note that the limit cycle will only start if there is a previous overflow. If no overflow occurs the system remains linear.

9

Digital signal processors

While the most demanding DSP applications will require custom designed VLSI devices to hard-wire the DSP algorithms, most common applications can be handled by the use of commercially available digital signal processors. For instance, it is likely that the desktop computer most people are using contains at least one DSP chip in it, most likely as part of the sound card. Owing to its common use, it is the purpose of this chapter to give a description of the common fundamental characteristics of these processors. We shall also attempt to describe some particular features of DSP chips by Texas Instruments and Analog Devices.

9.1 Common features

A few semiconductor manufacturers produce a range of DSP chips with different capabilities. Most of the DSP chips are single processor devices. There exist chips that integrate multiple DSP processors on the same chip such as the Texas Instruments TMS320C8x. Others combine a DSP processor with a microcontroller such as the Motorola DSP568xx.

Some manufacturers offer DSP cores. They are intended to be used as building blocks in creating a semi-custom chip. This allows the designer to integrate a programmable DSP and other custom circuitry onto a single application-specific integrated circuit (ASIC). The DSP core cuts design time and it is most useful for high volume production designs for specific applications in areas such as telecommunications. In some cases, the vendor providing the core is also the foundry fabricating the ASIC. In other cases, the vendor simply licenses the core design to the customer, who then selects an appropriate foundry.

Most of these processors share some common features, which facilitate the efficient computation of DSP algorithms.

9.1.1 Fast multiply-accumulate

It has been demonstrated in our discussion in the previous chapters that the multiply-add or multiply-accumulate (MAC) operation is encountered in all major DSP functions. These functions include filtering, FFT and correlation. Therefore, all DSP chips are designed to perform the MAC operation very efficiently. In fact, they are able to complete the MAC operation in one single instruction cycle. To achieve this, the multiplier and accumulator are integrated into the same data path of the processor. The accumulator register can usually store the extra bits resulting from the arithmetic operations to avoid overflow.

9.1.2 Multiple-access memory architecture

In most conventional microprocessors, the random access memory (RAM) can be used to store both the program instructions and the data. For example, to perform an addition, the microprocessor will have to fetch the instruction and then the data to be added. So the whole operation has to be performed in several instruction cycles. Most DSP chips, however, keep the instruction memory and data memory separate. This allows the processor to fetch an instruction and the associated data simultaneously.

To support simultaneous access to multiple memory locations, the DSP chips have to provide multiple on-chip buses, independent memory banks and/or multi-port on-chip RAM. This means that somehow the program and data must be transferred from external memory to on-chip internal memory.

9.1.3 Special addressing modes

DSP algorithms typically require the summation over a range of indices. For instance, the FIR filtering equation

$$y(n) = \sum_{m=0}^{N-1} h(m)x(n-m)$$

is a summation of product terms with the indices n and n–m. The indices refer to memory locations. To address them efficiently, DSP chips often incorporate dedicated address generation units. These units operate in the background, generating the next addresses in parallel with the execution of the current program instruction. Special addressing modes are therefore possible to perform such things as bit-reversed addressing for FFT and circular addressing for circular buffers.

9.1.4 Special program control

Special instructions are built-in for efficient looping that is often required by DSP algorithms. Other special instructions include those that move data from external to internal memories as a block, and low-overhead interrupts for fast input/output.

9.1.5 Peripheral interfaces

Most DSP chips incorporate serial and parallel I/O interfaces to other devices such as ADC and DAC. The DSP processor is often used as a coprocessor to another microprocessor, which acts as a host. Some DSP chips have special registers for communicating with the host processor. Some versions of the DSP chips have other peripheral devices integrated on-chip for special applications.

The hardware architecture, including data path design and the memory architecture, is discussed in more detail in the next section. Examples of how special instructions and addressing modes can be used are provided in section 9.3

9.2 Hardware architecture

In this section, we shall expand on some aspects of the hardware architecture of DSP chips. General design issues related to data path and memory organization are first discussed. They are then illustrated by describing the architectures of two specific families of DSP chips.

9.2.1 Data path

Data path refers to the complete arithmetic processing path, including multipliers, accumulators, other registers, and specialized units such as an address generation unit. The data paths of a fixed-point DSP (Texas Instruments TMS320C6x) and a floating-point DSP (Lucent DSP32C) are shown in Figures 9.1 and 9.2 respectively.

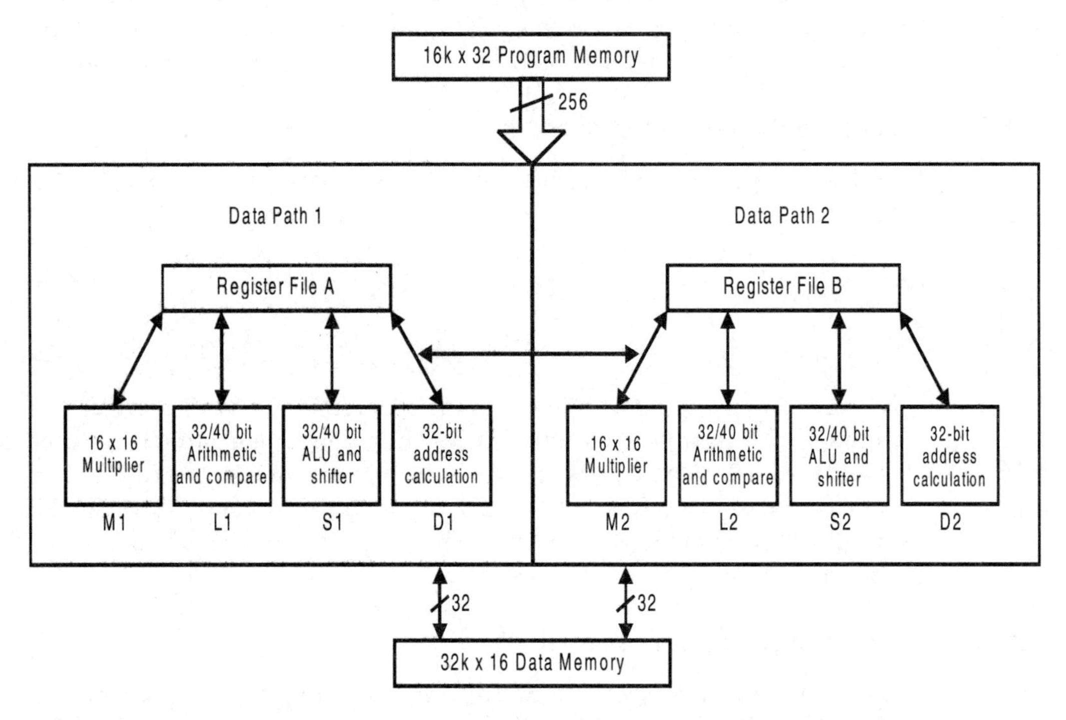

Figure 9.1
Data paths of TMS320C6x

Note that TMS320C6x family is one of the latest families of DSP chips. The fixed-point devices consist of two parallel data paths instead of the one data path that exists in other DSP processors.

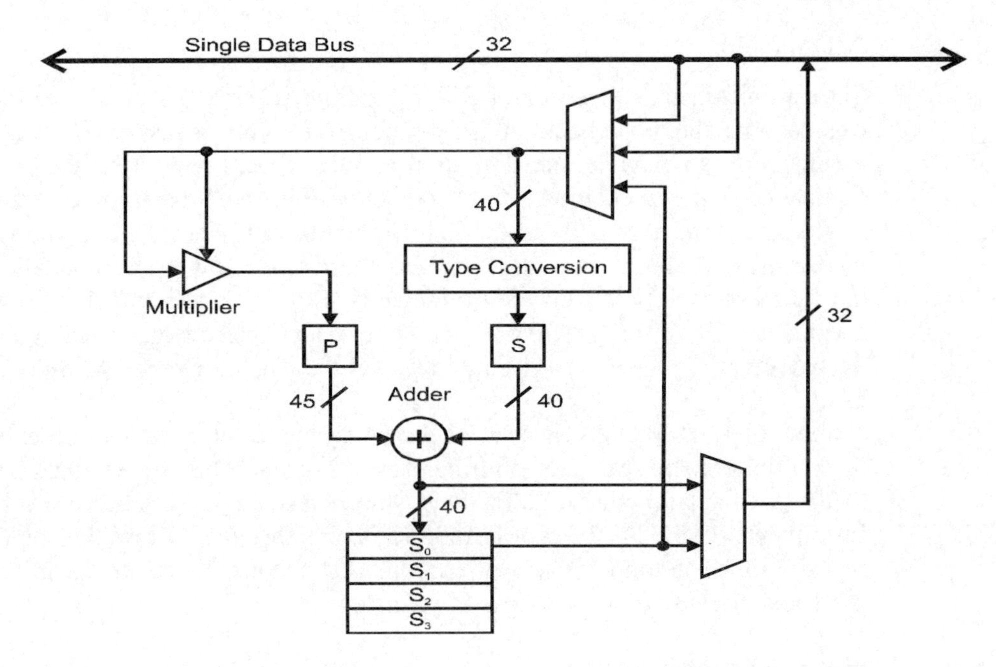

Figure 9.2
Data paths of Lucent Technologies' DSP32C

9.2.1.1 Multipliers

Multiplication is the central operation of DSP algorithms. Hence all DSP chips have a multiplier that can multiply two native-sized data in a single instruction cycle. But different designs lead to different characteristics. Some of them that are relevant to fixed-point DSP chips are listed below.

In the Motorola DSP5600x, the multiplier is integrated with an adder to form the MAC unit. Some other processors, such as the Lucent DSP16xx, the multiplier and adder are separate. The result of the multiplier is first kept in a product register before it is sent to the adder for accumulation. The result of the MAC operation will therefore be delayed by one instruction cycle before it can be used by the next instruction.

We know that the product of two n-bit fixed-point numbers will need $2n$ bits to store the result in order to avoid any loss of accuracy. Most fixed-point multipliers produce a result that is twice the word-length of their operands. So the multiplier itself does not introduce any error. But some multipliers produce results that are truncated. Examples of the latter design include the Zilog Z893x, which uses 16-bit operands and produces 24-bit results.

Some multipliers use pipelining to increase speed. Pipelining is a technique that allows two or more operations to overlap during execution. The task is broken down into a number of distinct sub-tasks, which are overlapped during execution. Thus the delay between time-inputs is presented to the multiplier to the time that the result may be available. It could be longer one instruction cycle even though the actual multiplication is done within that time. This delay is called latency. The advantage of pipelined multipliers is that if a long series of multiplications are to be performed, they are more efficient than the ones without pipeline. But latency is worst when only one multiplication is to be performed. The Clarkspur Design CD2450 DSP core uses a pipelined multiplier.

9.2.1.2 Accumulator

Since a fundamental operation of DSP algorithms is the MAC operation, the accumulator can become the bottleneck in the architecture. This is especially true if only one accumulator is available and it is used as one of the source operands and also as the destination of the calculation. Many DSP chips offer more than one accumulator.

The size of the accumulator should be larger than the size of the multiplier output word by several bits. These extra bits, known as guard bits, allow the accumulation of a number of results without overflow. Accumulators with *n* guard bits have the capacity to accumulate 2^n values without the need for intermediate scaling. The Lucent Technologies' DSP16xx has 4 guard bits and the Analog Device ADSP21xx has 8 guard bits.

Some other DSP chips, instead of providing guard bits, provide the output register of the multiplier to be scaled, by shifting it by a few bits. This is performed before adding it to the accumulator and usually done within the single instruction cycle. The Texas Instruments TMS320C2x and TMS320C5x, for instance, allow the multiplier product register to be automatically shifted right by 6 bits. However, guard bits are more preferable because there is no loss of precision.

9.2.2 Memory architecture

While the data path is important in speeding up the computation, a good memory architecture keeps the data path fed with data is equally important. Most DSP chips implement what is known as the Harvard architecture. Figure 9.3 illustrates typical microprocessor architecture known as the Von Neumann architecture and Figure 9.4 shows a general Harvard architecture.

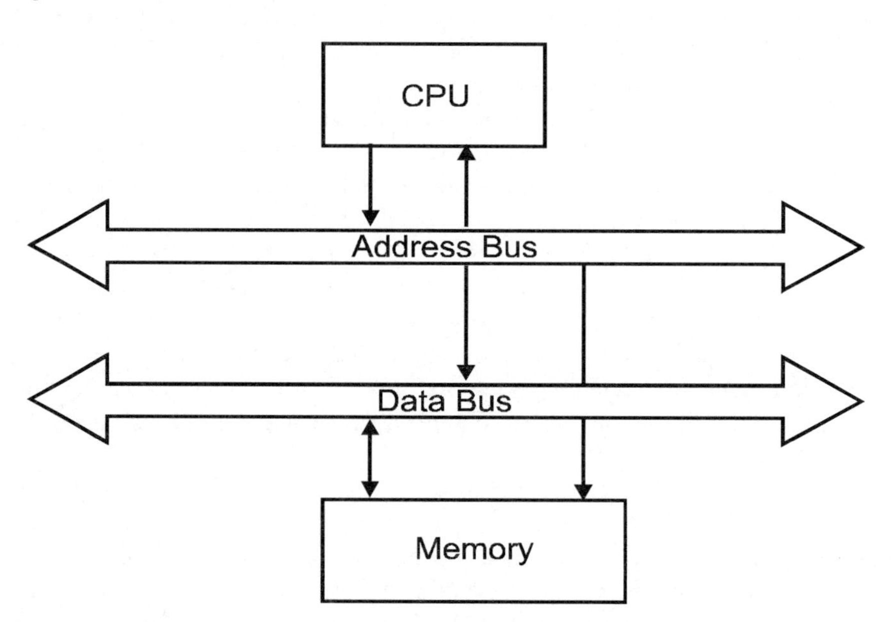

Figure 9.3
Von Neumann architecture

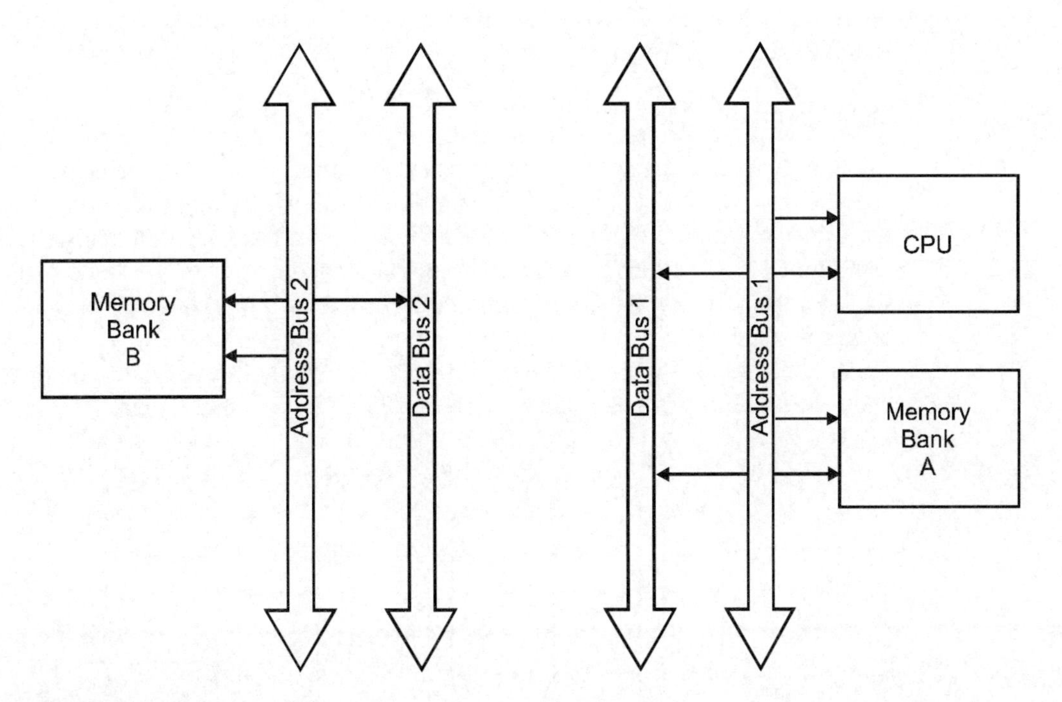

Figure 9.4
Harvard architecture

While the general microprocessor architecture has only one bus for both data and instructions, the Harvard architecture provides one for program instructions and two for data. The program and data memories are separate. Thus overlapping of instruction fetch (getting the next instruction from memory) and execution (involves reading and writing data to memory) is possible.

Most DSP chips implement some form of the Harvard architecture. They include the Texas Instruments TMS320 family, the Analog Devices ADSP2100 family and the Lucent Technologies DSP16xx family. It is interesting to note that for the DSP16xx processors, the full potential of dual bank of memories is not realized and writing to memory takes two instruction cycles. There are other processors that implement three banks of memory instead of two. Thus three independent memory accesses per instruction are possible. Processors in this category include the Zilog Z893x, the SGS-Thomson D950-CORE, and the Motorola DSP5600x, DSP563xx and DSP96002.

This multiple bus structure is too expensive to be extended to external (outside of the chip) memory. Usually only one address and one data bus are available off-chip. So it is important that data can be moved from external memory to on-chip internal memory efficiently.

9.2.2.1 Multiple-access memories

Another way to achieve multiple memory access in one instruction cycle is to use multiple-access memories. These memories can be accessed in a fraction of an instruction cycle, allowing multiple sequential accesses to be made on a single bus. The Lucent Technologies DSP32xx can complete four sequential memory accesses to the on-chip memories in a single instruction cycle.

Multiple access memories can be combined with the Harvard architecture to give even better performances. Zoran's ZR3800x processors have single-access program memory and dual-access data memory.

9.2.2.2 Multi-port memories

Another type of memory that can be used is called multi-port memory. It has multiple independent sets of address and data lines, allowing multiple independent memory accesses in parallel. So in this case we do not need to have separate banks of program and data memory since they can be accessed simultaneously from the same bank. Figure 9.5 shows a Harvard architecture combined with dual-port data memory and single-port program memory.

This architecture is used in the Motorola DSP561xx processors. The disadvantage of multi-port memory is that it takes up more silicon area to implement.

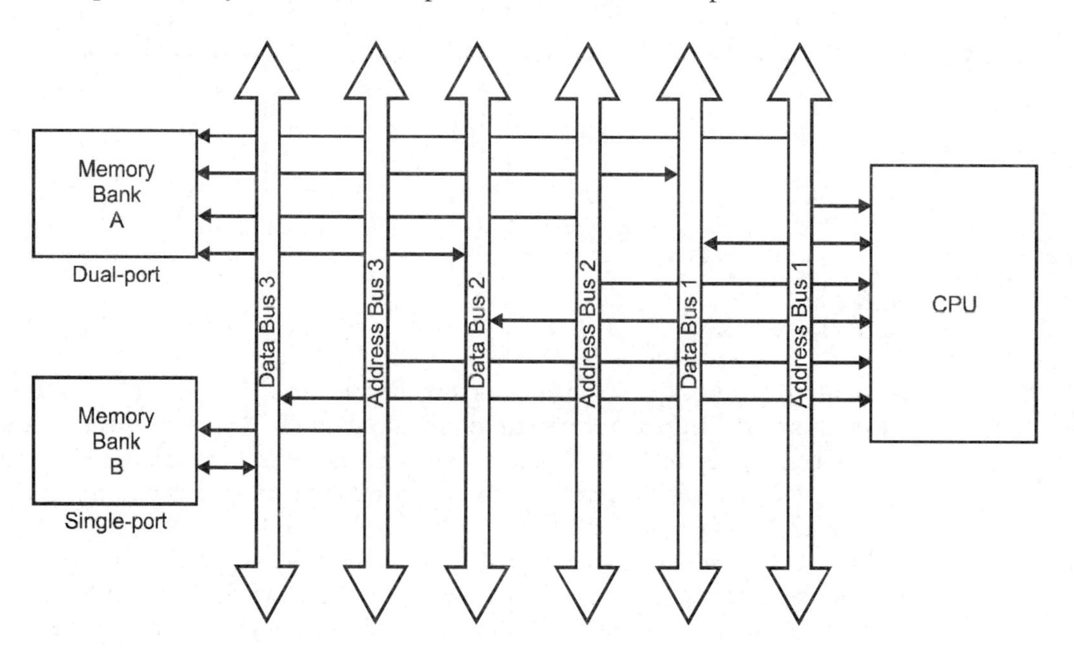

Figure 9.5
Harvard architecture with dual-port data memory

9.2.2.3 Program caches

A program cache is a small amount of memory for storing program instructions within the processor core. It reduces the need to fetch instructions from the program memory, thus speeding up operations. They are usually much simpler than those caches found in some advanced general-purpose microprocessors.

The simplest type of program cache is a single instruction repeat buffer. It is used in conjunction with the repeat instruction. The instruction that is to be repeatedly executed a number of times is loaded into this buffer. Subsequently, the same instruction is fetched from the cache instead of the program memory. This is implemented in the Texas Instruments TMS320C2x and TMS320C5x families of processors. Since program memory accesses are not required during repeat execution, the program memory can be used for data read or write access. During this time, the processor effectively has one more data bus available.

The repeat buffer can be designed to store more than a single instruction. In this case, a block of instructions can be loaded into the cache and repeated, freeing up the program memory bus for data access. This is very useful for algorithms containing loops with a few instructions. Such loops are often used in transforms, block data moves and filtering.

A more general form of multi-instruction repeats buffer is the single-sector instruction cache. It stores the number of the most recently executed instructions. If the program flow jumps back to one of the instructions in the cache (called a cache hit), the instruction is executed from the cache. The effectiveness of this type of cache obviously depends on the number of cache hits, which in turn depends on the algorithm. In some cases, the software designer can tailor the code to achieve more cache hits and so speeding up the execution of the algorithm. This type of cache can be found in the Zoran ZR3800x.

Multiple-sector instruction cache can also be found in some DSP chips. It works like the single-section variety except that two or more independent code segments can be stored. The Texas Instruments TMS320C3x processors have two sectors of 32 words each. Each sector stores instructions from different regions of program memory. The cache is updated when a cache miss (as opposed to cache hit) occurs. In this case, if the external address is from one of the two sectors currently associated with the cache, then the instruction is stored at the appropriate location in the cache. If the address is outside of that monitored by the cache, then the entire content of the sector is discarded and a new set of addresses will be monitored. The algorithm, which decides which cache sector will be discarded, is called the least recently used (LRU) algorithm. As the name implies, the cache with the most recent hit is kept and the other one is discarded.

Some DSP chips allow the programmer more control over the use of the cache. In some cases the programmer can lock the contents of the cache at some point in the program or disable the cache. Allowing manual control over the use of cache helps the developer to ensure that their programs will meet critical time constraints.

Even in cases where a physical cache is not present, the programmer can often manually move a section of program code from slower external memory to the faster internal memory for execution. This is called manual caching and often speeds up program execution significantly.

9.2.2.4 Direct memory access

Direct memory access (DMA) is the process of transferring data without the involvement of the processor itself. It is often used for transferring data to/from input/output devices. A separate DMA controller is required to handle the transfer. The controller notifies the DSP processor that it is ready for a transfer. Then the processor relinquishes control of its external memory bus and grants the control of the bus to the DMA controller. The DMA controller then transfers the specified amount of data and signals the processor upon completion of the transfer.

The Texas Instruments TMS320C3x, TMS320C4x, the Motorola DSP96002, and the Analog Devices ADSP2106x family of more sophisticated DSP chips all have an on-chip DMA controller.

9.2.3 Architecture of TMS320C5x

Figure 9.6 shows the functional block diagram of the Texas Instruments TMS320C5x family of DSP chips.

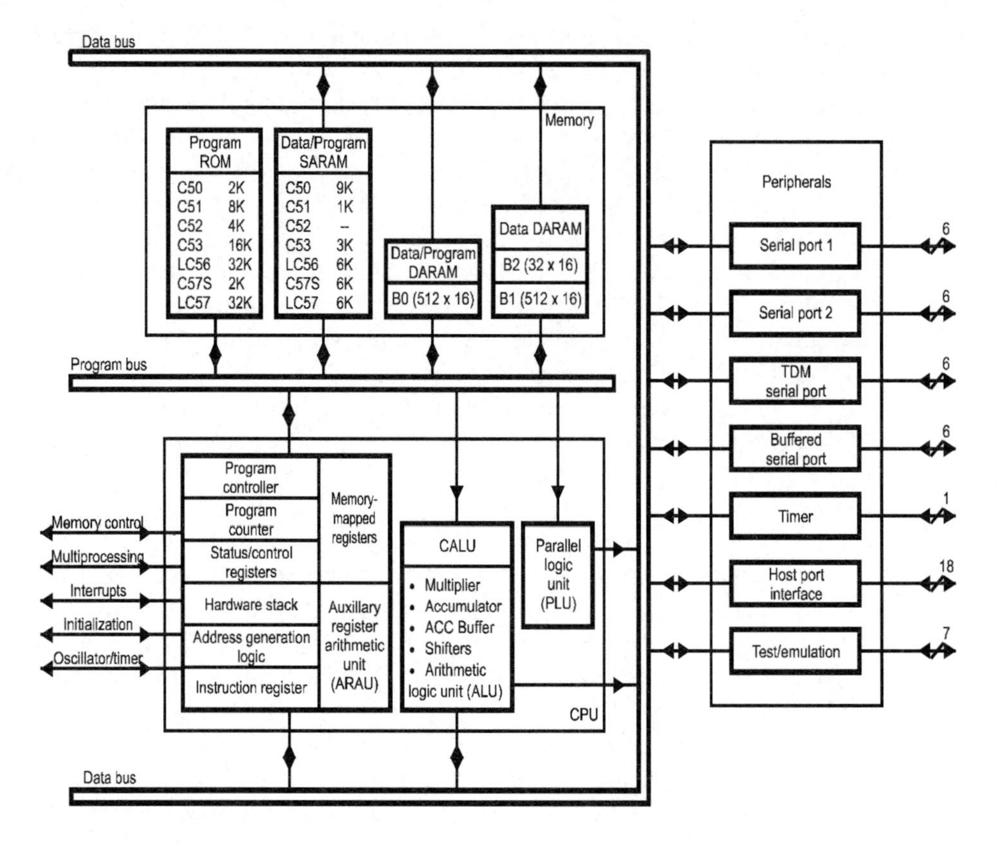

Figure 9.6
Block diagram of theTMS320C5x family

The central processor (CPU) consists of two arithmetic logic units (ALU), a parallel logic unit (PLU) and registers. The first ALU is called the central ALU (CALU). It is used for two's complement arithmetic and consists of the following:

- A 16×16 bit multiplier producing a 32-bit product
- A 32-bit accumulator
- A 32-bit accumulator buffer
- Shifters at the outputs of both the accumulator and the product register

The second ALU is called the auxiliary register arithmetic unit (ARAU). It is an unsigned 16 bit arithmetic unit that calculates indirect addresses by using inputs from the auxiliary registers, index register and the auxiliary register-compare register.

The scaling shifter is used for prescaling. It has a 16-bit input that is connected to the data bus and a 32-bit output connected to the ALU. It provides a left shift of 0 to 16 bits on the input data. Shifters are also connected to the output of the product register and the accumulator for post-scaling. They allow the CALU to perform numerical scaling, bit extraction, extended-precision arithmetic and overflow prevention.

The PLU operates independently and in parallel with the ALU. It performs boolean and bit manipulations. It can set, clear, test or toggle bits in a status register, control register or any data memory location. Its operation does not affect the contents of the accumulator or product register. There are eight memory-mapped auxiliary registers that can be used for indirect addressing.

Two circular buffers are available. The circular buffer control register controls them. The start and end addresses of the two buffers are stored in separate registers and the buffers can be enabled or disabled. They can be used with either increment or decrement type of updates.

Four internal buses allow simultaneous program and data access. They are:

- Program bus (PB)
- Program address bus (PAB)
- Data read bus (DB)
- Data read address bus (DAB)

The PAB provides addresses to program memory space for both reads and writes. The PB carries the instruction code and immediate operands from program memory to the CPU. The DB interconnects various elements of the CPU to data memory space. The program and data buses can work together to transfer data from on-chip data memory and internal or external program memory to the multiplier for single instruction cycle MAC operations.

The processor has a 4-deep pipeline for delayed branch, call and return instructions. For a given instruction sequence, the second instruction could be reading data at the same time the first instruction is writing data. It also provides a bit-reversed index-addressing mode for radix-2 FFTs.

The C5x DSPs carry a 1056-word 16-bit on-chip dual-access RAM. The memory space is divided into 3 individually selectable memory blocks:

- 512 word data or program block
- 512 word data block
- 32 word data block

The dual-access RAMs are intended for data storage only. But it can also be used to store program instructions. The two data buses (DB and DAB) allow the CPU to read and write to the dual-access RAM in the same instruction cycle.

There also on-chip single-access RAM. It can be configured as data or program memory or both. These RAMs are divided into 1 K or 2 K blocks. Each block can be accessed in parallel with the other blocks. However, only one access is allowed per cycle for a particular block. If the CPU requests multiple accesses to the same block, the RAM control logic schedules the accesses in multiple cycles.

9.2.4 Architecture of ADSP21xx

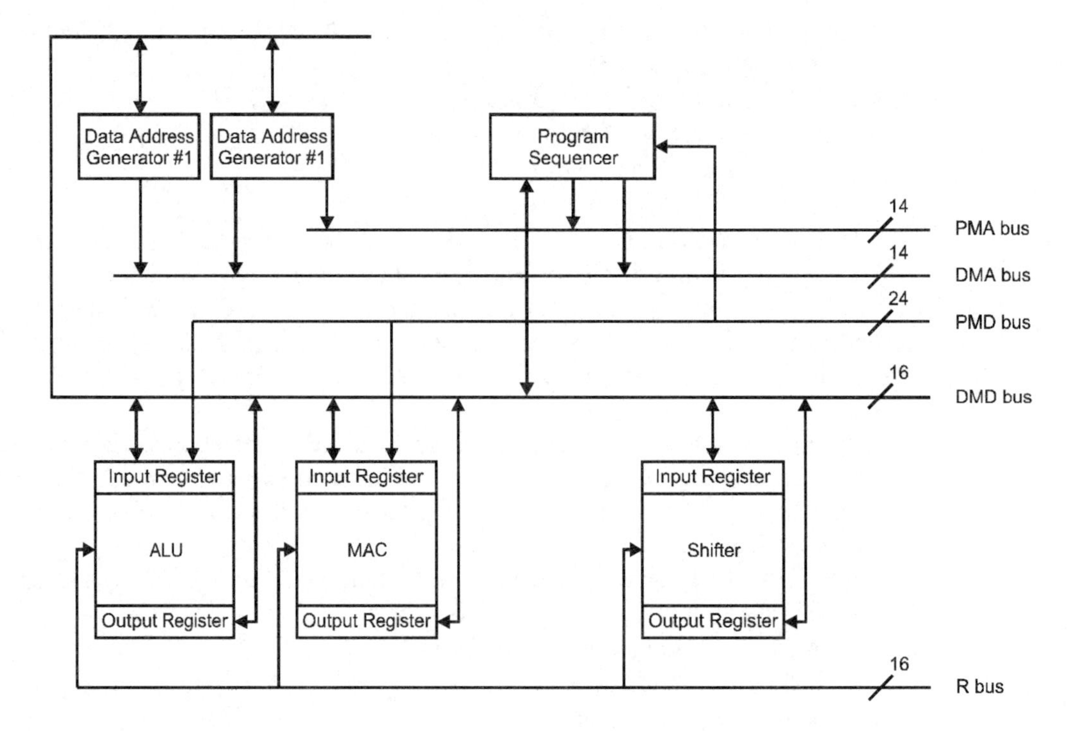

Figure 9.7
Internal architecture of the ADSP21xx family

The internal architecture of the Analog Device ADSP21xx family of DSP chips is shown in Figure 9.7. There are minor variations in the internal organization of functional blocks within the devices in the family. But generally they possess three independent units: an arithmetic logic unit (ALU), a multiplier-accumulator unit, and a barrel shifter. These units process 16-bit data. There are also two data address generators and a program sequencer.

Five internal buses are present:

- **Program data (PMD) bus**
  The PMD bus is used for transferring instructions from off-chip memory to the internal instruction register. Instructions are fetched and loaded into the register during one process cycle and they are executed during the following cycle while the next instruction is being fetched. It has the same width as the processor's instruction words, which is 24 bits.
- **Program address (PMA) bus**
  The address of the next instruction is generated by the program sequencer and is dependent on the current instruction and internal processor status. The program sequencer handles branching loop counters and zero overhead looping. This address is then placed on the PMA bus. The PMA bus is 14 bits wide. This allows direct addressing of 16 K words of program code.
- **Data memory address (DMA) bus**
  The DMA bus is also 14 bits wide, allowing direct addressing of 16 K words of data.

- **Data memory data (DMD) bus**
  The DMD bus transfers contents of a register to another register, or to an external memory location in a single cycle. There is also a PMD–DMD bus exchange unit that allows data to be passed from one bus to another.
- **Result (R) bus**
  It connects the computational units.

Each computational unit has its own set of input and output registers. These registers serve as stopover points for data between external memory and the computational units. This effectively introduces a single pipeline level on the input as well as on the output. The computational units are arranged side by side rather than in cascade. To avoid excessive pipeline delays when a series of different operations are performed, the R bus allows any of the output registers to be used directly without delay as the input to another computation.

An instruction cache is also present. It holds 16 words. Instructions loaded into the instruction register are also written into the cache memory. As additional instructions are fetched, they overwrite the current contents of the cache in a circular way. When the current instruction does a program memory data access, the cache automatically sources the instruction register if its contents are valid. The cache is most effective when executing a program loop where the instructions within that loop can be fully stored in cache memory. Operation of the cache is completely transparent to the user.

9.3 Special instructions and addressing modes

9.3.1 Circular buffers

To obtain an output sample from an order N FIR filter, $(N+1)$ MAC operations will be required. This corresponds to $N+1$ instruction cycles if each MAC operation takes one cycle to complete. The overhead of this filtering computation includes shifting the input sample from the input port to an internal register, the time required to update the registers and the time it takes to output the filtered output sample to memory. Apart from optimizing the MAC operation through hardware design, the overhead should also be minimized to achieve maximum throughput.

Consider a generic DSP architecture as shown in Figure 9.8.

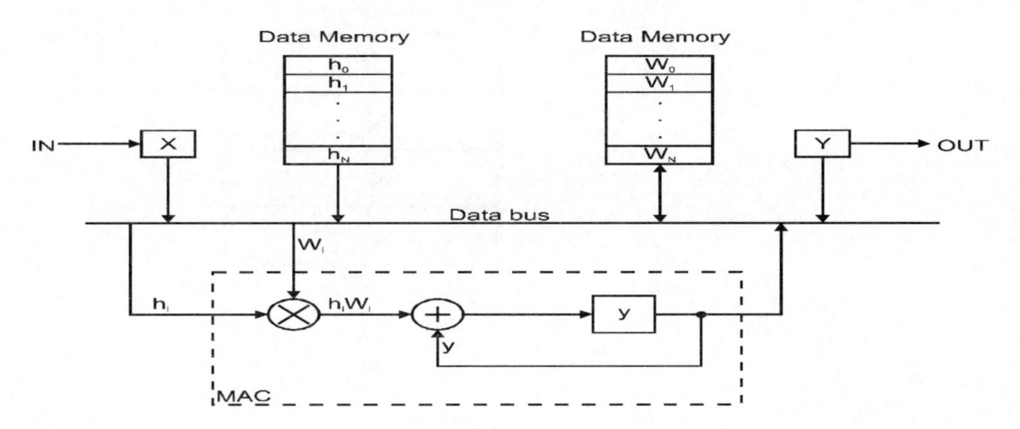

Figure 9.8
A generic DSP architecture

The sample processing algorithm for an N-th order FIR filter is
For each input sample x do:

$$w_0 = x$$
$$y = h_N w_N$$
$$for\ i = N-1,...,1,0 \text{ do:}$$
$$w_{i+1} = w_i$$
$$y = y + h_i w_i$$

Notice that the registers w_i are used to store the past input samples. These registers will need to be updated as the algorithm progresses. In the first generation of DSP chips, the operations:

$$w_{i+1} = w_i$$
$$y = y + h_i w_i$$

were carried out using two instructions, one for data shifting and the other for the MAC operation. Considering that so much effort has been put into optimizing the MAC operation, the extra instruction cycle required for updating a register seems like a waste. Therefore, in more modern DSP chips, these two operations can be carried out with one single instruction.

Another way to perform an internal update efficiently is by using circular buffers. Some DSP chips have built-in hardware to facilitate the implementation of circular buffers. Examples include the Analog Devices ADSP2101-21020, the Texas Instruments TMS320C30-C60, and the Motorola DSP56001 and DSP96002. The basic concept is that previously we moved the data from one location to another to perform the update. Now the data remain fixed at their respective memory locations but the addresses are updated instead.

Consider a third order FIR filter, with 4 input samples that need to be buffered. Instead of putting them in a straight line like we have done in figure 9.9, they are now arranged in a circle as shown in Figure 9.10.

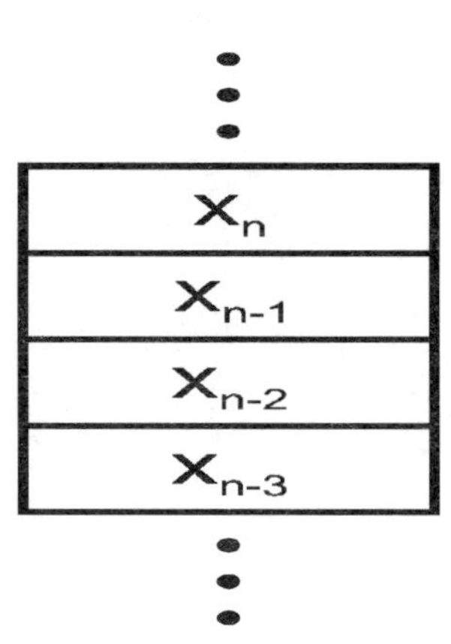

Figure 9.9
A linear buffer for data

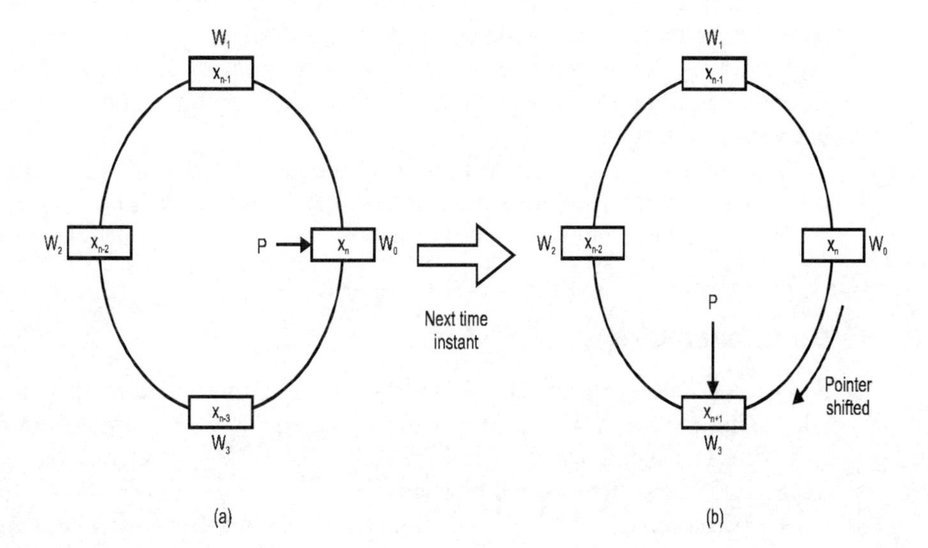

Figure 9.10
A circular buffer

In this diagram, the pointer p points to the first MAC operation. Figure 9.10(a) shows how the 4 input samples are stored at a certain time instant n. At the next time instant, a new input sample is available. Instead of storing it in w_0 as we have done previously, it is now stored in w_3 and the pointer moves clockwise by one location. This is shown in Figure 9.10(b). This means that the data in the other three registers remain where they are.

Figure 9.11 shows the pointer location and register contents for 8 successive time instants starting from $n = 0$.

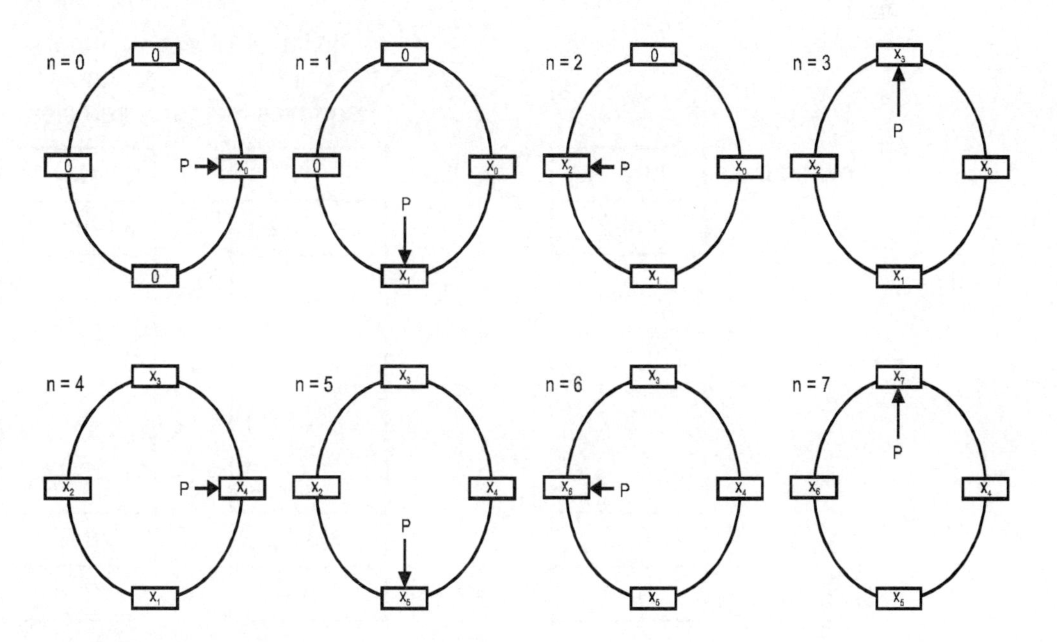

Figure 9.11
Pointers and register contents in 8 successive time instants

By using circular buffers, the amount of data movement is minimized. This is the reason why the register update and MAC operation could be performed in one single instruction cycle. So the total number of instruction cycles required for an order N FIR filtering operation remains at $N+1$. The DSP chip must have hardware to support circular or modulo addressing.

Circular buffers are very useful in implementing digital audio effects. A 100 msec reverberation with a sampling rate of 44.1 kHz corresponds to a circular buffer with 4410 samples. We have already seen in Chapter 5 how circular buffers can be used for wave table sound synthesis.

9.3.2 Code examples

Some real code examples will serve to illustrate certain special instructions and addressing modes available. We shall discuss the codes for implementing convolution or FIR filtering using circular buffers and the FFT butterfly. The Texas Instruments TMS320C5x instructions will be used.

Complete sections of codes are included here for completeness sake. The seemingly long programs should not deter the reader. There is no need to completely understand every part of the code.

9.3.2.1 FIR filtering

FIR filtering will be performed using circular addressing that implement a circular buffer. The register BK is initialized to the length of filter N. The locations for the data buffer and the filter coefficients must start from memory locations with addresses which are multiples of the smallest power of 2 that is greater than N. For example, if $N = 7$, the first address for the data buffer must be a multiple of 8 (2^3). Thus the least significant four bits of the beginning address must be zero. The data memory organization is illustrated in Figure 9.12.

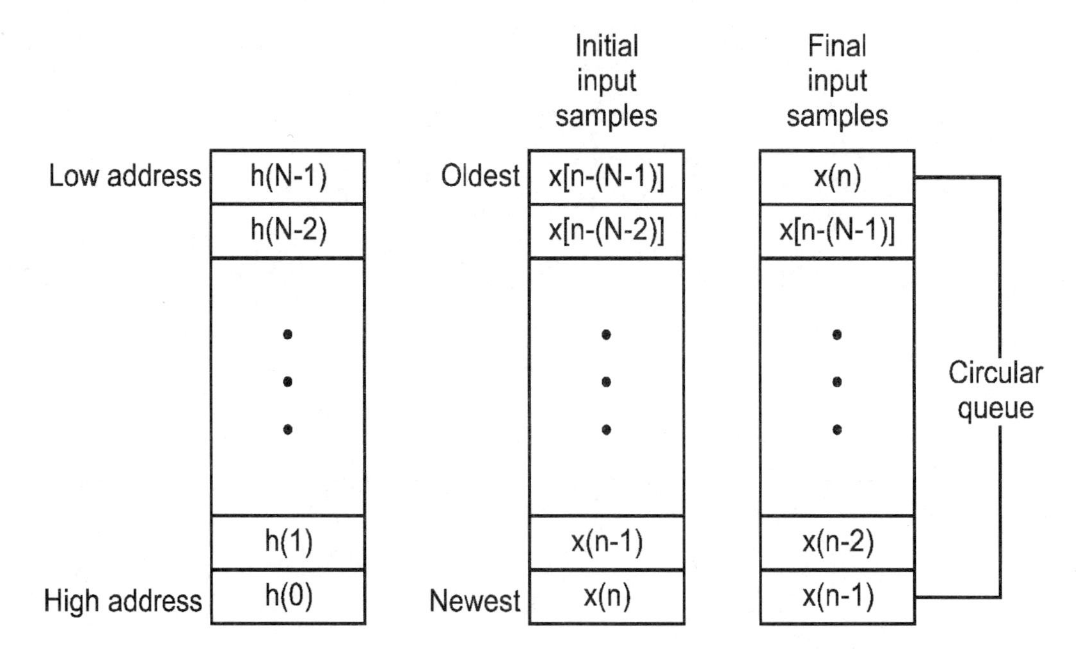

Figure 9.12
Data memory organization for the FIR filtering program

The codes for N=16 are given below.

```
; Implements a 16-th order FIR filter
        .mmregs
        .include  'main.inc'
; The filter coefficients
COEFF_START .sect     'coeff_fir'  ;filter coefficients
        .word 6Fh
        .word 0F3h
        .word 269h
        .word 50Dh
        .word 8A9h
        .word 0C99h
        .word 0FF8h
        .word 11EBh
        .word 11EBh
        .word 0FF8h
        .word 0C99h
        .word 8A9h
        .word 50Dh
        .word 269h
        .word 0F3h
        .word 6Fh
COEFF_END
FIR_DP          .usect   'fir_vars',0
d_filin         .usect   'fir_vars',1
d_filout        .usect   'fir_vars',1
fir_coeff_table .usect   'fir_coeff',20
d_data_buffer   .usect   'fir_bfr',40        ;buffer size
                         .def                fir_init  ;initialize filter
; This routine initializes circular buffers for both
; data and coefficients.
        .asg    AR0, FIR_INDEX_P
        .asg    AR4, FIR_DATA_P
        .asg    AR5, FIR_COEFF_P
        .sect   'fir_prog'
fir_init:
        STM     #fir_coeff_table,FIR_COEFF_P
        RPT     #K_FIR_BFFR-1                ;move coeffs from
        MVPD    #COEFF_FIR_START,*FIR_COEFF_P+;program to data
        STM     #K_FIR_INDEX,FIR_INDEX_P
        STM     #d_data_buffer,FIR_DATA_P    ;load cir_bfr address
                                                     ;for recent samples
        RPTZ    A,#K_FIR_BFFR
        STL     A,*FIR_DATA_P+                       ;reset the buffer
        STM     #(d_data_buffer+K_FIR_BFFR-1),FIR_DATA_P
        RETD
        STM     #fir_coeff_table,FIR_COEFF_P
;
; This subroutine performs FIR filtering using MAC instruction.
; Accumulator A (filter output) = h(n)*x(n-i) for i=0,1,...,15
        .asg    AR6,INBUF_P
        .asg    AR7,OUTBUF_P
        .asg    AR4,FIR_DATA_P
        .asg    AR5,FIR_COEFF_P
```

```
            .sect     'fir_prog'
            fir_task:
            LD        #FIR_DP,DP
            STM       #K_FRAME_SIZE-1,BRC            ;repeat 256 times
            RPTBD     fir_filter_loop-1
            STM       #K_FIR_BFFR,BK                 ;circular buffer size
            LD        *INBUF_P+,A           ;load the input value
fir_filter:
            STL       A,*FIR_DATA_P+%                ;replace oldest sample with
                                            ;new
            RPTZ      A,(K_FIR_BFFR-1)
            MAC       *FIR_DATA_P+0%,*FIR_COEFF_P+0%,A          ;filtering
            STH       A,*OUTBUF_P+          ;replace oldest buffer value
fir_filter_loop
            RET
```

Note that there is a special instruction FIRS that facilitates the implementation of exact linear phase FIR filters with symmetric impulse responses.

9.3.2.2 FFT

We shall now consider the coding of a 256-point real FFT. It is a radix-2, in-place algorithm. Memory allocation for this program is shown in Figure 9.13.

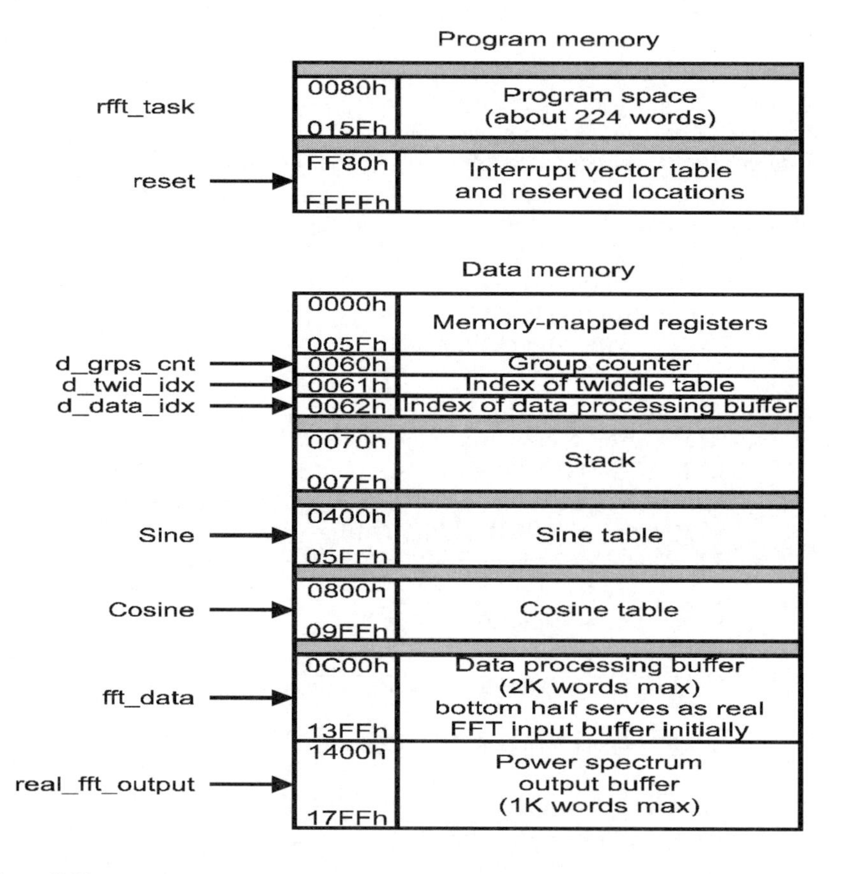

Figure 9.13
Memory allocation for the FFT program

There are four major parts:

- **Packing and bit-reversal of input**
  The input is bit reversed so that the output is in natural order. The original 2*N*-point real input sequence is copied into contiguous sections of memory (real_fft_input) and interpreted as an *N*-point complex sequence *d*(*n*). The even indexed real inputs form the real part of *d*(*n*) and the odd indexed ones form the imaginary part. This process is called packing. The complex sequence is then bit reversed and stored in the data processing buffer (fft_data).

```
;Bit Reversal Routine
        .asg    AR2,REORDERED_DATA
        .asg    AR3,ORIGINAL_INPUT
        .asg    AR7,DATA_PROC_BUF
        .sect 'rfft_prg'
bit_rev:
        SSBX    FRCT                        ;turn fractional mode on
        MVDK    d_input_addr,ORIGINAL_INPUT ;AR3->1st original input
        STM     #fft_data,DATA_PROC_BUF     ;AR7->data proc buffer
        MVMM    DATA_PROC_BUF,REORDERED_DATA;AR2->1st bit-reversed data
        STM     #K_FFT_SIZE-1,BRC
        RPTBD   bit_rev_end-1
        STM     #K_FFT_SIZE,AR0             ;AR0->half size of cir buf
        MVDD    *ORIGINAL_INPUT+,*REORDERED_DATA+
        MVDD    *ORIGINAL_INPUT-,*REORDERED_DATA+
        MAR     *ORIGINAL_INPUT+0B
bit_rev_end:
        RET
        .end
```

- ***N*-point complex FFT**
  An *N*-point complex FFT is performed in-place in the data processing buffer. The twiddle factors are stored in two separate tables, pointed to by sine and cosine. Each table contains 512 values, corresponding to angles ranging from 0 to almost 180 degrees. The indexing scheme used here permits the same twiddle tables for inputs of different sizes. Circular addressing indexes the table and the starting address of each table is required to start at an address with zeros in the eight least significant bits.

```
; There are log(N)-1 stages
        .asg    AR1,GROUP_COUNTER
        .asg    AR2,PX
        .asg    AR3,QX
        .asg    AR4,WR
        .asg    AR5,WI
        .asg    AR6,BUTTERFLY_COUNTER
        .asg    AR7,DATA_PROC_BUF           ;for stages 1 and 2
        .asg    AR7,STAGE_COUNTER           ;for remaining stages
        .sect   'rfft_prg'
fft:
;*****Stage 1*****
        STM     #K_ZERO_BK,BK               ;BK=0 so that
                                            ; ARn+0% == *ARn+0
        LD      #-1,ASM                     ;outputs div by 2 at each stage
        MVMM    DATA_PROC_BUF,PX            ;PX->PR
```

```
        LD      *PX,A                           ;A := PR
        STM     #fft_data+K_DATA_IDX_1,QX               ;QX -> QR
        STM     #K_FFT_SIZE/2-1,BRC
        RPTBD   stage1end-1
        STM     #K_DATA_IDX_1+1,AR0
        SUB     *QX,16,A,B                      ;B := PR-QR
        ADD     *QX,16,A                        ;A := PR+QR

        STH     A,ASM,*PX+                      ;PR' := (PR+QR)/2
        ST      B,*QX+                  ;QR' := (PR-QR)/2
        LD      *PX,A                           ;A := PI
        SUB     *QX,16,A,B                      ;B := PI-QI
        ADD     *QX,16,A                        ;A := PI+QI
        STH     A,ASM,*PX+0             ;PI' := (PI+QI)/2
        ST      B,*QX+0%                        ;QI' := (PI-QI)/2
        LD      *PX,A                           ;A := next PR
stage1end:

;*****Stage 2*****
        MVMM DATA_PROC_BUF,PX                   ;PX->PR
        STM     #fft_data+K_DATA_IDX_2,QX               ;QX->QR
        STM     #K_FFT_SIZE/4-1,BRC
        LD      *PX,A                           ;A := PR
        RPTBD   stage2end-1
        STM     #K_DATA_IDX_2+1,AR0
; 1st bufferfly
        SUB     *QX,16,A,B                      ;B := PR-QR
        ADD     *QX,16,A                        ;A := PR+QR
        STH     A,ASM,*PX+                      ;PR' := (PR+QR)/2
        ST      B,*QX+                  ;QR' :=(PR-QR)/2
        LD      *PX,A
        SUB     *QX,16,A,B
        ADD     *QX,16,A
        STH     A,ASM,*PX+
        STH     B,ASM,*QX+
; 2nd butterfly
        MAR     *QX+
        ADD     *PX,QX,A                        ;A := PR+QI
        SUB     *PX,*QX-,B                      ;B := PR-QI
        STH     A,ASM,*PX+                      ;PR' := (PR+QI)/2
        SUB     *PX,*QX,A                       ;A := PI-QR
        ST      B,*QX
        LD      *QX+,B
        ST      A,*PX
        ADD     *PX+0%,A
        ST      A,*QX+0%
        LD      *PX,A
stage2end:
;*****Stage 3 thru Stage logN-1*****
        STM     #K_TWID_TBL_SIZE,BK             ;BK=twiddle table size
        ST      #K_TWID_IDX_3,d_twid_idx        ;init index of table
        STM     #K_TWID_IDX_3,AR0
        STM     #cosine,WR                              ;initial WR pointer
        STM     #sine,WI
        STM     #K_LOGN-2-1,STAGE_COUNTER
```

```
        ST      #K_FFT_SIZE/8-1,d_grps_cnt
        STM     #K_FLY_COUNT_3-1,BUTTERFLY_COUNTER
        ST      #K_DATA_IDX_3,d_data_idx      ;init index for data
stage:
        STM     #fft_data,PX                ;PX->PR
        LD      d_data_idx,A
        ADD     *(PX),A
        STLM    A,QX                          ;QX->QR
        MVDK    d_grps_cnt,GROUP_COUNTER
group:
        MVMD    BUTTERFLY_COUNTER,BRC                 ;# of butterflies
        RPTBD   butterflyend-1
        LD      *WR,T
        MPY     *QX+,A
        MACR    *WI+0%,*QX-,A           ;A := QR*WR+QI*WI
                                          ;QX->QR
        ADD     *PX,16,A,B                ;B :=(QR*WR+QI*WI)+PR
        ST      B,*PX
        SUB     *PX+,B
        ST      B,*QX
        MPY     *QX+,A
        MASR    *QX,*WR+0%,A
        ADD     *PX,16,A,B
        ST      B,*QX+
        SUB     *PX,B
        LD      *WR,T
        ST      B,*PX+
        MPY     *QX+,A
butterflyend:
;Update pointers for next group
        PSHM    AR0
        MVDK    d_data_idx,AR0
        MAR     *PX+0
        MAR     *QX+0
        BANZD   group,*GROUP_COUNTER-
        POPM    AR0
        MAR     *QX-
;Update counters and indices for next stage
        LD      d_data_idx,A
        SUB     #1,A,B
        STLM    B,BUTTERFLY_COUNTER
        STL     A,1,d_data_idx
        LD      d_grps_cnt,A
        STL     A,ASM,d_grps_cnt
        LD      d_twid_idx,A
        STL     A,ASM,d_twid_idx
        BANZ    D,stage,*STAGE_COUNTER-
        MVDK    d_twid_idx,AR0
fft_end:
        RET
        .end
```

- **Separation of even and odd parts**
  Separates the FFT output into four independent sequences: RP, RM, IP and IM, which are the even real, odd real, even imaginary and odd imaginary parts, respectively.

- **Generation of final output**
  One more set of butterflies are needed to generate the $2N$-point complex output, corresponding to the DFT of the original $2N$-point real input sequence. The output resides in the data processing buffer.

Codes for the last two parts do not involve any new instructions and are not included.

9.4 General purpose microprocessors for DSP

General-purpose microprocessors are becoming increasingly powerful. Quite often, a substantial amount of spare capacity is available in personal computers or embedded systems. This spare capacity can be harnessed for use in less demanding DSP applications. At the same time, some more recent microprocessors, such as the Motorola/IBM PowerPC 6xx, the MIPS R10000, the Sun UltraSPARC and the Hewlett-Packard PA-7100 and the Intel Pentium MMX, have adopted some DSP chip features. Some are able to perform a floating-point MAC operation in one single instruction cycle in some circumstances. They also have special instructions to perform multimedia signal processing that are required in modern computers.

Intel has adopted a native signal processing (NSP) initiative that seeks to use the host processor in personal computers to perform such tasks as audio compression and decompression, sound synthesis as so on. As the host processor becomes more powerful and operates at increasingly higher clock speeds, more DSP functions can be performed.

9.5 Choosing a processor

Although many DSP chips are similar, there are many subtle differences between them. We have briefly described some of the differences between designs. It is not easy to compare different processors. Hopefully this background knowledge will help the user to choose the right device for the application at hand.

Many vendors state their processor performs using MIPS (million instructions per second). But this is a very subject value that depends on the device architecture. A MIPS rating will not necessarily reflect the performance of your algorithm running on different processors. When using benchmark programs for comparison, make sure that algorithms similar to the actual application are used. The most important matter is the performance of the entire DSP system for the particular application. The criteria of choice should include system complexity, cost and development time. Some other considerations include the following:

- The amount of internal RAM: Since the access time for internal RAM is typically shorter than that for external RAM, the performance of your programs may vary according to the amount of internal RAM.
- If high data throughput is required, then the number of DMA (direct memory access) controllers available for handling movement of data through the system without loading the CPU should be considered.
- Some DSP chips have high-speed communication ports available for data transfer to and from other devices in the system.

The relatively high cost of floating-point devices has prevented their widespread use. Most systems use fixed-point DSP chips. However, more application specific versions of these devices are becoming available. They target specific application areas such as mobile cellular radio, video and speech coding, control, etc.

The development time of the whole system depends very much on the quality of software and hardware tools available. This is the topic of the next chapter.

9.6 To probe further

It is not the purpose of this chapter to compare the performances of the DSP chips available. A very detailed performance report has been produced by Berkeley Design Technology, Inc. Their contact details can be found in Table 9.1. Other comparisons between particular processors can also be found on the world wide web.

Details of the DSP chips can be obtained from the manufacturers.

Texas instruments	http://www.ti.com
Analog devices	http://www.analog.com
Motorola	http://www.mot.com

10

Hardware and software development tools

In this final chapter we shall take a brief look at some of the development tools for both hardware and software DSP systems. Some of these tools are free and in the public domain while most others are commercial. Obviously we cannot list all of them exhaustively. But certainly some of the ones we will describe here are very popular. It should be emphasized that we are not endorsing any of the vendors or their products.

The right tool for the right job is so important that it cannot be over-emphasized. Some of the DSP chips and systems may look impressive in terms of their performance figures. But in the end the DSP designer still needs to rely on good and efficient development tools to build the system on time and within budget.

We shall first briefly review the DSP system design flow. Then the tools available from the DSP chip manufacturers, software simulation tools, and other third party development tools will be briefly discussed.

10.1 DSP system design flow

A simplified DSP system design flow is depicted in Figure 10.1.

It is simplified because we assume that the final product will be implemented with a DSP chip on purpose-designed hardware. As we have mentioned in previous chapters, there are many other approaches that can be taken. For instance, the whole system can be a piece of software running on a general-purpose microprocessor. At the other end of the spectrum, we may need to design a specific VLSI device to implement the algorithm for a time-critical application.

The design flow is composed of several main areas. They are:

- System requirement definition.
- Development of algorithm.
- Selection of DSP chip.
- DSP hardware development.
- DSP software development.

- System integration.
- System debugging and testing.

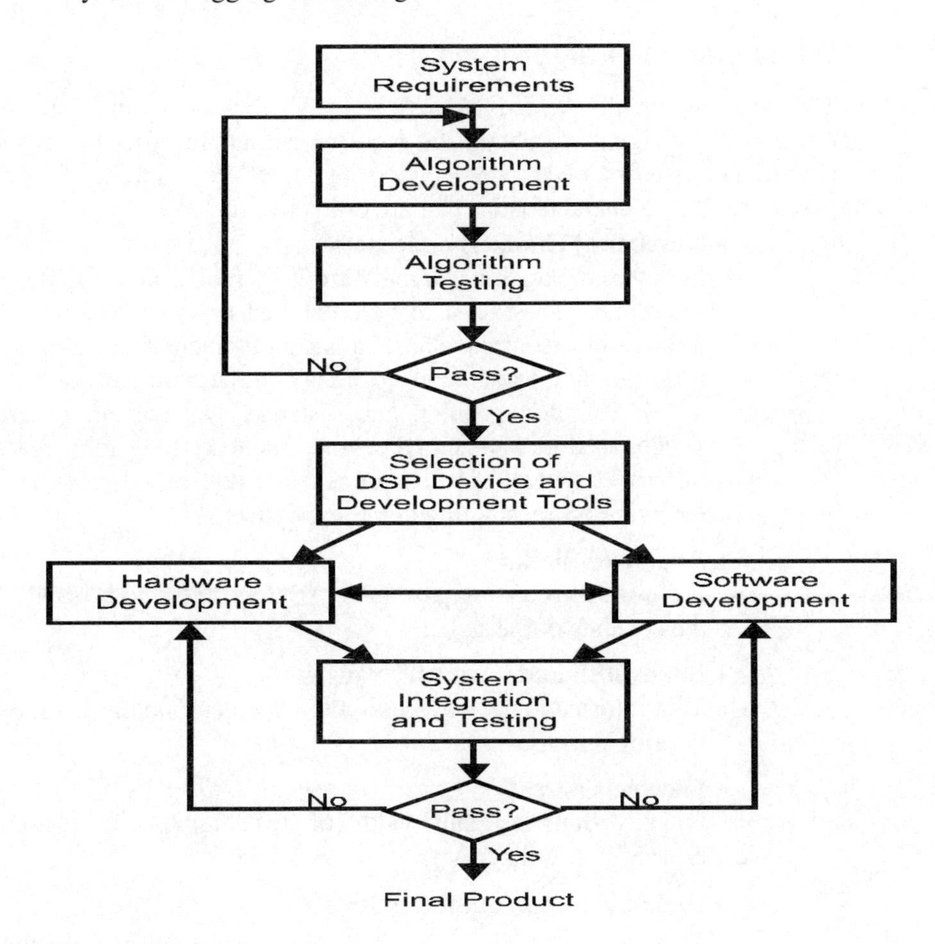

Figure 10.1
DSP system design flow

10.1.1 System requirement definition

Good engineering designs require a thorough knowledge of the problem at hand. It is essential that at the start of the project all requirements be defined. This process is often overlooked, however. The success of the final system is judged primarily by satisfying and perhaps even exceeding the basic requirements documented. The definition typically includes at least the following:

- The input and output signal or data requirements
- The interface design specifications (IDS)
- Prime item development specification (PIDS)

The IDS specifies all interface characteristics such as data rates, data lengths, control, and message protocols. It may also include the types of connectors used and electrical characteristics. The PIDS specifies all signal and non-signal processing requirements. The signal processing requirements may include the bandwidth, throughput delay, and modes of processing. The non-signal processing requirements may include the size and form factor of the final product, weight, power requirements, reliability and cooling requirements.

10.1.2 Development of algorithms

The next step in the design process is to develop or adopt a suitable algorithm for processing the signals to obtain the required output. In order to decide on the suitable algorithm(s), we first need a good understanding of the characteristics of the signals to be processed. Signal characteristics that are of interest include:

- **Number of channels or sensors**
  In some applications, signals are gathered from a number of sensors at the same time. They need to be combined in some way in order to extract the information required. These applications include sonar in submarines, which receives underwater acoustic signals to determine if other vessels are present and if so, their position and distance. Antenna arrays are also becoming widely used in commercial cellular phone systems to increase the capacity. In this case, several antennas attached to the base station of a cell are either receiving or transmitting at the same time.
- **Analog or digital**
  If the signal is analog, a suitable ADC will need to be used in order to capture the contents of the signal.
- **Bandwidth and frequency ranges**
  This information is obviously used for determining a sampling rate and for filtering purposes.
- **Spectral contents**
  Even within the bandwidth of the signal, the phase and magnitude characteristics may vary.
- **Dynamic range (number of bits)**
  This may affect our choice of fixed point or floating point implementations.
- **Steady-state or transient or both**
  The duration of the whole signal is usually too long to be processed all at one time. The signal is usually processed either sample by sample or block by block. If the signal is in steady state, the boundary of these blocks may not affect the outcome of the processing too much. However, for transient signals, the transient duration is usually relatively short and the boundaries of the block may have implications on the outcome.
- **Deterministic or random**
  Random signals are those which obey certain statistical properties such as distribution of amplitude. For digital signals with a finite alphabet, the probability of occurrence of each alphabet may be different or we may assume this probability is uniformly distributed. Most signals, especially digital signals, are modeled as random signals. Signals that are not random are known as deterministic. A signal may be entirely deterministic or entirely random or both. It also depends on which level we are modeling the signal. For instance, the DTMF signal of a certain key on the telephone keypad can be considered as deterministic because each time that key is pressed, this particular signal will be generated. During transmission, random noise will be added and the received signal is now a combination of both a deterministic

and a random signal. The transmitter knows which key on the keypad has been pressed but the receiver does not. So as far as the receiver is concerned, the signal it receives can be any of the DTMF signals that are possible. The received signal can therefore be modeled as an entirely random signal. Random signal processing is a very important and interesting area within the broad field of DSP.

- **Type of noise**
  Noise is that component of a signal that is not wanted. Noise can be additive or multiplicative, meaning that it is either added to the desired signal or multiplied with the signal. Most noise are additive. Examples include the noise added to a transmitted signal through the channel. On television, we can sometimes see some speckle noise showing up as white or black dots randomly on the screen. Multiplicative noise can result from camera shake or out of focus blur.
- **Data format, multiplexing and codes**
  For digital signals, especially those that have been encoded or multiplexed, we need to know the format in order to recover the original symbol sequence.
- **Data rate**
- **Desired information**
  Sometimes it is easy to forget what information we are trying to extract from a signal. The desired information may simply be the original signal that is as noise free as possible. We may also be extracting the digital signal symbol sequence. In speech recognition systems, the desired information is the string of words that have been spoken.

The above list is obviously not exhaustive. But once the fundamental characteristics of the signals to be processed are available, we can then search for a suitable algorithm. In most cases, algorithms are already available and they only need to be customized to the present needs. For instance, FIR and IIR filtering algorithms are widely studied as we have discussed in previous chapters. All that is needed is to design the suitable filter, which depend on the frequency characteristics of the signal. If a suitable algorithm is not readily available, one has to be developed, which may take anything from a few days to a few years.

In some cases, the performance of the algorithm is already known. If not, then its performance has to be characterized as well. The performance criteria depend on the particular application. Some performance criteria include the throughput delay, the signal to noise ratio, and the accuracy of the extracted information. Criteria can also be subjective. For instance, the quality of an image after processing often needs to be assessed by human subjects. Also, the quality or clarity of computer generated speech also need to be assessed subjectively.

10.1.3 System implementation

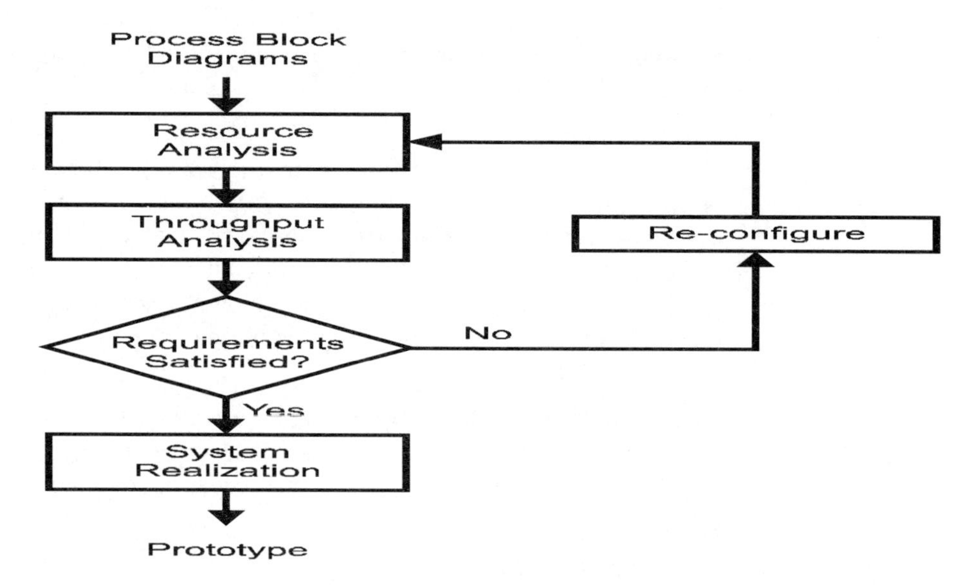

Figure 10.2
DSP system implementation procedure

The system implementation steps are illustrated in Figure 10.2. The whole process is first drawn as a block diagram, from the point where signal or data is captured by the system to where the data leaves the system. This diagram can be derived with the help of the system requirement definitions. An analysis of the resources required by each part of the whole process is then carried out.
This analysis includes:

- **Processor engine resource analysis**
  The amount of time required by the DSP processor to process a unit of data (maybe a block or a single sample). This will depend on the algorithm.
- **Memory resource analysis**
  The amount of internal and external memory required.
- **Data communication resources analysis**
  The amount of time needed to move data from external memory to internal memory and vice versa. It should also take into account the communication between the DSP chip and the control processor if one is present.
- **Control processor resource analysis**
  If a separate control processor is in the system, then the computational demand on this processor needs to be assessed. For instance, in some communication systems, the DSP chip may handle signal modulation and demodulation while a separate processor takes care of the protocol and external interface.
- **Input/output processor resource analysis**
  In some systems, there are input and/or output processors that handle the external interfaces. These processors include those that handle the parallel or serial interface.

The throughput of the system can then be analyzed. This is compared to the system requirements and if certain requirements are violated, then the system configuration will need to be adjusted.

10.1.4 System debugging and testing

Once the system has been implemented, it can be tested against specifications. Setting the appropriate set of tests according to specifications requires insight and thorough understanding of the specifications and operating conditions. Ideally debugging is carried out by the design team while testing is done by a separate team that is not involved directly with the detailed design. In this way, there is a greater likelihood that some bugs, which are overlooked by the design team, can be picked up by the testing team.

Debugging is usually carried out with the help of in-circuit emulators, logic analyzers, and software debuggers. All vendors of DSP chips offer some kind of in-circuit emulators and other software tools for development purposes.

10.2 Development tools

A search through the Internet will reveal that there are numerous resources for the development of DSP systems. Some of these are available from the chip manufacturers for their specific chips. Others are either third-party commercial or shareware. There is also some design software available in the public domain, although they are mainly for non-commercial educational uses.

10.2.1 High-level language tools

The first generation of DSP chips is programmed primarily by using their assembly languages. High-level language software tools (compilers, etc) are virtually non-existent. However, the use of high-level language such as 'C' for software development in DSP applications has become more common recently. There are several reasons:

- **Productivity**
  Writing programs in high-level languages are much easier than writing programs in assembly language. Quite often, algorithms are developed and tested in a high-level language. These programs can then be compiled to the appropriate processor's machine code directly and in most cases, without change.
- **Maintainability**
  Anyone who has developed programs in assembly languages and high-level languages will agree that high-level language codes are much easier to maintain.
- **Portability**
  It is not difficult to re-compile a high-level language program for a different target DSP chip if necessary.
- **Efficiency concerns**
  There are obviously concerns regarding execution speed. The most efficient codes are hand-coded in assembly language. But with the speed of DSP chips increasing, in most applications some inefficiencies are quite acceptable. In fact, the most time-critical parts of the program can be hand-coded in assembly language with the rest developed in a high-level language. Most

linkers will be able to link object codes generated by assemblers and compilers to produce the executable program.

In order to create high-level language compilers that are efficient for DSP chips, there are several areas of concern that need to be addressed:

- **Memory usage**
  As we have mentioned in the previous chapter, most DSP chips have on-chip and external memories. The way they are used will often determine the efficiency of the program. The memory space of a DSP system is also typically partitioned in some special way. There must be some ways in which the programmer can tell the compiler where to put certain variables or program codes. Compilers also tend to produce codes that require more memory compared with hand coding. This may be a concern for processors with a small addressing space.
- **Special instructions usage**
  DSP chips have special instructions to perform some tasks such as the MAC operation very efficiently. The compiler will need to recognize such constructs and use these special instructions instead of the more general ones. These concerns are more pronounced in fixed-point DSP systems. Firstly, fixed-point DSP chips are often used in applications that have more cost effectiveness concerns. So efficient memory usage is an even more important. Secondly, fixed-point algorithms have their special needs for scaling and rounding as discussed previously. Most high-level languages do not support fixed-point data types, thus making programming fixed-point algorithms much more difficult.

  Generally, efficiency concerns for floating-point DSP systems are much less severe.
- **Languages available**
  Among the high-level languages, C is probably the most popular. All vendors that support high-level language development on their DSP chips or cores have a C compiler for their processors. So far, only the Lucent Technologies' DSP16xx and the Zoran AR3900x do not have a C compiler available.

  There is a very high quality general purpose C compiler called GNU C widely available on the Internet. It is developed by the Free Software Foundation. Some people have made modifications so that some versions of it can generate codes for DSP chips. Most notably, there is one version for the Motorola DSP5600x family.

  C is, after all, a general purpose programming language. It lacks some important features that will simplify the coding of DSP algorithms. For instance, a fixed-point data type would be very useful for fixed-point processors. Many vendors have added their own specific extensions to the standard C language to support some of these constructs. These extensions include providing ways to assign certain variables to certain areas of memory and the inclusion of assembly language segments into the C source code.

  The ANSI Numerical C Extension Group (NCEG), which is an ANSI standards committee is working towards standardizing extensions to the C language to support numerical computations. It is anticipated that once the standard has been agreed upon, vendors will provide at least a subset of the extensions for use with their compilers.

Another high-level language that is gaining popularity is C++. The ability to create new data types and operations using C++ makes it much more flexible compared with C. The use of these user-defined data types and operators gives the compiler more room for optimization. Additionally, object-oriented programming techniques can be used which makes program more maintainable. At present, only the Texas Instrument's TMS320C3x and TMS320C4x families of DSP chips have C++ compilers available for them from a third-party vendor.

Some run-time libraries useful for DSP applications are available from vendors. They include libraries for mathematical functions, signal-processing functions, vector processing and application libraries for speech codecs, modems and image processing.

Source-level debuggers are usually bundled with the compilers. Most of them make use of the windowing systems available on most computers for visual user interface.

10.2.2 Assembly language tools

A typical assembly language development environment for DSP systems will include the following:

- Assembler
- Linker
- Instruction-set simulator
- Debugger
- Development boards
- In-circuit emulator
- Software libraries

Figure 10.3 shows the relationships between the various parts of the development environment.

- **Assemblers and linkers**

  Assemblers for DSP chips are no different from assemblers for general-purpose processors. Most of them are macro assemblers with 'standard' features such as conditional assembly.

  The common object file format (COFF) is the standard format for assembled object code files. COFF allows the annotation of object codes with debugging information if necessary. This common format also simplifies the integration of third-party libraries.

- **Simulators**

  Instruction-set simulators simulate the execution of a processor at the instruction level on a host computer. The user can see changes in the various registers, memory and flags as the program is executed. The user can single-step through a program. It is very useful for debugging and algorithm optimization. A limitation of such simulators is that it simulates the processor alone and does not support the simulation of peripherals or other I/O processors.

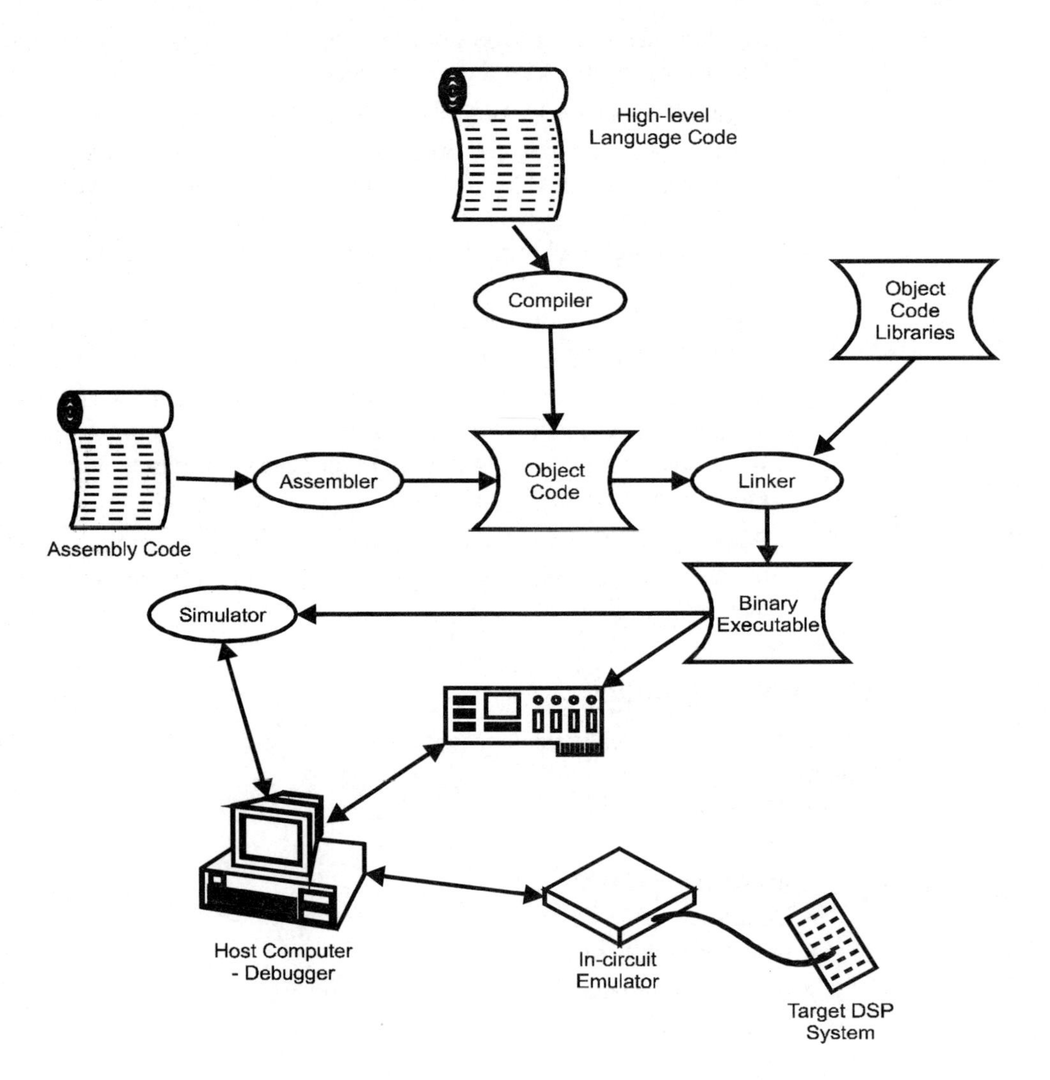

Figure 10.3
DSP system development environment

- **In-circuit emulation**

  In-circuit emulators (ICE) are hardware systems that sit between a host computer and the DSP system. Software on the host computer allows the user to monitor and control the processor in the target system as it executes programs. The user may single-step through a program and optionally modify the contents of memory locations or registers. The ICE is an important debugging tool because it allows the developer to see what's happening when the programs are executed in real-time on the target system.

  There are three basic types of ICE:

- **Pod-based emulator.**

  The DSP chip is removed from the target system. The ICE has a pod that is plugged into the socket for the DSP chip in the target system. The emulator contains a special version of the target DSP processor that is being emulated. This special processor has additional hardware for controlling it. The disadvantage is its price; they are usually quite expensive and are processor specific. Also, the emulator pod changes the electrical characteristics of the circuit and may cause electrical time problems. For this reason, pod-based emulators usually do not allow full-speed operation.

- **Scan-based emulator.**
  This type of emulation is becoming popular in recent years. This is because the new generation of DSP chips has built-in debugging logic together with a special serial port to access this logic by external devices. Some of these processors use an IEEE standard 'JTAG' (Standard 1149.1) compatible port. Debugging features of the chip can be accessed by simply connecting a host processor to this port using a special adapter. The on-chip debugging logic monitors the processor operations in real-time, halting it when a break point is reached. Debugging information can then be accessed via the serial port.

  There are 3 main advantages of scan-based emulation compared with the pod-based one. First one is that the processor does not have to be removed from the target system. So the electrical characteristics are not affected. Another advantage is that the number of signal lines coming out of the target system is small (5 for JTAG) because communication is serial. These signals can operate at a much lower speed than the processor's other signals. Third advantage is that it supports full-speed operation of the DSP chip.

- **Monitor-based emulator.**
  This is the most inexpensive type of emulation. A supervisory program, called the monitor, is run on the DSP processor. The monitor using one of the serial ports on the DSP chip handles communication between the target system and the host computer. The debugging program is run on the host computer. The advantage of this approach is that no special hardware is needed. It is the cheapest of the three types of emulation. But it is usually not possible to set real-time break points. Also, since the monitor runs on the DSP processor itself, the state of the processor is changed before it can be examined by the user. This approach is normally used with low-cost development boards or evaluation boards.

- **Development boards**
  Most DSP chip manufacturers produce low-cost development boards or evaluation boards. They are very useful for learning about the specific chip and its capabilities, experimenting with algorithm implementations, and simple system developments. There also exist a whole range of stand-alone or plug-in boards for personal computers or workstations. They are usually more powerful, typically with more memory and some of them can be plugged into the expansion slots of PCs. They are useful for program development. Cost of these systems varies from a few hundred US dollars to several thousand dollars depending on the processor and memory configurations. Some of these cards even have multiple processors on them for parallel processing.

10.2.3 Other software tools

There are a number of software available that supports DSP algorithm development and programming. Some of these incorporate very good visual programming environments and programming can be done by dragging and dropping block diagrams and interconnecting them. They provide tools for simulation; fixed-point analysis and some even have DSP code generators.

Some of them are listed below:

- **MATLAB**
  Produced by Mathworks Inc. It is very popular with academic institutions and as a result has become an 'industry standard' for simulation and signal processing algorithm development. Many new DSP textbooks use MATLAB code as examples. It also has options for fixed-point analysis and DSP code generation. There is also a visual block-diagram-programming environment called Simulink. Details can be obtained from

  http://www.mathworks.com

  It is available for both PCs and UNIX workstations. A more limited student version is available to full-time students.

- **DADisP**
  DADiSP is an interactive graphics worksheet – a visually oriented software package for the display, management, analysis and presentation of scientific and technical data. It can collect, manipulate, edit, reduce, transform, display and analyze data. Details are available at the DSP development corporation web site

  http://www.dadisp.com

  There is a free student version and a 30-day commercial trial version that can be downloaded from their web site.

- **MatrixX**
  MatrixX is marketed by Integrated Systems Inc. It is designed to be a complete system modeling and simulation environment using building blocks in a graphical design environment. Mathematical analysis and visualizing tools block model development is supported. Automatic software code generation and push-button configuration of the real-time hardware are available. Details are available at

  http://www.isi.com

- **Signal processing worksystem (SPW)**
  SPW from the Alta Group is a simulation and design tool that bridges the gap between system level simulations and the realization of the system components. SPW supports realizations with digital signal processors as well as hardware realizations. Details at

  http://www.altagroup.com

  This software is only available for UNIX workstations.

- **COSSAP**
  The COSSAP digital signal processing development system provides a unified environment throughout the design process. It makes use of stream-driven simulation (SDS), which operates on the natural flow of data through the system so designers do not have to define architecture before they develop their algorithms. Since this is essentially a self-timed circuit, defining multi-rate and asynchronous systems is no more difficult than defining a single-rate system. This approach also results in considerably lesser overhead for the simulation engine, typically allowing it to run 8–16 times faster than clock-cycle-based simulation, according to Synopsis Inc.

The COSSAP product family includes signal processing libraries and HDL code generator for behavioral and RTL design in VHDL or verilog, DSP code generator for optimal C generation and DSP developer kits for DSP co-simulation with TI and Lucent Technologies DSP chips. The synopsis web site is located at

http://www.synopsis.com

This software is only available for UNIX workstations.

- **Hypersignal**
  Hypersignal is a graphical DSP environment for developing real-time DSP applications using DSP and data acquisition boards. It also includes an automatic ANSI C source code generator for generating C source code from visual designs. Details can be found at the web site of Hyperception Inc.

  http://www.hyperception.com

- **Scilab**
  Scilab is a high-level language for numerical computations in a user-friendly environment. Scilab is developed at INRIA and is available free via anonymous ftp in source and binary formats. Scilab runs on Windows 95/NT, linux and most UNIX workstations. Scilab can be obtained by anonymous ftp from 'ftp.inria.fr' (internet 192.93.2.54), in directory '/INRIA/Scilab'.

Naturally this list is not exhaustive. Note that we are not recommending any of these products. It is simply provided for reference purposes.

10.2.4 Real-time operating system

An operating system controls accesses to system resources and manages the order of execution of programs on a processor. Real-time operating systems (RTOS) are ones that guarantees the latency (time delay) of interrupts. In other words, the time delay between the instant when an interrupt occurs and the time when a special interrupt service routine is activated to service that interrupt is bounded. An RTOS will be most useful when multiple interrupts may occur at unspecified times.

Real-time operating systems are more popular with general-purpose processors. A few are now available for DSP processors. They are usually adapted from the ones for general purpose processors. So far only one is designed for DSP chips right from the start. It is called SPOX from Spectron Microsystems. There are versions of a range of floating-and fixed-point DSP chips from Texas Instruments, Motorola and Analog Devices.

Appendix A

Binary encoding of quantization levels

Consider an n-bit binary number representing a full-scale range R. In other words, the range R is being quantized into 2^n quantization levels. If R is unipolar, the quantized value x_Q lies in the range $[0,R]$. If it is bipolar, x_Q lies in the range $[-R/2,R/2]$.

We shall denote the n-bit pattern as a vector $b = [b_{n-1}, b_{n-2}, \ldots, b_1, b_0]$ where b_{n-1} is called the most significant bit (MSB) and b_0 is the least significant bit (LSB). There are many ways in which this n-bit pattern can be used to encode x_Q. Three most common ways are:

- **Unipolar natural binary**

$$x_Q = R\left(b_{x-1}2^{-1} + b_{x-2}2^{-2} + \ldots + b_1 2^{-(x-1)} + b_0 2^{-x}\right)$$

- **Bipolar offset binary**

$$x_Q = R\left(b_{x-1}2^{-1} + b_{x-2}2^{-2} + \ldots + b_1 2^{-(x-1)} + b_0 2^{-x} - 0.5\right)$$

- **Bipolar two's complement**

$$x_Q = R\left(\overline{b}_{x-1}2^{-1} + b_{x-2}2^{-2} + \ldots + b_1 2^{-(x-1)} + b_0 2^{-x} - 0.5\right)$$

Here b_{n-1} denotes the complement of b_{n-1}.

Example A.1

For $R = 2$ V and 3-bit (8-level) quantization, the correspondence between the binary representations and the quantized value are given in the following table.

$b_2b_1b_0$	Natural binary	Offset binary	2's complement
111	1.75	0.75	–0.25
110	1.50	0.50	–0.50
101	1.25	0.25	–0.75
100	1.00	0.00	–1.00
011	0.75	–0.25	0.75
010	0.50	–0.50	0.50
001	0.25	–0.75	0.25
000	0.00	–1.00	0.00

The unipolar natural binary representation encodes levels in the range 0 to 2 V. Offset binary and 2's complement encodes –1 V to 1 V.

Appendix B

Practical sessions

The practical sessions associated are designed to enhance the readers' understanding of the lecture materials. Some sessions may go beyond what is being taught for those who are interested. Thus they not only enhance but also extend the knowledge that has been gained from the lectures.

Most of the practical sessions are software based. They make use of the widely used MATLAB software from Mathworks, Inc. Other sessions use the Texas Instruments DSP boards for experimentation. The readers are encouraged to explore as far as time allows.

Below is a summary of the practical. Detailed notes on each practical are covered later.

- **Introduction to MATLAB**
  The objective of this practical is to introduce the MATLAB software: how to use it and what it can do, etc. It should be done before the other practical that makes use of the software.
- **Introduction to SIMULINK**
  Simulink is a software that provides a drag-and-drop interface and other graphical facilities to perform visual programming. No knowledge of MATLAB is required as the interface is meant to be intuitive. The objective of this practical is to be able to use SIMULINK to simulate simple DSP operations.
- **FIR filter design**
  The objective of this practical is to deepen the understanding of the FIR filter design process. Various techniques introduced in the lectures will be used. Readers can attempt to design filters that are relevant to their area of work.
- **IIR filter design**
  Similar to FIR filter design practical, the objective of this practical is to enhance the understanding of the design of IIR filters. IIR filters can then be compared to FIR designs. Insights into their relative merits can be gained.

- **Filter realization**
  Simulink and the DSP block set will be made use of in this practical to experiment with different digital filter realization methods discussed in the lectures. The effects of finite word length and coefficient quantization can also be experimented with.
- **Image processing**
  Image processing is inherently a 2-dimensional signal processing operation. But often it is treated as 1-dimensional (1-D). In this practical some of the insight gained in 1-D DSP will be applied to image processing. The image processing toolbox integrated with MATLAB will help in this area.
- **Sampling and quantization**
  The objective of this practical is to observe the effects of sampling and quantization on the frequency spectrum of signals. A simulation model of the ERMES transmitter discussed in Chapter 5 will be built. Various sampling and quantization effects will be introduced and the result observed.
- **DSP implementation**
  The TMS320C5x DSP chip will be made use of to perform some speech processing and analysis tasks. The objective is to gain some first hand experience in compiling, assembling, downloading and debugging programs written for the chip. No actual programming will be required. Insight into the effectiveness of some development tools discussed in the lectures will be gained.

Introduction to MATLAB

Objectives

To provide a brief overview of the functionality and basic features of MATLAB.

Equipment required

A 486/Pentium PC running Windows95 with MATLAB version 5.x and the signal processing toolbox installed.

Notation

The commands that user needs to enter into the appropriate window on the computer are formatted with the typeface as follows:

```
plot(x,y)
```

Brief description of MATLAB and the signal processing toolbox

MATLAB

MATLAB is a powerful collection of tools for algorithm expression, computation and visualization. It provides much of the control and flexibility of a traditional high-level programming language. However MATLAB is extremely easy to learn and is very compact. This allows you to express algorithms in concise and readable code. MATLAB also provides an extensive set of ready-to-use functions including mathematical and matrix operations, graphics, color and sound control.

MATLAB is an ideal software tool for studying digital signal processing (DSP). Its language has many functions that are commonly needed to create and process signals. The plotting capability of MATLAB makes it possible to easily view the results of calculations and to visualize what is happening.

This is a collection of toolboxes built on the MATLAB numeric-computing environment. The toolbox supports a wide range of specific operations, from signal processing to control system, image processing and economic modeling.

Exercises on basic features

These simple exercises will take you through the basic command structures of MATLAB so that you will be able to understand the commands that we will use later on in other exercises.

Start MATLAB by clicking on the MATLAB icon on the desktop. The MATLAB command window should be opened with a prompt '>>'.

Simple math

Just like a calculator, MATLAB can do simple maths. Type

```
>> 4+2+5
```

(without the command prompt). The answer is given a name.

Q1: What is the variable name of the answer? ___________

Try to do multiplication (*) and division (/).

Vector and matrices

MATLAB stands for 'MATrix LABoratory'. All of MATLAB's calculations are performed on matrices. A scalar value is a 1×1 matrix.

Let's build a vector. At the MATLAB prompt, enter:

This command creates a vector A containing 201 elements between 0 and 100 (inclusive). The 'linspace' command makes the vector in a 'linear' fashion, e.g. 201 evenly spaced points with the beginning value of 0 and the end value of 100. The single quote after the command line is the transpose command. 'linspace' generates a row matrix. The use of the single quote transforms it into a column matrix.

Notice that the result of this command is output to the screen. Now try to place a semicolon (;) at the end of the line and execute it again. (Hint: you may use the up-arrow key to scroll back to the previous command). This time the output is not displayed.

Create another column vector B, with 201 linearly spaced points between 25 and 75 without displaying the results. Then type B at the command prompt to view your vector.

Addition of vectors is straightforward.

Q2: What is the result of entering the following command?

```
>> C=A+B
```

Q3: What is the result of entering the following command?

```
>> D=A*B
```

Remember that A and B are matrices. Attention should be paid to the dimension of the matrices when doing multiplication. Both A and B are 201 × 1 matrices. The size of all variables in the current session can be checked by entering:

```
>> whos
```

The following multiplication should work:

```
>> D=A'*B
```

Element-by-element multiplication can also be done. The command format is as follows:

```
>> E=A.*B
```

Q4: What are the dimensions of matrix E?

Here are two ways to build a matrix:

(1) `>> M=[1 2 3;4 5 6;7 8 9];`

(2) `>> N=[A B];`

Q5: What are the dimensions of M and N?

Note that the values of A and B are copied to N. So if the values of A and B are changed after N is created as above, N will still hold the old values in A and B.

Help

Probably the most useful command in MATLAB is 'help'. Enter the following:

```
>> help
```

A list of topics MATLAB has help files on is returned. Try entering the following:

```
>> help elfun
```

A list of elementary math functions in MATLAB is returned. For more help on a certain function (for example, `for`) type

```
>> help for
```

Alternatively, you may click on the '?' icon on the command window. A new window appears and you may now click on the item of interest to show the respective help information. Try this out now.

File execution

You may extend the available MATLAB commands by creating your own. These commands or functions are usually stored in what is called M-files. The syntax of these files is simply a sequence of statements, which could execute from the MATLAB prompt put into a single file, where each line ends with a semicolon.

Under the File drop-down menu in the command window, select New→M file. A new window (MATLAB editor/debugger window) will appear. Enter the following lines into that window:

```
t=linspace(0,2*pi,100);
x=sin(t);
plot(t,x);
title('Sine Function');
xlabel('radians');
ylabel('amplitude');
```

Save this to a file by selecting File→Save in the drop-down menu of the MATLAB editor/debugger window. Enter a filename of your own choice (say, test1). The file will be saved with extension .m appended.

Now go back to the MATLAB command window and enter this filename at the command prompt. The commands in this file are executed and a plot (of one period of a sine function) is created.

This type of M-files are called script M-files. Another type of M-file is called function M-files. They are different from script M-files in that they take input arguments and the output are placed in output arguments.

In the editor/debugger window, select File→New to create a new M-file. Enter the following and save it with the filename 'flipud'.

```
function y = flipud(x)
% FLIPUD Flip matrix in up/down direction
% FLIPUD(X) returns X with columns preserved and rows
% flipped in the up/down direction. For example,
%
%
% X = 1 4 becomes 3 6
% 2 5 2 5
% 3 6 1 4
%

if ndims(x)~=2, error('X must be a 2-D matrix.'); end
[m,n]=size(x);
y=x(m:-1:1,:);
```

After saving this file, go back to the command window. Create a matrix X as in the example given in the file by entering

```
>> X=[1 4;2 5;3 6];
```

Then apply the 'flipud' function.

```
>> Y=flipud(X)
```

Check that the matrix returned is as described. Note that the filename of a function M-file is always the same as the name of the function itself.

Optional exercises

These exercises should be taken if you have time and an interest in understanding more about MATLAB.

- Enter tour at the command prompt. A separate 'MATLAB tour' window will appear.
- Move the cursor to 'Intro to MATLAB' on the left-hand side of the window and click on it.
- Go through the introduction by clicking on the 'Next>>>' button when ready.
- In a similar way to (c) above, go through the following categories one by one: matrices, numeric, visualization, and language/graphics.
- You may choose to go through the examples in each of these categories passively by clicking on the appropriate button when prompted by the text that appears.

- Alternatively and this is recommended, that when you see the commands shown in the text on the left, go back to the MATLAB command window by a single click on that window. Then type in those commands as shown and see MATLAB work. Commands are shown with the prompt '>>' in the text box in the 'slideshow player' window. You will need to go back and forth between the 'slideshow player' window and the MATLAB command window.
- When you have finished all the categories and their associated examples as listed in (a), click on the 'main window' button at the bottom of the window to return to the 'MATLAB Tour' main window.
- When you have finished all the examples, click on the 'main window' button at the bottom of the window to return to the 'MATLAB tour' main window.
- Then click on the 'exit' button on the bottom right-hand corner of the 'MATLAB tour' main window to exit the tour.
- Type quit in the MATLAB command window to exit MATLAB.

Introduction to SIMULINK

Objective

To provide:

- A brief overview of the functionality and applications of SIMULINK.
- A tutorial on the use of SIMULINK to generate simulation models.
- A tutorial on the DSP Block Set.

Equipment required

A 486/Pentium PC running Windows95 with MATLAB version 5.x, SIMULINK version 2.1 and DSP block set installed.

Notation

The commands that users need to enter into the appropriate window on the computer are formatted with the typeface as follows:

```
plot(x,y)
```

Brief description of SIMULINK and the DSP blockset

SIMULINK

SIMULINK is a software package for modeling, simulating and analyzing dynamical systems. It supports linear and non-linear systems, modeled in continuous-time, discrete-time, or a combination of the two. Systems can also be multi-rate, i.e. having different parts that are sampled or updated at different rates.

SIMULINK provides a graphical user interface (GUI) for building models as block diagrams, using click-and-drag mouse operations. With this interface, you can draw the models just as you would with pen and paper. It has a set of 'standard' block library consisting of sinks, sources, linear and non-linear components and connectors. User created and defined blocks are also possible.

It runs under the MATLAB environment.

DSP blockset

The DSP blockset is a collection of block libraries for use with Simulink dynamic system simulation environment. These libraries are designed specifically for DSP applications. They include operations such as classical, multi-rate, and adaptive filtering, complex and matrix arithmetic, transcendental and statistical operations, convolution, and Fourier transforms.

Building models

We shall now attempt to build three SIMULINK models, starting with a simple one.

Sine wave integrator SIMULINK model

Procedure

We shall now attempt to build a simple model using SIMULINK. You should close all the demo windows with only the MATLAB window on. The model we shall be building will simply integrate a sine wave and display the input and output waveforms.

- To start SIMULINK, type simulink (followed by the enter key) at the MATLAB prompt. A window titled 'library: simulink' will appear.
- In the 'library: simulink' window, in the 'file' drop-down menu, choose 'new->model'. A new window will now appear with a blank screen. You might want to move this new model window to the right side of the screen so you can see its contents and the contents of the block libraries at the same time.
- In this model, you need to get the following blocks from these libraries:
  Source library: the sine wave block

  Sinks library: the scope block

  Linear library: the integrator block

  Connections library: the mux block
- Open the source library to access the sine wave block. To open a block library, double-click on the library's icon. Simulink then displays all the blocks in that library. In the source library, all the blocks are signal sources.
- Now add the sine wave block to your model by positioning your cursor over that block, then press and hold down the mouse button. Drag the block into the model window. As you move the block, you can see the outline of the block and its name move with the pointer.
- Place the block in your model window by releasing the button when it is in the position you want. In the same way, copy the other three blocks into the model window.
- The > symbol pointing out of a block is an output port. If the symbol points to a block, it is an input port. A signal travels out of an output port and into an input port of another block through a connecting line.
- The mux block has 3 input ports; we need only 2 of them in our model. To change the number of input ports, open the mux block's dialog box by double clicking on the block. Change the 'number of inputs' parameter value to 2. Then click on the 'close' button.

- Now we need to connect the blocks. Connect the sine wave block to the top input port of the mux block: position the pointer over the output port of the sine wave block, hold down the mouse button and move the cursor to the top input port of the mux block. The line is dashed while the mouse button is down. Release the mouse button. The blocks are connected.
- Connect:
  The output port of the integrator block to the other input of the mux.

  The output of the mux to the scope.
- The only remaining connection is from the sine wave block to the integrator. We shall do so by drawing a branch line from the line connecting sine wave to the mux. Follow the steps:
  Position the cursor on the line.

  Press and hold down the CTRL key on the keyboard. Press the mouse button.

  Drag the cursor to the Integrator block's input port.

  Release the mouse button and the CTRL key.
- Open the scope block to view the simulation output. Keep the scope window open.
- Set the simulation parameters by choosing the 'parameters' from the 'simulation' drop-down menu. In the dialog box that appears, set the 'stop time' to 15.0. Close the dialog box.
- Choose 'start' from the 'simulation' menu. Watch the traces of the scope block's output.
- Simulation stops when it reaches the time specified or when you choose 'stop' from the 'simulation' menu.
- You may save the model by choosing 'save' from the 'file' menu.

Questions

(a) Explain the phase shift between the integrated and sine wave form?
(b) Do you expect to see this phase shift in practice? Why?

Audio effects – reverberation

The second simulation will demonstrate the interaction between MATLAB and SIMULINK. We shall simulate audio reverberation. The simulation model is shown below.

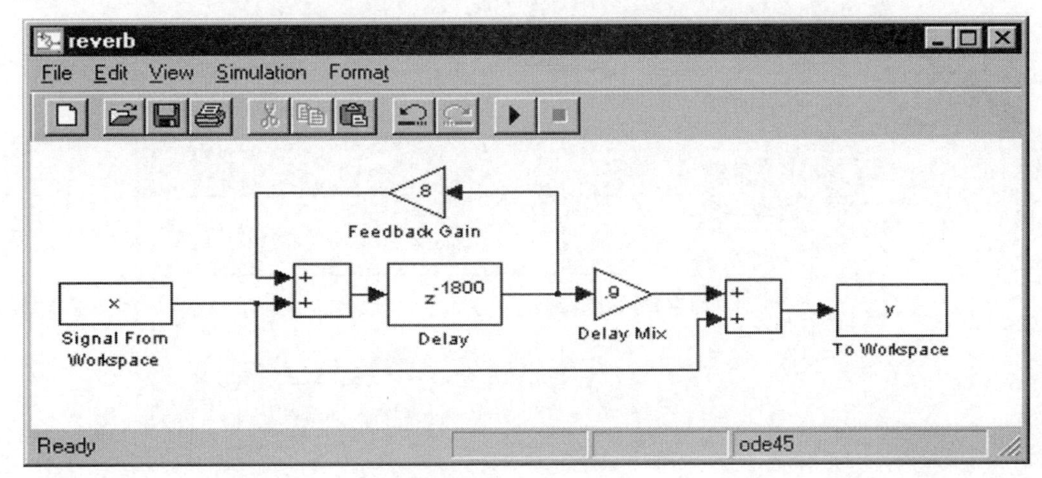

Create a new SIMULINK model as shown in the figure. Note that the blocks entitled 'feedback gain' and 'delay mix' are actually 'gain' blocks in the linear library. Blocks can be renamed simply by clicking on the titles and editing them. Change the gain values in the gain blocks to that shown in the figure.
Also set the following block parameters:
(1) Signal from workspace
 variable name = x
 sample time = 1/fs
(2) To workspace
 variable name = y
 max. no. of rows = inf
 decimation = 1
 sample time = -1
(3) Delay
 integer sample delay = 1800
 initial condition = 0

Once the model has been setup, save it to a filename of your choice.

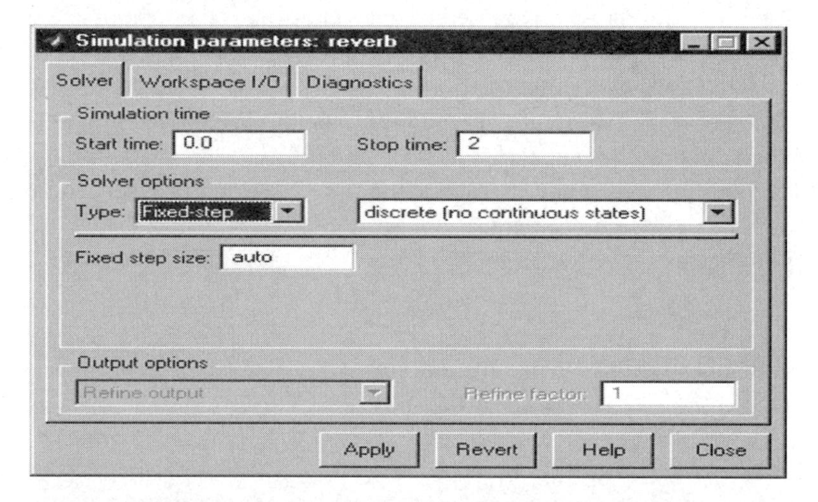

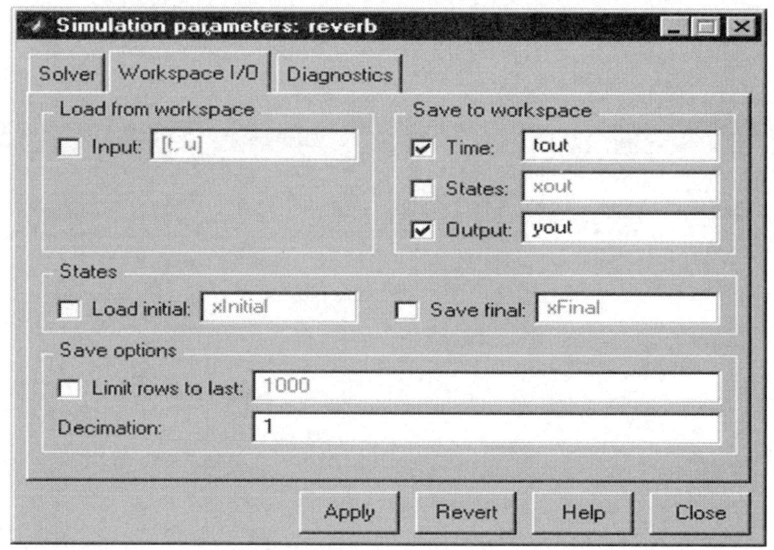

Select simulation→parameters and set up the simulation parameters the same as that shown in the above figures.

Now go back to the MATLAB command window, enter the following:

```
>> load reverbsrc
>> fs = 16000;
```

Data from a file called 'reverbsrc.mat' has been loaded into the workspace. Check that data has been loaded to a variable called x (using the `whos` command). You can also hear the sound using the command:

```
>> sound(x,fs)
```

Now run the simulation. The result can be heard by using the command in MATLAB:
>> sound(y,fs)

Adaptive noise cancellation

An adaptive noise cancellation system has been described briefly in Chapter 1. We shall now attempt to build a simulation model to study its operation. This model will need to include blocks from the DSP blockset

The simulation model is shown in the figure below.

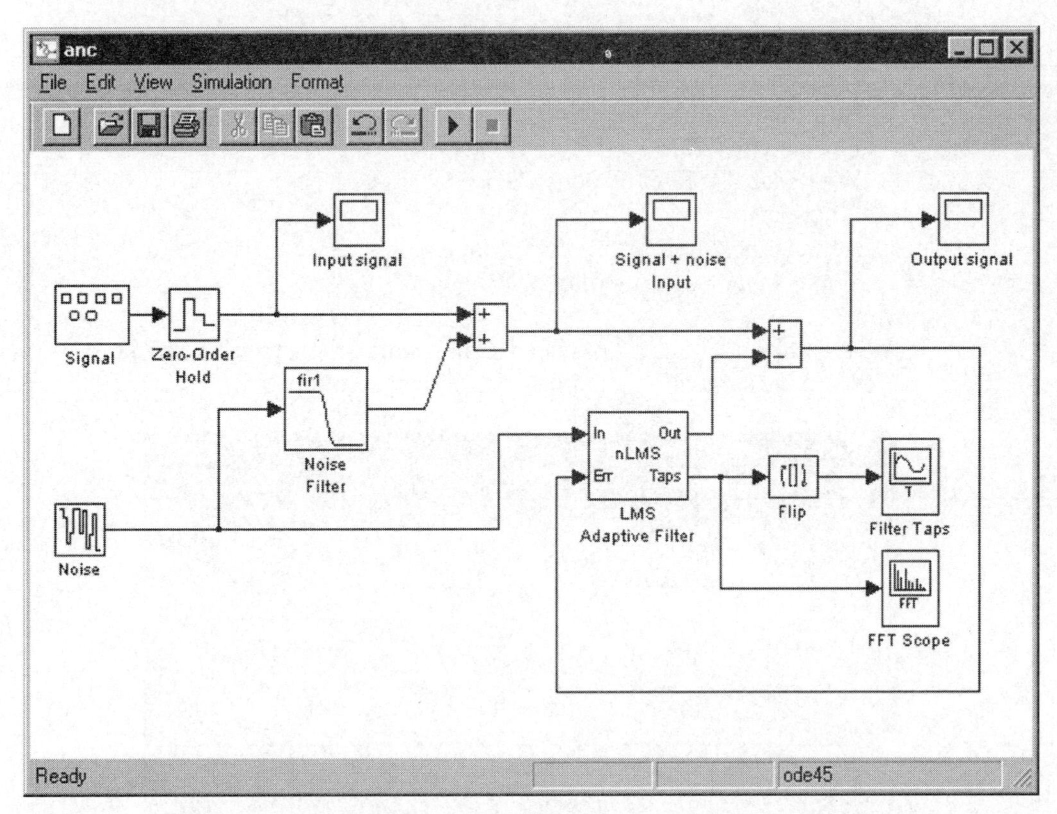

Details of the blocks in this model and their parameter settings are given below:

(1) Signal
   Actual block used: signal generator
   waveform = sine
   amplitude = 1.0
   frequency = 0.345573
   units = rad/sec

(2) Noise
Actual block used: bandlimited white noise
noise power = 1
sample time = 1
seed = [23341]

(3) Noise filter
Actual block used: digital FIR design
method = classical FIR
type = lowpass
order = 31
lower bandedge = 0.5
upper bandedge = 0.6

(4) LMS adaptive filter
FIR filter length= 32
step size, mu = 0.5
initial condition = 0.0
sample time = 1

(5) FFT scope
Frequency units = hertz
Frequency range = half
Amplitude scaling = dB
FFT length = 256
Y-axis label = filter response, dB

(6) Filter taps
Actual block used: time vector scope
Y-axis label: adaptive filter coefficients

Save the model once it has been setup using a filename of your choice. Set up the simulation parameters as shown below.

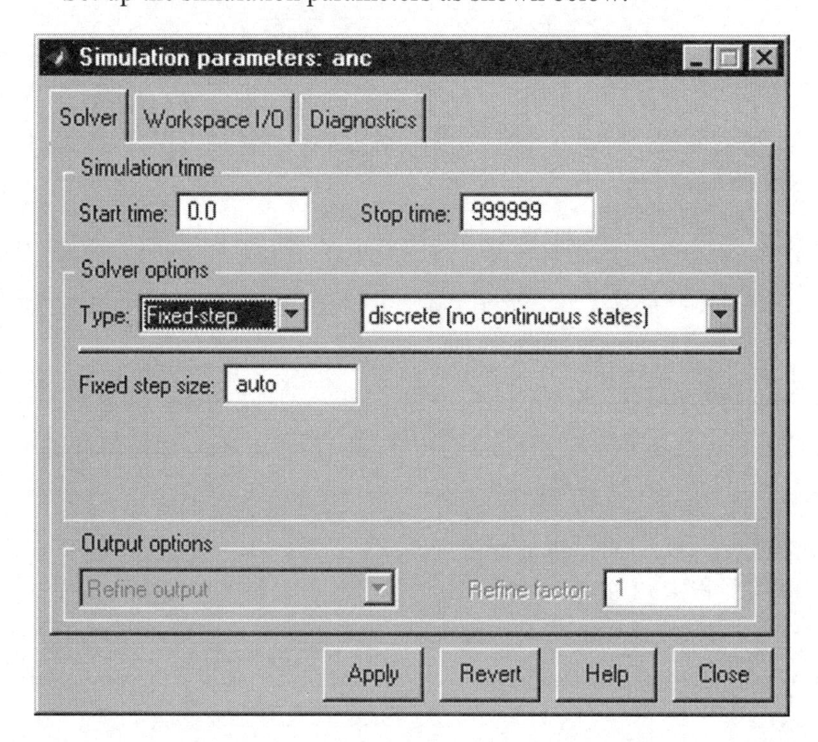

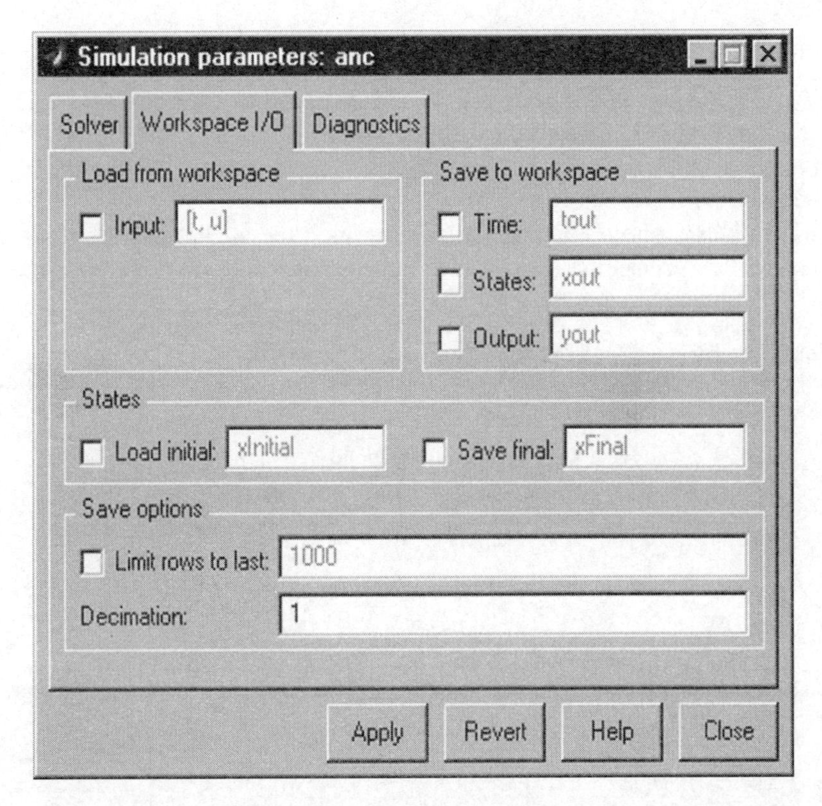

Now run the simulation and compare the input, input + noise, and output.

Questions:

(1) Does the system perform better if the LMS adaptive filter length is changed to 64?
(2) What if the LMS adaptive filter length is shortened to 24?

Discrete Fourier transform and digital filtering

Objectives

To reinforce concepts learnt in the lectures in the following areas:

- DFT and FFT
- Aliasing
- Convolution and filtering
- Overlap-add and overlap-save methods

Equipment required

A 486/Pentium PC running Windows95 with MATLAB version 5.x, SIMULINK 2.x and the signal processing toolbox installed.

Notation

The commands that users need to enter into the appropriate window on the computer are formatted with the typeface as follows:

```
plot(x,y)
```

DFT, windowing and aliasing

(a) Start MATLAB by clicking on the MATLAB icon on the desktop. The MATLAB command window should be opened with a prompt '>>'.

(b) Enter

```
>> sigdemo1
```

(c) The screen below should appear, containing the time and frequency representation of a sinusoid. You are seeing the discrete samples of a sine wave and the absolute value of its DFT, obtained using the FFT algorithm.

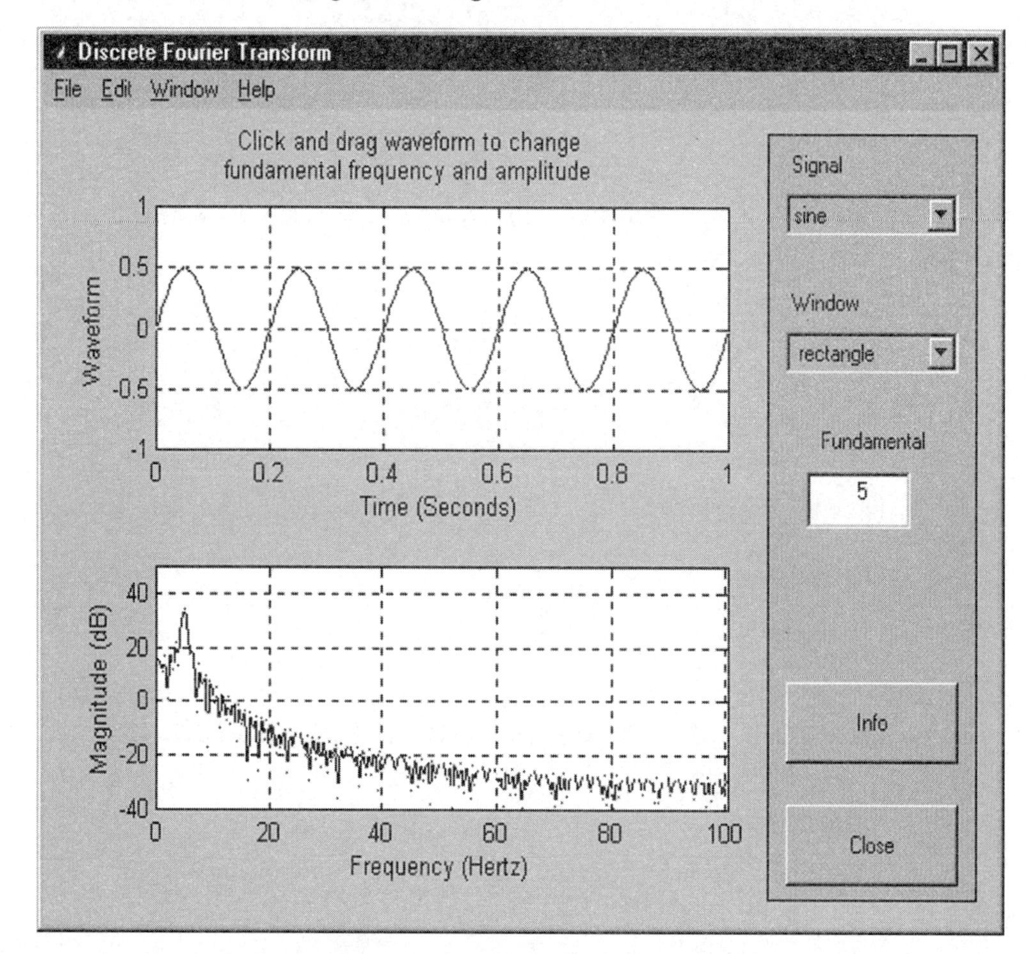

Q1: Does the peak of the frequency spectrum correspond to the frequency of the sinusoid?

Q2: Draw the theoretical spectrum of a sinusoidal signal. Is what is shown here correspond to what you expect? If not, why not?

(d) To increase the frequency of this sinusoid, click on the curve in the top window and while holding the mouse button down drag the mouse towards the left margin. Upon releasing the mouse button we observe that the fundamental frequency of the sinusoid has increased and is displayed in the window called 'Fundamental'.

Q3: How does the spectrum change when frequency of the sinusoid is increased?

Q4: What happens to the spectrum when the frequency of the sinusoid exceeds 100? Explain what happened.

Q5: What is the sampling frequency for this demonstration?

(e) The original window applied to the selected signal is a rectangular window. This means that the sinusoids are cut off abruptly at both ends of the signal. A different window may be applied by selecting a window from the 'window' drop-down menu. Select the Hamming window.

Q6: Does the peak of the frequency spectrum correspond to the frequency of the sinusoid?
Q7: How does the spectrum of the signal differ from the one obtained using the rectangular window?
Q8: See a plot of the Hamming window function in Figure 6.20 of the manual. Compare this to a rectangular window (Figure 6.13). Can you guess what contributes to the difference in the resulting spectrum?

(f) Try all the available windows and compare the resulting spectra.

Q9: Which window gives the smallest side-lobes (the artefacts at both sides of the peak)?

(g) Change the waveform by opening the drop-down menu called 'Signal' and clicking the mouse on 'square'.
Observe the corresponding time and frequency representations.

Q10: In changing the signal from sine wave to square wave, what do you notice about the harmonics (the peaks in the spectrum) ?

(h) Click on the CLOSE button to end this session.

Filtering a signal

Here's an example of filtering with the signal processing toolbox.

(a) First make a signal with three sinusoidal components (at frequencies of 5, 15, and 30 Hz).

```
Fs=255;
t=(0:255)/Fs;
s1=sin(2*pi*t*5); s2=sin(2*pi*t*15); s3=sin(2*pi*t*30);
s=s1+s2+s3;
```

(b) The sinusoids are sampled with a sampling period of 1/Fs and 256 points are included. Now plot this signal.

```
plot(t,s);
xlabel('Time (seconds)');
ylabel('Time waveform');
```

(c) To design a filter to keep the 15 Hz sinusoid and get rid of the 5 and 30 Hz sinusoids, we create a 50-th order FIR filter with a passband from 10 to 20 Hz. The filter was created with the FIR1 command.

```
b=fir1(50,[20/Fs 40/Fs]);
a=1;
```

Use the command help fir1 to see how the function fir1 is used. The filter coefficients are contained in the variables b. To see their values simply type b at the command prompt. b is also the impulse response of the filter.

(d) Display its frequency response.

```
[H,w]=freqz(b,a,512);
plot(w*Fs/(2*pi),abs(H));
xlabel('Frequency (Hz)');
ylabel('Mag. of frequency response');
grid;
```

(e) Filter the signal using the filter command. The filter coefficients and the signal vector are used as arguments.

```
sf=filter(b,a,s);
```

(f) Display the filtered signal (`sf`).

```
plot(t,sf);
xlabel('Time (seconds)');
ylabel('Time waveform');
axis([0 1 -1 1]);
```

Q11: Does it look like a single sinusoid?

(g) Finally, display the frequency contents of the signal before and after filtering.

```
S=fft(s,512);
SF=fft(sf,512);
w=(0:255)/256*(Fs/2);
plot(w,abs([S(1:256)' SF(1:256)']));
xlabel('Frequency (Hz)');
ylabel('Mag. of Fourier transform');
grid;
```

Q12: Which frequencies have been removed from the composite signal?

Linear and circular convolution

The above digital filtering operation is performed using the `filter` function.
Q13: Using the command `whos`, find the dimensions of `b`, `s` and `sf`.

Filtering is basically a linear convolution between the impulse response of the filter and the input signal. So the output of the filter can also be obtained by the linear convolution function `conv`.

Q14: From the dimensions of `b` and `s`, what is the length of the sequence resulting from the linear convolution of `s` and `b`?

Check your answer by performing the linear convolution and checking the dimension of the result:

```
sc = conv(b,s);
whos
```

Q15: Is your answer to Q14 correct?

Q16: Is `sf` the truncated version of `sc`? Check the values of both sequences.

Circular convolution of `b` and the first 51 elements of s (s1) to obtain c1:

```
B=fft(b);
s1=s(1:51);
S1=fft(s1);
C1=B.*S1;
c1=ifft(C1);
```

Note that `C1` is obtained by element-by-element multiplication of `B` and `S1`.

Q17: Compare the values of `c1` with that of `sc`. Do you notice any differences?

Now perform the circular convolution of `b` and the second 51 elements of `s(s2)`.

```
s2=s(52:102);
S2=fft(s2);
C2=B.*S2;
c2=ifft(C2);
```

Q18: Do you expect the values in `c2` to be the same as the elements 52 to 102 of `sc`? Give your reason.

Overlap-add and overlap-save methods

We cannot obtain the correct linear convolution results by simply putting the circular convolution results together. To obtain the correct linear convolution results, we need to use overlap-save or overlap-add methods as described in section 4.7.2 in the manual.

We shall divide the signals into 2 blocks of length 128 each. Enter the following:

```
s1 = s(1:128);
s2 = s(129:256);
```

Q19: What are the values of L and M (refer to section 4.7.2 of the manual) in this case? We shall start with the overlap-add method.

Q20: How many zeros will need to be appended after each block?

Create a vector of this many zeros.

```
nz= % set this to the value of your answer in Q20
z = zeros(1,nz);
```

Now append the zeros to `s1` and `s2`:

```
sz1 = [s1 z];
sz2 = [s2 z];
```

Perform the circular convolutions using DFT:

```
B=fft(b,128+nz);
SZ1=fft(sz1);
SZ2=fft(sz2);
R1=B.*SZ1;
R2=B.*SZ2;
r1=ifft(R1);
r2=ifft(R2);
```

Now align the two results and add them.

```
z2=zeros(1,128);
r1 =[r1 z2];
r2 =[z2 r2];
r = r1+r2;
```

Q21: Is the resulting vector `r` the same as that obtained by linear convolution (`sc`)?
Based on the MATLAB codes above, implement the overlap-save method. Also breaking the signal `s` into two 128-element blocks.
Write your MATLAB code here:

Q22: Are the results of overlap-add and overlap-save the same?

FIR filter design

Objective:

To provide:

- Deeper understanding of the characteristics of FIR filter.
- An understanding in the use of software tools in the design of filters.
- Verification of the examples in the lecture.

Equipment required:

A 486/Pentium PC running Windows95 with MATLAB version 5.x and signal processing toolbox version 4 installed.

Notation:

The commands that users need to enter into the appropriate window on the computer are formatted with the typeface as follows:

plot(x,y)

Exercises:

Starting sptool.

'sptool' is a graphical environment for analyzing and manipulating digital signals, filters and spectra. Through sptool, you can access 4 additional tools that provide an integrated environment for signal browsing, filter design, analysis and implementation.

- Start MATLAB by clicking on the MATLAB icon on the desktop. The MATLAB command window should be opened with a prompt '>>'.
- Enter sptool at the command prompt. A separate 'SPTool' window will appear.

- `sptool` has now started. We shall use it to design some filters and use them for filtering signals. We shall also make use of the signal browser and spectrum viewer to examine the properties of the unfiltered and filtered signals.

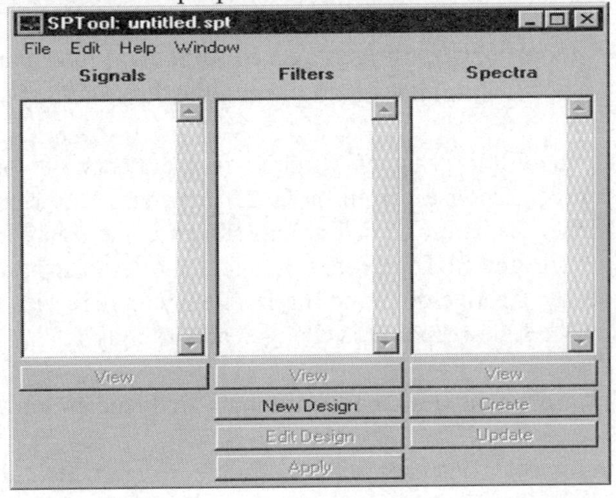

The SPTool window

Using the filter designer

Using the filter designer you can design IIR and FIR filters of various lengths and types, with standard frequency band configurations.

(a) Open the filter designer by pressing the button new design on the SPTool window. The filter designer is now activated with a separate window appearing as below.

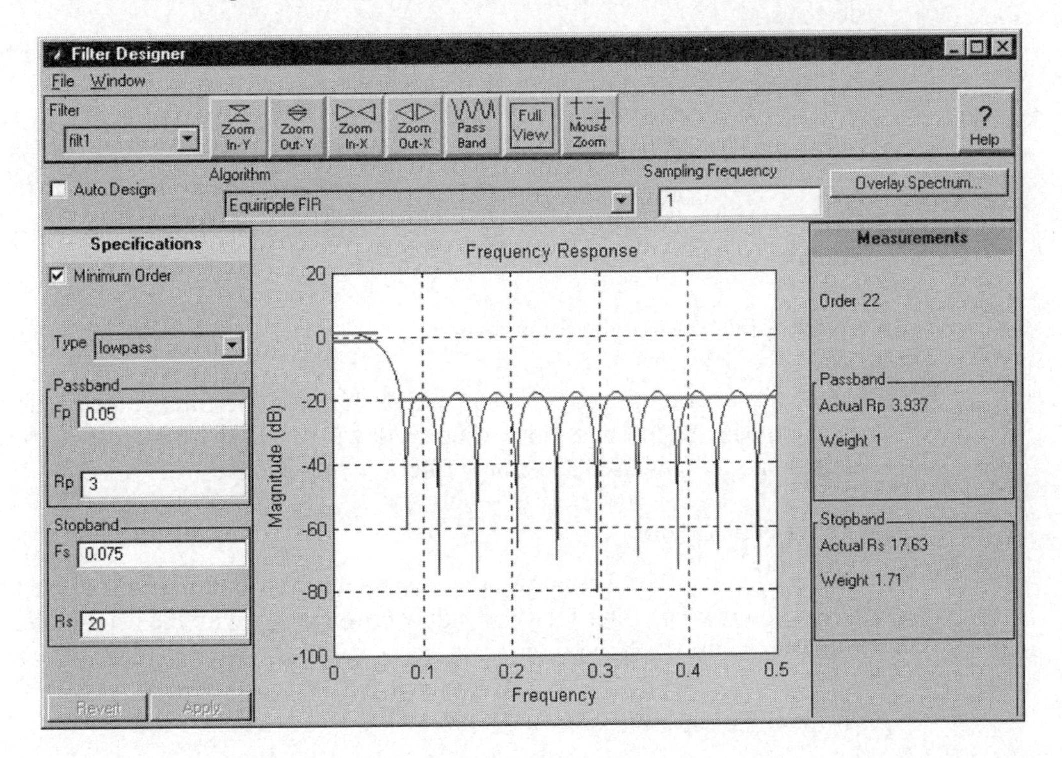

The filter designer window

(b) The filter designer window has the following components:

- A magnitude response area.
- A design panel for viewing and modifying design parameters of the current filter.
- Zoom controls for getting a closer look at filter features.
- Specification lines for adjusting the constraints.

(c) When the filter designer window first appears, it contains the specifications and magnitude response for an order 22, low-pass, equiripple FIR filter (designed using the Remez exchange algorithm), as shown in the figure above.

(d) Go back to the SPTool window, under the 'edit' drop-down menu, choose sampling frequency. Change the sampling frequency to 7418 Hz. Notice now that the frequency axis of the filter response is changed accordingly.

Q1: What are the maximum and minimum frequencies shown in the frequency axis?

(e) Go to the filter designer window, in the design panel, click on the 'down-arrow' next to the word lowpass. A list of frequency configurations is shown.

Q2: What frequency configurations are available?

(f) Select bandpass by clicking on the word.

(g) Then set fs1 to 1200, fp1 to 1500, fp2 to 2500 and fs2 to 2800. These fields define the width for the passband to stopband transition, in hertz.

(h) Set Rp (passband ripple) to 4. Set Rs (stopband attenuation) to 30. The units are in decibels.

(i) Now specify the filter design method. Click on the 'down-arrow' next to the word equiripple FIR. A list of design methods is shown.

Q3: Which of the design methods shown are for FIR filters?

(j) Click on Kaiser window. The new filter with the new specifications should now be designed and the results shown.

Q4: What is the order of the filter designed?

(k) Place the cursor on the constraints (straight lines) in the filter response diagram. Press on the mouse button and move it up or down. When the button is released, the new constraints are now used and a new filter is computed.

Using the filter viewer

(a) Go back to the SPTool window and click on the view button (right above the new design button). The filter viewer window now appears with the magnitude and phase responses of the designed filter.

Q5: Is the filter linear phase?

Q6: What happened to the stopband ripples as shown in the filter designer window?

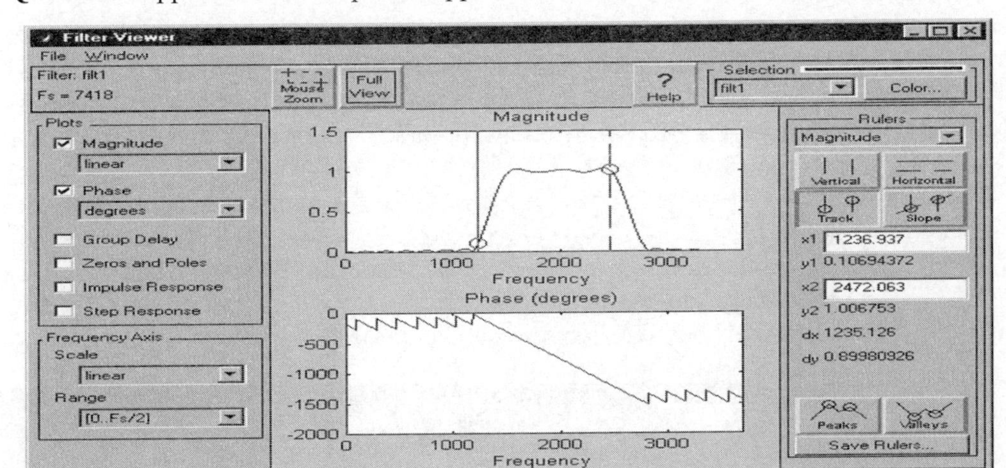

The filter viewer

(b) The filter viewer has the following components:

- A main plots display area for viewing one or more frequency domain plots of the selected filter.
- A plots panel for selecting which subplots to display.
- A frequency axis panel for specifying x-axis scaling in the main plot area.
- A filter identification panel which displays information about the current selected filter.
- Zoom controls for getting a closer look at the plots.

(c) Go to magnitude in the plot panel. Click on the 'down-arrow' next to the word linear. This determines the scaling appearing on the y-axis of the magnitude response plot. Now click on the word decibel.

Q7: How does the magnitude response plot look using decibels as magnitude units compared with the previous linear scale?

(d) It is not easy to tell whether the phase response is linear because of wrapping of the angles. The group delay response makes it clearer. Click on the 'tick box' next to group delay.

Q8: Is the filter linear phase?
Note: Group delay is defined as the derivative of the phase with respect to frequency. So a linear phase filter has constant group delay response.

(e) Click on the tick-box next to phase and group delay to remove those plots.
(f) Click on the tick-box for impulse response to see a plot of the impulse response of this filter.
(g) Then click on the tick-box for step response to see a plot of the response of this filter when unit step input is applied.

Using the signal browser

(a) We shall now get a signal from a file stored previously using MATLAB. Go back to the SPTool window. Click on file to obtain the drop-down menu. Then click on import.
(b) An 'Import to SPTool' window appears. Click on from disk. Then click on the browse button.
(c) In the 'select file to open' window, under the toolbox\signal directory double click on mtlb.
(d) Now in the file contents panel of the 'import to SPTool' window, the variable names mtlb and Fs can be seen. First click on mtlb. Then click on the right arrow leading to the data text box. The window should be as shown below:

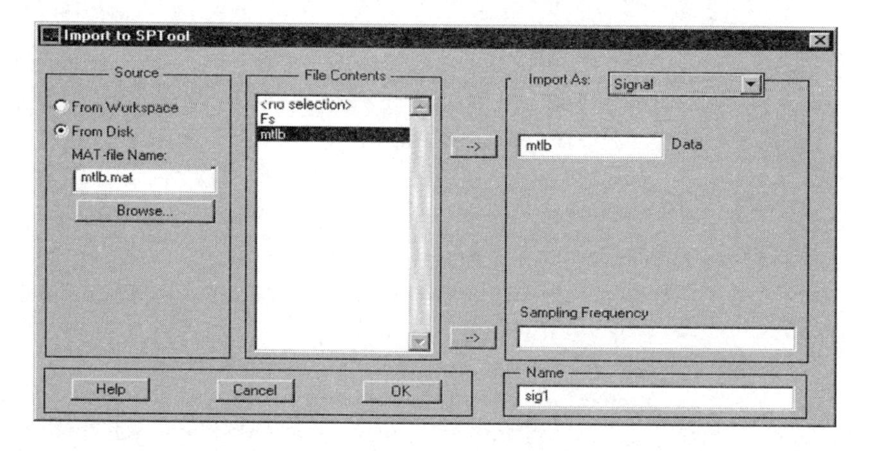

Importing signal to SPTool

(e) Now click on Fs and then click on the right arrow leading to the sampling frequency textbox. Click OK.
(f) Click on the view button under the signals textbox. The signal browser is now activated.

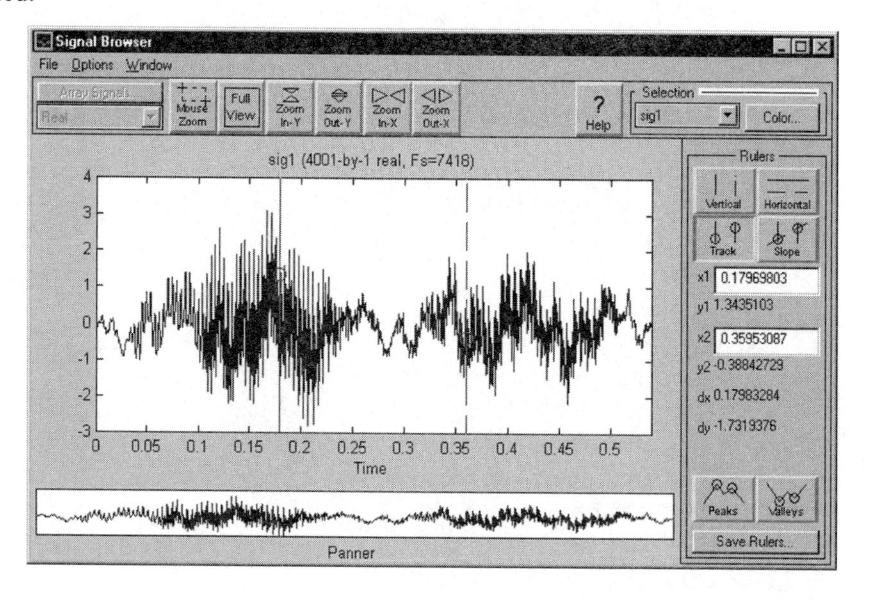

The signal browser window

(g) The signal browser window has the following components:
- A main display area for viewing signals graphically.
- A panner for seeing which part of the signal is currently displayed.
- Display management control (at the top left) with array signals and real.
- Zoom controls for getting a closer look.
- Rulers and line display controls for making signal measurements and comparisons.

(h) Move the cursor over one of the vertical lines in the signal display. The cursor now changes into the shape of a hand. While holding the mouse button down, move the vertical line back and forth. Notice that the numbers in the rulers and line display controls change values reflecting the position of the vertical line. You can do the same with the other vertical line.

(i) Click on the Zoom in X button 2 to 3 times. Notice the changes in the signal display. Also notice a box appears in the panner below indicating the position of the currently displayed portion of the signal.

(j) Clicking on Zoom out X will have the opposite effect.

Using the spectrum viewer

(a) Go back to the SPTool window. Click on the create button under the spectra textbox. The spectrum viewer is activated.

(b) In the spectrum viewer window, click on the apply button on the lower left. The following display should be obtained.

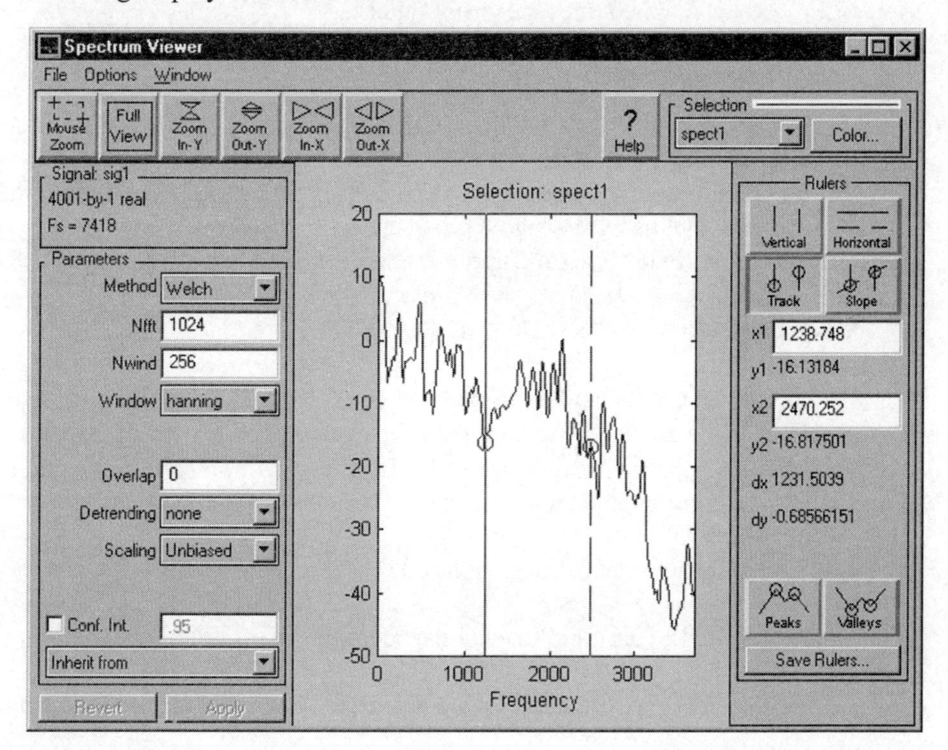

The spectrum viewer window

(c) The spectrum of the signal mtlb is displayed.

(d) The spectrum viewer window has the following components:
- A main display area for viewing spectra.

- A parameter frame for viewing and modifying the parameters or method for computing the spectrum.
- Zoom controls.
- Rulers and line display controls for making spectral measurements and comparisons.
- Spectrum management buttons: inherit from, revert and apply.
- A signal identification panel.

(e) The ruler and line display controls are similar to that in the Signal Browser. The 2 vertical lines for measurement can be dragged back and forth.

Q9: What is the frequency of the spectral peak that is closest to 2 kHz? (Use the ruler to make measurements).

(f) In the parameter frame, pull down the menu for window (click on the 'down-arrow').

Q10: Which windows are available in this menu?

(g) Choose Hamming window.
(h) In the overlap textbox, enter 100. There is now 100 samples overlap between successive windows of the signal for FFT. Click on apply.

Q11: Are there any difference between the current spectrum and the one displayed earlier? Which ones looks smoother? ____________________

Applying the filter to the signal

(a) Go back to the SPTool window. Click on the apply button under filters. In the 'apply filter' window, click OK.
(b) The signal sig1 has been filtered filt1 to produce sig2. Highlight sig2 and view the filtered signal using the signal browser.
(c) Create a new spectrum and apply it to sig2, which has already been selected.

Q12: Is the spectrum what you would expect?

Design the low-pass and high-pass filters for the loudspeaker crossover network as specified in the manual. The crossover frequency is 1 kHz with passband ripple of 0.1 dB and stopband attenuation of at least 60 dB. The transition band starts and ends at ±200 Hz from the crossover frequency. Use Kaiser window design.

Q13: What is the order of filter required?

Apply this filter to sig1 and display the spectrum of the filtered signal.

Design the crossover filters using the Remez exchange algorithm (equiripple design).

Q14: What is the order of filter required?
Apply this filter to sig1 and display the spectrum of the filtered signal.

Q15: Is the spectrum significantly different from the one obtained in the previous delivery. If so, in what way?

IIR filter design

Objective:

To provide:

- Deeper understanding of the characteristics of IIR filters.
- An understanding in the use of software tools in the design of filters.
- Verification of the examples in the lecture.

Equipment required:

A 486/Pentium PC running Windows95 with MATLAB version 5.x and signal processing toolbox version 4 installed.

Prerequisite:

You should have completed the discrete Fourier transform and digital filtering service before attempting this one. We assume that you are already familiar with `sptool`.

Exercises:

(1) IIR filter design using the filter designer.

- The filter designer lets you design digital filters based on classical functions including Butterworth, Chebyshev (Chebyshev I), inverse Chebyshev (Chebyshev II), and elliptic filters.
- Start `sptool` and create a new design. (Refer to FIR filter design exercise if you are not sure how to do this.)
- Select a Chebyshev type 1 IIR filter and high-pass as configuration.
- Set the sampling frequency to 2000 Hz using sampling frequency ... from the edit menu in SPTool window.
- In the filter designer, set fs (stopband edge frequency) to 700. Set fp (passband edge frequency) to 800.
- Set Rp (passband ripple) to 2.5. Set Rs (stopband attenuation) to 35. The unit is decibel.
- Uncheck the minimum order tick-box and enter 7 in the textbox for an order 7 filter. Click on the apply button.
- The new filter should now be computed.

Q1: Does this filter satisfy the specifications?

- Enter 6 in the order textbox.

Q2: Does this filter satisfy the specifications?

- Try an even lower order filter.

Q3: What is the lowest order that still satisfies the specification?

- Click view under filters in the SPTool window to activate the filter viewer. Look at the phase response. Click on the tick box for group delay to give you a better picture.

Q4: Is the phase response linear? (Or, equivalently, is the group delay constant?)

Q5: If the answer to Q4 is no, then in which frequency region does the group delay change the most?

- Now select a Butterworth IIR filter. Let the specifications remain the same as before.

Q6: What is the lowest order Butterworth filter that satisfies the specification?

- Go to the filter viewer.

Q7: Is the phase response linear? (Or, equivalently, is the group delay constant?)

Q8: If the answer to Q4 is no, then in which frequency region does the group delay change the most? ____________________________

- Select a Chebyshev type 2 IIR filter.

Q9: What is the lowest order Chebyshev II filter that satisfies the specification?

- Go to the filter viewer.

Q10: Is the phase response linear? (Or, equivalently, is the group delay constant?)

Q11: If the answer to Q4 is no, then in which frequency region does the group delay change the most?

- Now select an elliptic IIR filter.

Q12: What is the lowest order elliptic filter that satisfies the specification?

- To the filter viewer.

Q13: Is the phase response linear? (Or, equivalently, is the group delay constant?)

Q14: If the answer to Q4 is no, then in which frequency region does the group delay change the most?

Verify the example designs in the manual.

(a) Design the Butterworth IIR filter as specified in the example of section 7.2.1 of the manual.

Q15: Is the filter response the same as in Figure 7.4?

(b) Design the Chebyshev I filter as specified in the example of section 7.2.2.

Q16: Is the filter response the same as in Figure 7.7?

(c) Design the inverse Chebyshev (Chebyshev II) filter as specified in the example of section 7.2.3.

Q17: Is the filter response the same as in Figure 7.9?

(d) Design the elliptic filter as specified in the example of section 7.2.4.

Q18: Is the filter response the same as in Figure 7.9?

IIR filtering

(a) Import the signal mtlb using the procedures in (4) of discrete Fourier transform and digital filtering exercise.
(b) Go back to the SPTool window, under the edit menu, choose sampling frequency. Change the sampling frequency to 7418 Hz.
(c) Design a band-pass filter using an elliptic response.
(d) Set fs1 to 1200, fp1 to 1500, fp2 to 2500 and fs2 to 2800.
(e) Set Rp (passband ripple) to 4. Set Rs (stopband attenuation) to 30.
(f) Click on Auto in the parameter panel of the filter designer to let the program select the appropriate filter order automatically.

Q19: What filter order is needed?

Q20: Compare with the FIR filters designed using the same specifications, which one has the lower order?

(g) Filter the signal sig1 by clicking on apply under filters in the SPTool window. Click OK to generate sig2.
(h) View sig2 using the signal browser.
(i) Click on create under spectra in the SPTool window to view the spectrum of sig2.

Q19: Is the filtered spectrum what you expected?

(j) Design an FIR filter using Kaiser window with the same specifications. Filter the signal sig1 using this FIR filter. Then view the spectrum of the output signal sig3.

Q20: How do the spectra of sig2 and sig3 compare?

Filter realization and wordlength effects

Objective:

To provide a deeper understanding and illustrations of:

- Wordlength effects
- Cascade realization of IIR filters

Equipment required:

A 486/Pentium PC running Windows95 with MATLAB version 5.x and signal processing toolbox version 4 installed.

Notation:

The commands that users need to enter into the appropriate window on the computer are formatted with the typeface as follows:

plot(x,y)

Exercises:

ADC quantization effects

Note that MATLAB computes everything in floating point. To simulate quantization effects we shall create two functions `fpquant` and `coefround,` which are not standard MATLAB functions.

(a) First start MATLAB. Under the command window, select `file→new→m-files`. Then enter the following into the editor/debugger window.

```
function X = fpquant(s,bit)
%FPQUANT simulated fixed-point arithmetic
%-------
% Usage: X = fpquant( S, BIT )
%
% returns the input signal S reduced to a
%     word-length of BIT bits and limited to the range
% [-1,1). Wordlength reduction is performed by
% (1) rounding to nearest level and
% (2) saturates when input magnitude exceeds 1.

if nargin ~= 2;
 error('usage: fpquant( S, BIT ).');
end;
if bit <= 0 | abs(rem(bit,1)) > eps;
 error('wordlength must be positive integer.');
end;

Plus1 = 2^(bit-1);
```

```
X = s * Plus1;
X = round(X);
X = min(Plus1 - 1,X);
X = max(-Plus1,X);
X = X / Plus1;
```

(b) Save it with the filename fpquant.m (which stands for fixed-point quantizer). Now enter the coefround function and save this one in coefround.m.

```
function [aq,nfa]=coefround(a,w)

% COEFROUND quantizes a given vector a of filter
% coefficients by rounding to a desired
% wordlength w.

f=log(max(abs(a)))/log(2); % Normalization of a by
n=2^ceil(f); % n, a power of 2, so that
an=a/n; % 1>an>=-1
aq=fpquant(an,w); % quantize
nfa=n; % Normalization factor
```

(c) Generate a linearly increasing sequence v and obtain its quantized values using a 3-bit quantizer:

```
v=-1.1:1e-3:1.1;
vq=fpquant(v,3);
dq=vq-v;
figure(1), plot(v,vq)
figure(2), plot(v,dq)
```

(d) The mean, variance and probability density of the quantization error can be obtained by

```
mean(dq)
std(dq)
[hi,x]=hist(e,20);
plot(x,hi/sum(hi))
```

Q1: What is the range of distribution of errors?

Q2: Is the distribution even?

The power spectrum of the error can be displayed with
spectrum(dq)

Q3: Does this power spectrum differ from the theoretical? If so, how do they differ?

Filter coefficient wordlength effects

(a) Design a linear phase FIR low-pass filter and display its frequency response:

```
f=[0 0.4 0.6 1];
m=[1 1 0 0];
h1=remez(30,f,m);
[H,w]=freqz(h1,1,256)
plot(w,20*log10(abs(H)))
xlabel('Normalized Frequency');
ylabel('Magnitude squared (dB)');
```

Q4: What are the specifications (order, passband, stopband, ripples, etc) of the filter being designed?

(b) Quantize the coefficients to 10 bits:

```
wlen=10;
h1q=coefround(h1,wlen);
```

(c) Now display the frequency response of the filter:

```
[Hq,wq]=freqz(h1q,1,256)
hold on, plot(wq,20*log10(abs(H)))
```

Q5: How does the response of the coefficient-quantized filter differ from the original?

Q6: Does coefficient quantization destroy the linear phase property of the filter?

(d) Change the number of bits (wlen) to 8 and repeat the above.

Q7: How do the stopband ripples of these three versions of the filter differ?

(e) Next, implement an IIR elliptic low-pass filter with passband edge at 0.4 Hz (normalized) and a stopband attenuation of 40 dB.

```
[b,a]=ellip(7,0.1,40,0.4);
[H,w]=freqz(b,a,512);
figure(2), plot(w,20*log10(abs(H)))
```

(f) Quantize the coefficients to 10 bits:

```
w0=10;
[bq,nb]=coefround(b,w0);
[aq,na]=coefround(a,w0);
[Hq,w]=nb/na*freqz(bq,aq,512);
hold on, plot(w,20*log10(abs(Hq)))
```

Compare the frequency response with that of the unquantized filter. Try using other wordlengths (such as 12, 13, 14 bits).

Q8: What is the minimum number of bits required so that the minimum stopband attenuation remains below 40 dB?

Cascade implementation of IIR filters

(a) Implement the IIR filter designed in the previous section using the cascade structure. First find the poles and zeros:

```
p=roots(a)
z=roots(b)
```

(b) `p` and `z` comes either in complex conjugate pairs or real. In this case there are three complex conjugate pairs and a real root for each polynomial. Choose the pair with the largest magnitude from `p` and the ones nearest to zero for `z`.

```
p1=[0.2841+0.9307i 0.2841-0.9307i];
z1=[-0.3082+0.9513i -0.3082-0.9513i];
b1=real(poly(z1));
a1=real(poly(p1));
[H1,w]=freqz(b1,a1,512);
figure(3), plot(w,20*log10(abs(H1)))
```

(c) Repeat the above for the remaining two pairs of `p` and `z`. Assign them to variables `p2, z2, b2, a2, H2` and `p3, z3, b3, a3, H3` respectively.

(d) The cascade frequency response can be obtained by

```
Hc=H1.*H2;
Hc=Hc.*H3;
sf=H(1)/Hc(1);
Hc=Hc*sf;
hold on, plot(w,20*log10(abs(Hc)))
```

(e) Quantize `b1, b2, b3` and `a1, a2, a3` to wordlengths of 10 bits. Compare the resulting cascaded frequency response with quantized coefficients to that obtained in the previous section.

Q9: How sensitive is the frequency response to coefficient quantization when the filter is implemented as a cascade of second order structures? Is the frequency response better or worse than the one obtained with coefficients quantized to the same number of bits?

DSP system development

Objective:

- To introduce some features of the TMS320C54x family of DSP chips.
- To show how a simple DSP board can aid in the development of a DSP system.
- To go through the process of assembling, loading and debugging a DSP assembly language program.

Equipment required:

- TMS320C5x DSP starter kit (DSK), with cables and power supply.
- PC with TMS320C5x development software installed.
- Disk of examples supplied by IDC.
- Microphone, speakers, signal generator, oscilloscope and spectrum analyzer (if available).

Hardware setup

Check that the DSK board has been connected as shown in the documentation provided in the Appendix. The DB25 printer cable is connected to the PC's parallel port on one end and to the DSK board on the other. The power supply is connected to the power supply connector.

Software setup

The appropriate software should already be loaded. Click on the 'C54x code explorer' icon to start the debugger. Note that the debugger will only start if the DSK board has been powered up and connected properly.

The directory **C:\DSKPLUS** is where the software required for this practical resides. Load the example files needed for this practical to this directory. It should come as a separate disk supplied by IDC.

Exercises:

Familiarization with the development board

Some relevant chapters from the TMS320C54x DSKplus user's guide and the TLC320AC01C analog interface circuit data manual have been extracted in the Appendix. Please refer to them if necessary. Your instructor should also have original copies of these manuals available.

(a) Take a close look at the DSKplus development board. Identify the 3 main devices on this board: the DSP chip, the programmable analog interface, and the PAL for host port interface.

Q1: Which one of the TMS320C54x family of chips does this development board use?

The table below shows the internal program and data memory sizes for various chips in this family of DSP chips.

Memory Type	**'541**	**'542**	**'543**	**'545**
ROM:	28K	2K	2K	48K
Program	20K	2K	2K	32K
Program/Data	8K	0	0	16K
DARAM	5K	10K	10K	6K

Q2: What are the configurations for the chip on the DSKplus development board?

Note: DARAM (dual access RAM) can be configured as data memory or data/program memory.

(b) The AC01 analog interface circuit provides a single channel of voice quality data acquisition and playback. Quantization resolution is 14 bits. The default sampling frequency is 15.4 kHz. The sampling frequency can be changed by programming the A and B registers of the AIC.
The master clock frequency is now 10 MHz.

$$f_5 = \frac{f_{\text{MCLK}}}{2 \times (\text{register A value}) \times (\text{register B value})}$$

Q3: What are some of the combinations of values for registers A and B that will produce a 15.4 kHz sampling rate? What about a 10 kHz sampling rate?

(c) The on-board 10 MHz oscillator provides a clock to the board. However, the C542 creates a 40 MHz internal clock.

Assembly language program structure

(a) The assembler that comes with the DSKplus is called an algebraic assembler. It enables users to program in assembly language without having extensive knowledge of the mnemonic instruction set.

(b) Start up the PC and open an MS-DOS window. Go to the directory **C:\DSKPLUS** by entering

CD C:\DSKPLUS.

(c) Start up the text editor by entering `EDIT FIR.ASM`. You are now looking at the source code for a simplified FIR filtering program. The function of this program will be discussed later.

(d) Find the assembler directives **.setsect** in the program file.

Q4: How many 'setsect' directives are there?

Q5: What addresses do these directives define?

The last number on the 'setsect' directive statement indicates whether program (0) or data (1) space is used.

(e) The .copy directive copies source code from the file with name enclosed in double quotes.

Q6: How many files in total does this program consist of?

Q7: Can you identify the data areas and the program areas in this program?

Q8: What are the starting addresses of the filter coefficients, the input data and thc output data?

(f) Try to understand roughly what the code does. The comments in the file should make it quite clear.

Q9: Find the file that initializes the analog interface chip. What sampling frequency is being used (Hint: find the values for A and B registers)?

(g) Refer to the TMS320C54x DSP algebraic instruction set manual to find out what the instructions repeat and macd do. These are at the heart of the FIR filtering program.

(h) When you feel you understand the program, exit the text editor by pressing ALT-F followed by **X**. You should now return to the MS-DOS prompt.

Using the assembler and debugger

(a) Assemble the file `FIR.ASM` by entering

```
dskplasm fir.asm -l
```

Note that the last letter in the above command is lowercase L.

Q10: What messages do you see as the file is assembled?

(b) Check that the file FIR.OBJ is created.

Q11: What other files are created?

(c) Go back to windows. Click on the C54x code explorer icon in the code explorer group to start the debugger or use the start menu.

(d) Click on file, followed by load program

(e) Go to the directory `C:\DSKPLUS`

Q12: How many `'.OBJ'` files are present in that directory?

(f) Double click on `FIR.OBJ`

(g) The code has been loaded onto the C54x board. You should be able to see the source code in the disassembly window on the left-hand side.

Note: If the program has not been loaded properly, the source code in FIR.ASM will not appear in the disassembly window. In that case, exit code explorer and reset the DSKplus board by unplugging the power connection and reconnect again. Then repeat (c) to (f) above. If problem persists, get the instructor to help do a self-test on the board.

(h) The debugger consists of 4 windows: disassembly, CPU registers, peripheral registers and data memory windows. The toolbar on top of the screen includes buttons for single stepping, running, and resetting the DSKplus board. These buttons allow you to step over or into functions. The animation button supports a graphical representation of a variable or buffer. The data can be viewed in either the time or frequency domain. The debugger's online help is accessed through a button on the interface. It can be helpful in providing answers to common questions you may have while you are using the tool.

(i) The first line of the program is highlighted in the disassembly window. Click on the step into button on the top to single-step through the program. Single-step through the first 3 lines of the program.

Q13: Which registers have been changed?

Q14: What color does the contents of these registers turn into?

(j) Open the data memory window. Examine the contents of the locations where the filter coefficients, the input data and the output data are stored.

Q15: What are the contents of the output data area before and after the filtering instructions?

(k) Reset the program by clicking on the reset button at the top.

(l) You can dynamically change the contents of registers and data memory. Try increasing the most recent input data by 100 h (hexadecimal). Execute the program again.

(m) You can also change the contents of these data memory locations during the execution of the program. Try reducing the third input data by 100 h after the macd instruction has been performed 5 times.

You have now gone through the basic steps in assembling and examining the operation of a program using the debugger. These are routine procedures when developing a DSP program for execution on a DSP chip.

(n) If an oscilloscope or spectrum analyzer is available, you may observe the output of the board. The program generates random noise, which is then filtered. The output signal should show the frequency response of the low-pass filter.

(o) If a microphone is available, see if you can modify the source code of the program to accept input from the input port (instead of the random noise sample generated within the program). Then run the program, speak into the microphone and listen to the filtered output.

Designing and implementing an FIR filter (optional)

We shall now go through the process of designing an FIR filter and putting the coefficients into the FIR filtering program. We shall perform the design using MATLAB, generate the filter coefficients and put that into our FIR filter program.

(a) Design an 80th order FIR filter with a cut-off frequency of 0.25 Hz (normalized) using the Hamming window method. The MATLAB function to be used is `FIR1`.

Q16: What is the real cut-off frequency?

(b) Quantize these coefficients to 15 (or 16) bits. Scale the resulting quantized coefficients by a factor of 2^{15} (or 2^{16}).

(c) You can now enter these values as filter coefficients into the coefficient file. Copy the original coefficient file and rename it to a filename of your choice. Copy, rename and change the main program source (`FIR.ASM`) to reflect the change in coefficient filename. Enter the coefficients into the appropriate locations in the file.

(d) Assemble and run the FIR filter program. If an oscilloscope or spectrum analyzer is available, check if the program is behaving as expected. Otherwise, listen to the filtered white noise using the speakers. The original filter has a cut-off frequency of around 970 Hz. It should sound quite different from the current one.

(e) If time permits, try some other filter cut-off frequencies.

Sigma-delta techniques

Objective:

To reinforce the concepts and techniques used in sigma-delta converters, namely,

- Oversampling
- Quantization noise spectral shaping.

Equipment required:

A 486/Pentium PC running Windows95 with MATLAB version 5.x and Simulink 2.1 installed.

Exercises:

Please refer to section 2.4.4 of the manual for concepts and techniques used in sigma-delta converters.

Oversampling

The simulation model that we will use in studying the effect of oversampling is shown in the figure below.
Note: It makes use of the function fpquant that we have defined in filter realization and wordlength effects exercise.

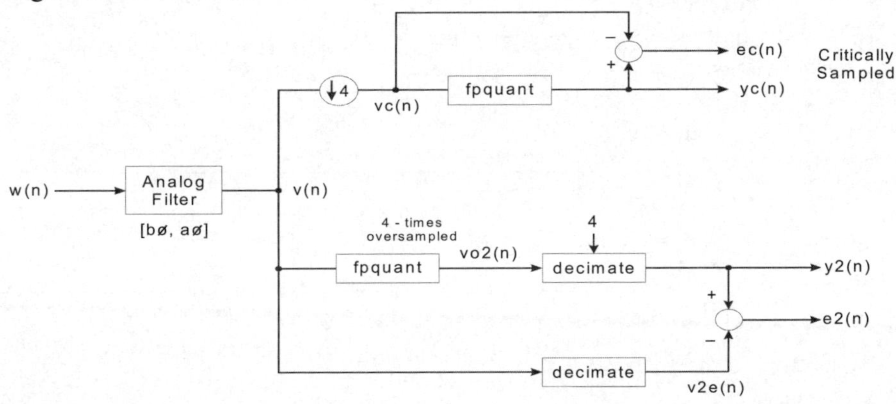

Figure C.1
Simulation model for studying oversampling effects

The signal source is random (white) noise *w*(*n*), which has a flat spectrum. This signal is filtered appropriately to produce a random signal *v*(*n*) with the desired bandwidth. The MATLAB code for doing that is:

```
[b0,a0]=ellip(7,0.1,60,0.195);
w=(rand(1,8000)-0.5)*2);
v=filter(b0,a0,w);
```

We have generated 8000 samples of the signal. The filter used is a 7-th order elliptic low-pass filter with 0.1 dB passband ripple and at least 60 dB attenuation in the stopband. The cut-off frequency is 0.195 Hz normalized.

v(*n*) is now the 4 times oversampled signal. The critically sampled signal *vc*(*n*) is generated by downsampling *v*(*n*) by 4 (taking 1 out of 4 samples).

```
n=1:4:length(v);
vc=v(n);
```

The signals *v*(*n*) and *vc*(*n*) are now quantized to 10 bits.

```
yc=fpquant(vc,10);
vo2=fpquant(v,10);
```

The quantization noise power (in dB) for *vc*(*n*) is calculated and stored as variable dbe1.

```
ec=yc-vc;
dbe1=10*log10(cov(e1));
```

In actual systems, the oversampled signal will be digitally filtered and then downsampled (see the manual for details), we shall do the same here. The quantization noise power (in dB) dbe2 is calculated using only the downsampled version.

```
y2=decimate(vo2,4);
v2e=decimate(v,4);
e2=y2-v2e;
dbe2=10*log10(cov(e2));
```

Q1: What is the difference between dbe1 and dbe2?

Q2: How many bits of quantization does the improvement (in Q1) represent? Is that roughly what is expected?

Quantization noise spectral shaping

The second technique that sigma-delta converters use is the reshaping of the quantization noise spectrum by using error feedback. Figure C.2 shows the simulation model that we will use.

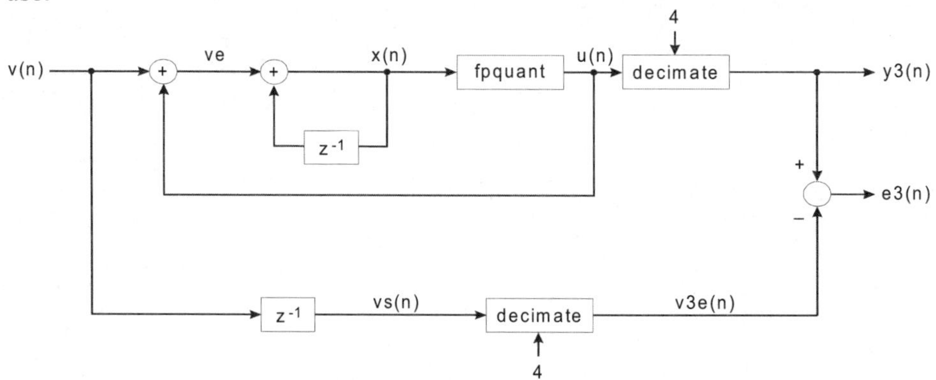

Figure C.2
Simulation model for studying quantization noise shaping effects

The upper portion of the simulation model is the sigma-delta system using error feedback. The lower portion provides us with the reference for calculating the noise power (in dB). We shall first generate the sequence of outputs $u(n)$.

```
x=0;
for n=1:length(v),
  u(n)=fpquant(x,10);
  ve=v(n)-u(n);
  x=ve+x;
end
```

The above code may take longer to execute. We have to compute the output sample-by-sample instead of operating on the whole vector/matrix for which MATLAB is optimized. Then the output signal is decimated (filtered and down sampled) to produce the actual output and error.

```
y3=decimate(u,4);
```

Before we decimate the input signal, we need to shift the sequence to the right by one place because of the one sample delay introduced by the integrator in the loop.

```
vs(2:length(v))=v(1:length(v)-1);
vs(1)=0;
v3e=decimate(vs,4);
```

The quantization error (in dB) can now be calculated.

```
e3=y3-v3e;
dbe3=10*log10(cov(e3));
```

Q3: What is the difference between dbe2 and dbe3?

Q4: How many bits of quantization does the improvement (in Q3) represent? Is that roughly what is expected?

Sigma-delta A/D converter

Start SIMULINK and construct a model as shown below:

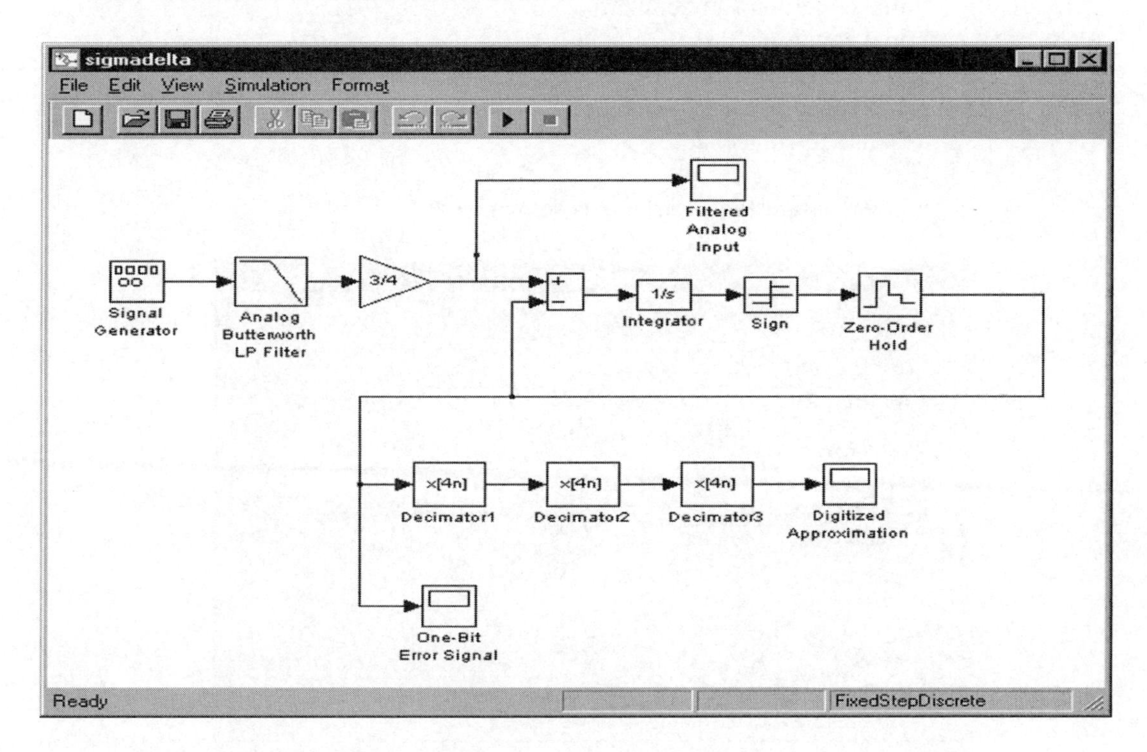

The following parameters are used for the blocks:

(1) Signal generator
  waveform: square
  amplitude: 1
  frequency: 80 Hz

(2) Analog Butterworth LP filter
  cutoff frequency: 2*pi*400
  order: 5

(3) Zero-order hold
  sample time: 1/512000

(4) Decimator 1
  Actual block used: FIR decimation
  FIR filter coefficients: fir1(31,0.15)
  decimation factor: 4
  input sample time: 1/512000

(5) Decimator 2
  Actual block used: FIR decimation
  FIR filter coefficients: fir1(31,0.15)
  decimation factor: 4
  input sample time: 1/128000

(6) Decimator 3
  Actual block used: FIR decimation

FIR filter coefficients: fir1(31,0.15)
decimation factor: 4
input sample time: 1/32000

(7) Integrator
external reset: none
initial condition source: internal
initial condition: 0
upper saturation limit: inf
lower saturation limit: -inf
absolute tolerance: auto

Simulation parameters are setup as shown below.

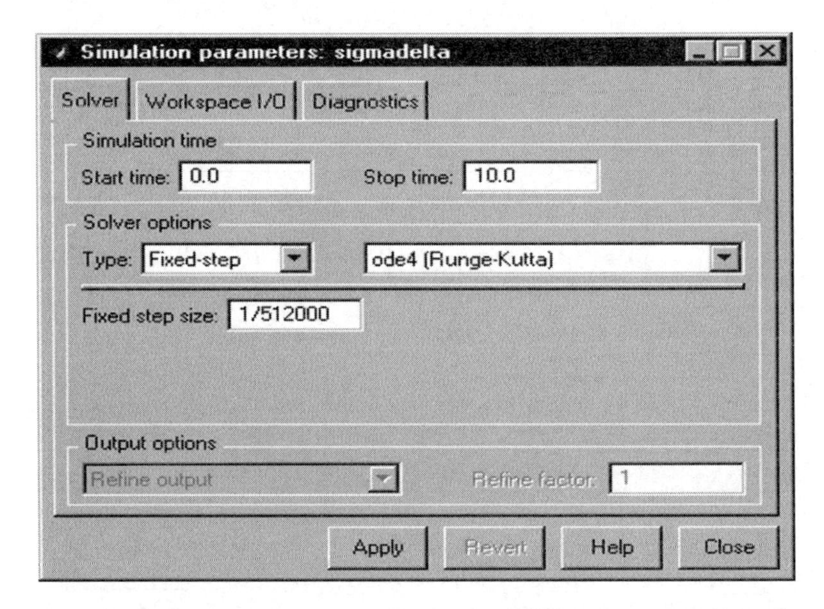

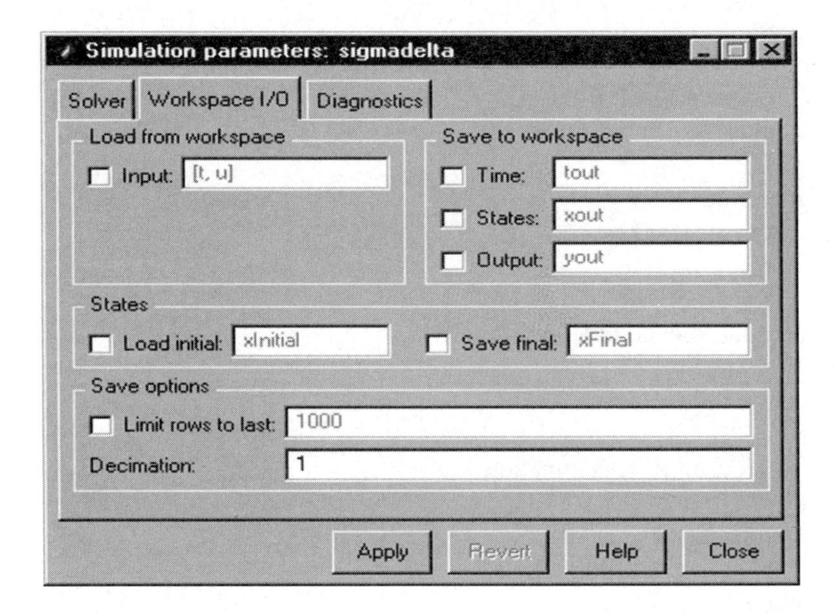

Start the simulation. The output of the converter is the quantized version of the input.

Q5: Can you understand how the converter works?

Digital image processing

Objective:

- To provide an introduction to the DSP area of image processing.
- To illustrate linear and non-linear filtering on images.

Equipment required:

A 486/Pentium PC running Windows95 with MATLAB version 5.x and image processing and signal processing toolboxes installed.

Notation:

The commands that users need to enter into the appropriate window on the computer are formatted with the typeface as follows:

plot(x,y)

Exercises:

Displaying images.

(a) Start MATLAB. Enter

```
I = imread('ic.tif');
J = imrotate(I,35,'bilinear');
imshow(I)
figure, imshow(J)
```

An image of an IC is displayed and it is rotated by 35° counterclockwise.

(b) To display a sequence of images,

```
load mri
montage(D,map)
```

Image analysis

(a) In image analysis, we typically want to obtain some pixel values or their statistics. Enter the following:

```
imshow canoe.tif
impixel
```

Click on two or three points in the displayed image and then press 'return'. The pixel values are displayed. Notice that since this is a color image, the RGB values are shown.

(b) To obtain the intensity values along a certain straight line:

```
imshow flowers.tif
improfile
```

The cursor is changed to a cross hair when it is over the image. Specify a line segment by clicking on the end points. Then press 'return'.

(c) Image contours can be obtained:

```
I=imread('rice.tif');
imshow(I)
figure, imcontour(I)
```

(d) Image histograms are useful. One use of histogram has been discussed in the lecture.

```
I=imread('rice.tif');
figure(1), imshow(I)
figure(2), imhist(I,64)
```

(e) Edge detection is also a very useful operation.

```
I=imread('blood1.tif');
BW=edge(I,'sobel');
figure(1), imshow(I)
figure(2), imshow(BW)
```

You may also run `edgedemo` for an interactive demonstration of edge detection.

Image enhancement

(a) Intensity adjustment:

```
I=imread('rice.tif');
J=imadjust(I,[0.15 0.9],[0,1]);
figure(1), imshow(I)
figure(2), imshow(J)
```

Compare this adjustment with the following:

```
J=imadjust(I,[0 1],[0.3 0.8])
imshow(J)
```

(b) Histogram equalization

```
I=imread('pout.tif');
J=histeq(I);
imshow(I)
figure(2), imshow(J)
```

Histograms of the two pictures can be compared:

```
figure(1), imhist(I)
figure(2), imhist(J)
```

(c) Median filtering

First, read in an image and add noise to it.

```
I=imread('eight.tif');
J=imnoise(I,'salt & pepper',0.02);
figure(1), imshow(I)
figure(2), imshow(J)
```

Now median filter the image:

```
K=filter2(fspecial('average',3),J)/255;
L=medfilt2(J,[3 3]);
figure(1), imshow(K)
figure(2), imshow(L)
```

The first figure uses linear filtering, and the second one uses median filtering. Which one is better?

(d) Adaptive filtering

```
I=imread('saturn.tif');
J=imnoise(I,'gaussian',0,0.005);
```

```
K=wiener2(J,[5 5]);
figure(1), imshow(J)
figure(2), imshow(K)
```

This filter is called Wiener filter.

Fourier transform

(a) Construct an artificial image:

```
f=zeros(30,30);
f(5:24,13:17)=1;
imshow(f,'notruesize')
```

(b) Compute the 256×256 DFT:

```
F=fft2(f,256,256);
F2=log(abs(F));
imshow(F2,[-1,5], 'notruesize');
colormap(jet); colorbar
```

The DC coefficient is displayed in the upper-left corner. It can be moved to the center by

```
F2=fftshift(F);
imshow(log(abs(F2)),[-1,5]);
colormap(jet); colorbar
```

These are just some of the operations provided by the image processing toolbox. Explore it further by going to the MATLAB demos for this toolbox in a similar way to introduction to MATLAB exercise.

Index

Note: Page numbers in *italics* refer to figures and tables

Accumulator, 208
Active noise control (ANC), 79, 94–7
Adaptive differential pulse code modulator (ADPCM), 6
Adaptive differential quantization, 5–6
Aliasing, 19–*20*
  effect of, *21*
All-pole filter, 185
Analog filter, 163–4
Analog signal reconstruction, 42–5
  ideal, 43
  image-rejection postfilters, 45
  staircase, 44–5
Analog-to-digital converter (ADC), 14, 23, 25, 34–49
  dual slope ADC, 36
  flash ADC, 36–7
  quantization noise shaping, 40–2
  sigma-delta ADC, 37–9
  successive approximation, 34–6
  typical DSP system, 14
Anti-alias filter, 21, *22*, 23
Application specific integrated circuit (ASIC), 101
Arithmetic logic unit (ALU), 212
Arithmetic operations, 189, 190
Autocorrelation coefficients, 57–60
  implementation, 60
  periodic sequences, 59–60

Band-pass filter, 3, 143
Blackman window, *117*–18
Butterfly operation, *68*, 69, *70*–2

Canonical form, *172*, 173–6, *177*
Cauer filters, 157
Central processor (CPU), 212
Chebyshev filters responses, 150–8
Chebyshev polynomial functions, 150–*1*
Chebyshev rational function, 155
Circular buffer, 215–18
Circular convolution, 65, 74–5, 172
Classification of systems, 54–5
  causal vs non-causal, 54
  linear vs non-linear, 54
  stable vs unstable, 55
  time-variant vs time-invariant, 54
Coefficient quantization model, *193*
Common object file format (COFF), 233
Companding, 29
  process, *5*
Companding or non-uniform quantization, 5
Complex frequency variable, 158
Contrast enhancement transformation, *92*
Convolution, concept of, 55–7
Convolution filters, 102
Convolution using DFT, computation of, 74–8
  circular convolution, 74–6
  long data sequences, 76–8
    overlap-add method, 76–7
    overlap-save method, 77–8
Crossover filter responses, 123–*4*

Design examples, 142–3
Design flow:
  development of algorithm, 226, 228

analog or digital, 228
bandwidth and frequency ranges, 228
data format, multiplexing and codes, 229
data rate, 229
desired information, 229
deterministic or random, 228
dynamic range, 228
number of channels or sensors, 228
spectral contents, 228
steady-state or transient or both, 228
type of noise, 229
DSP hardware development, 226
DSP software development, 226
selection of DSP chip, 226
system debugging and testing, 227
system integration, 227
system requirement definition, 226
input and output signal or data requirements, 227
interface design specifications (IDS), 227
prime item development specification (PIDS), 227
Digital filter, 164–5
Digital filter realizations
cascade form, 179–81
FIR filters, 180
IIR filters, 180–1
direct form, 171–9
FIR filters, 178–9
IIR filters, 171–8
finite word-length effects, 191–203
coefficient quantization errors, 191–5
limit cycles, 200–3
overflow errors, 196–200
rounding and truncation, 195–6
other structures, 183–6
fast convolution, 186
frequency sampling structure, 186
lattice structure, *183*–6
vwave digital filter, 186
parallel form, 181–3
representation of numbers, 187–90
fixed-point representation, 187–9
floating-point representation, 189–90
software implementation, 186–7
sample processing algorithms, 186–7
Digital filters, classification of, 98–9
Digital signal processors:
architecture, *215*
choosing a processor, 224
circular buffers, 215–18
code examples, 218–24
FFT, 220–4
FIR filtering, 218–20
common features, 204–5
fast multiply-accumulate, 204–5
multiple-access memory architecture, 205
peripheral interfaces, 205
special addressing modes, 205
special program control, 205
frequency modulator using, *87*
general purpose microprocessors for DSP, 224
hardware architecture, 206–15
architecture of ADSP21xx, 214–15
architecture of TMS320C5x, 211–13
data path, 206–8
memory architecture, 208–11 *see* Harvard architecture; Von Neumann
oversampling method, 39
special instructions and addressing modes, 215–24
Digital-to-analog (D/A) conversion, 14, 46–8
bit stream DAC, 47–8
multiplying DAC, 46–7
Digital video discs (DVD), 8
Digital waveform generation using digital filters, *81*
Dirac delta function, 23
Direct form structure, *173*, 178, 193
Direct memory access (DMA), 211
Discrete Fourier transform (DFT), properties of, 13, 61, 63, 64–7
convolution, 65
correlation, 66
even and odd functions, 65
frequency shifting, 66
linearity, 65
modulation, 66–7
Parseval's relation, 65
periodicity, 64–5
real sequences, 65
time delay, 66
Discrete frequency spectrum, effect of, 62, *66*
Discrete-time Fourier series (DTFS), 62
aperiodic signals, 63–4
periodic signals, 62–3
Discrete-time signals operations, 52–3
block diagram representation, 53
delay or shift, 52
scalar addition and multiplication, 52–3
vector addition and multiplication, 53
Discrete-time system, *3*, 54, 57, 90, 98, 102
Dither, amplitude distribution of, *32*
DSP, application areas of, 4–12
adaptive filtering, 8–10
channel equalization, 9–10
echo cancellation, 9
noise cancellation, 8–9
control applications, 10–11

DSP, application areas of (*Continued*)
digital communication receivers and transmitters, 11–12
image and video processing, 7–8
image compression and coding, 8
image enhancement, 7
image restoration, 7–8
sensor or antenna array processing, 11
speech and audio processing, 4–7
coding, 4–7
recognition, 7
synthesis, 7
DSP system, 3–4
application, 79–90
block processing, 4
design flow, 226–31
development of algorithms, 228–9
system implementation, 230–1
system requirement definition, 227–8
development environment, *234*
implementation procedure, 230
control processor resource analysis, 230
data communication resources analysis, 230
input/output processor resource analysis, 230
memory resource analysis, 230
processor engine resource analysis, 230
sample-by-sample processing, 3
Dual-tone multi-frequency (DTMF), 80

Electrical/electronic engineering (EE), 2
Elliptic filter *see* Cauer filters; Chebyshev filters
Equiripple FIR filters, 134
ERMES (European radio message system), 83, 84, 85, 86, 100, 128
premodulation filter in, *85*
specifications in, 99–*100*
ETSI (European Telecommunications Standards Institute), 83

Fast Fourier transform (FFT), 13, 61, 64, 67–71
computational savings, 69
decimation-in-frequency algorithm, 69–70
other fast algorithms, 71
Filter coefficients, 98
Filter design process, 99–106
approximation, 99–101
implementation, 101
performance analysis, 101
synthesis and realization, 101
Filter impulse response, 107, *110*
Filter response, 134
Filter structures, 171
Filters, comparison of, *157*
see also Magnitude response

Finite impulse response (FIR), 98
characteristics of, 102–6
frequency response, 102–3
linear phase filters, 104–6
designed using frequency sampling, *131*
structure, 178
structure with overflow non-linearity added, 202
FIR filters *see* Non-recursive filters; Convolution filters
First order sigma-delta ADC, *40*–1
Formants, 89
Frequency, definition of, 2
Frequency-domain:
discrete-time signals, representation of, 61–78
interpretation, *114*
model of coefficient quantization, 191–4
Frequency sampling design, 166–7
Frequency sampling method, 128–34
design formulas, 129–30
transition region, 130–4
Frequency warping, *161*

Graphical user interface (GUI), 245
Gray-scale modification, 91
Guard bits, 72, 208

Hamming window, *115*–17, 120
magnitude response, *116–17*
Hardware and software development tools, 231–7
assembly language tools, 233–5
assemblers and linkers, 233
development boards, 235
in-circuit emulation (ICE), 234
monitor-based emulator, 235
pod-based emulator, 234
scan-based emulator, 235
simulators, 233
high-level language tools, 231–3
efficiency concerns, 231
languages available, 231–2
maintainability, 231
memory usage, 231
portability, 231
productivity, 231
special instructions usage, 231
other software tools, 235–7
COSSAP, 236
DADisP, 236
hypersignal, 237
MATLAB, 236
MatrixX, 236
scilab, 237
signal processing worksystem (SPW), 236
real-time operating system, 237

Harvard architecture, 208–*9*
  dual-port data memory, *210*
High-pass filter, 3

Ideal low-pass filter:
  and its impulse response, 43
  response, *107*
IIR filter response, calculation of, 166
Image enhancement, 91–4
  contrast enhancement, 91–3
  noise reduction, 93–4
Infinite impulse response (IIR), 98, 145
  lattice filter, *185*
Infinite impulse response (IIR) filter design:
  approach, *147*, 157, 162–3
  characteristics of IIR filters, 146–7
  direct design methods, 165–9
    frequency sampling, 165–7
    least squared equation error design, 167–9
  FIR vs IIR, 169–70
  IIR filters from analog filters, 157–65
    bilinear transformation method, 160–2
    frequency transformation, 162–5
    impulse invariant method, 157–60
  review of classical analog filter, 147–50
    Butterworth function, 148–50
    Chebyshev approximation, 150–2
    elliptic function, 155–7
    inverse Chebyshev approximation, 153–4
  see Recursive filters
In-phase and quadrature signals, *87*
Institute of Electrical and Electronic Engineers
    (IEEE), 189
Interchanging the cascade, *172*
Internal buses:
  data memory address (DMA), 214
  data memory data (DMD), 215
  program address (PMA), 214
  program data (PMD), 214
  result (R), 215
Inverse DFT and its computation, 64

Jacobian elliptic functions, 155

Kaiser windows, *120*–2, 191

Latency, 207
Least recently used (LRU) algorithm, 211
Least significant bit (LSB), 188
Limit cycle, 200
Linear buffer for data, *216*
Linear convolution, 57, 60, 65, 74, 75
Linear phase response, *104*–6, 169
Linear prediction, 6–7
Linear predictive model, 90
Linear programming method, 141–2
Linear time-invariant (LTI), 54
Low-pass filter, 2–3, 117, 142

Magnitude and phase responses, 99, *103*, 165
Magnitude response:
  Butterworth filters, *148*, 150
  Chebyshev filters, 150–8
  five filters in the graphic equalizer, 126
  Kaiser window designed FIR filter, 122
  length-21 bandpass filter, 140
  low-pass filter, 118
  rectangular window, 114
  rectangular windowed filter, 111
Manual caching, 211
Maximally flat filters, 148
METEOR, *142*–3
Most significant bit (MSB), 188
Multi-channel ANC system, *97*
Multiple-access memories, 209–10
Multiply-and-accumulate (MAC), 74, 172, 204
Multi-port memory, 210

Native signal processing (NSP), 224
Natural and man-made signals, 78
Non-recursive filters, 99, 134
Non-uniform quantization, 5, *28*
Notation, 50
Numerical C Extension Group (NCEG), 232
Nyquist interval, 33, 43, 82, 106, 137
Nyquist rate, 17, 21, 37

Odd length filter response, *139*
Optimum and minimax filters *see* Equiripple
    FIR filters
Original and quantized signal, *31*
Overlap-add method, 76–*7*
Overlap-save method, 77–*8*
Oversampling ratio, 37, 39, 40
Oversampling stage of a bit stream
    DAC, *47*–8

Parallel form realization, *181*
Parallel logic unit (PLU), 212
Parks-McClelland method, 134–41
  approximation problem, 135–6
  design formulas, 138–41
  equiripple solution, 136
  Remez exchange algorithm, 137–8
Perfect reconstruction filter banks
    (PRFB), 144
Periodic signal generation, 80–3
  digital waveform generation, 81
  DTMF example, 83
  generating arbitrary frequencies, 82–3

Periodic signal generation (*Continued*)
generating integer multiples of fundamental frequency, 81–2
sampling frequency, 81
Phonemes, 89
POCSAG, 83, 84, 87
Practical implementation issues, 71–4
bit reversal, 71
computation of real-valued FFTs, 72–3
computational complexity, 73–4
fixed point implementations, 72
Primal problem, 142
Processing signals digitally, benefits of, 1–2
Program and data access, 213
data read address bus (DAB), 213
data read bus (DB), 213
program address bus (PAB), 213
program bus (PB), 213
Program caches, 210–11
Pulse code modulator (PCM), *6*
Pulse density modulated (PDM) waveform, *48*

Quantization, 24–33
dithering, 30–3
non-uniform, 28–30
sample-and-hold, 24–5
uniform, 25–8
Quantization noise, model of, 27
sampled and oversampled systems, *39*
spectrum shaped by the sigma-delta technique, 41
spectrum with dithering, 33
Quantized filter response, 194
Quantized signal spectrum, *31*

Real-time operating systems (RTOS), 237
Rectangular window, 108–*10*
impulse response, *110*
Recursive filters, 99, 102, 145
Reflection coefficient, 184, 186
Relative merits, 176

Sample and hold circuit, *24*, 34
Sampling, 15–23
aliasing, 20–1
anti-aliasing filters, 22
frequency domain interpretation, 18–20
mathematical representation, 23–4
practical limits on sampling rates, 22–3
theorem, 16–17
Shannon's sampling theorem, 16, 17
Shape parameter, 119, 120, 127
Sigma-delta converter model, 37, 39, *40*
Signal reconstruction, *42*
Signals, definition of, 2
Single channel ANC system, *94*
Snapshots, 15
Software radio architecture, 12
Spectrum, 2
Speech production model, *6*, *90*
Speech synthesis, 88–90
classification of sounds, 89
production mechanism, 88–9
production model, 90
Staircase reconstructors, 42, 44–5
Successive approximation register (SAR), 34
Superposition principle, 54

Transfer functions:
cascade of, *179*
of overflow non-linearity, *203*
Transposed FIR direct structure, *179*
Transposed structures, 177–8
Transversal structure or tapped delay line structure *see* Direct form structure
Typical discrete-time signals, 50–1
random, 51
unit impulse, 50–1
unit step, 51

Uniform sampling, 16
Unit impulse sequence, *50*, 52
Unit step sequence, *51*–3

Voltage source multiplying DAC, *47*
Von Neumann architecture, *208*

Wave table synthesis, 81
Window method, 106–28
Blackman window, 117
Hamming window, 115–16
Kaiser window, 118–28
Bessel functions, 127–8
design, 119–20
design steps, 121–2
high-pass filter design, 122–7
rectangular window, 108–15
another interpretation, 114–15
performance evaluation, 110–13
Winograd Fourier transform (WFT), 71
Wireless transmitter implementation, 83–8
DSP implementation, 86–7
other advantages, 88
specifications, 84–6
WKS sampling theorem, 17

Zero order holds (ZOH) *see* Staircase reconstructors
Zero padding, 60